VOLUME ONE HUNDRED AND TWO

VITAMINS AND HORMONES

Thymosins

VOLUME ONE HUNDRED AND TWO

VITAMINS AND HORMONES

Thymosins

Series Editor

GERALD LITWACK, PhD

Toluca Lake, North Hollywood, California

ELSEVIER

AMSTERDAM • BOSTON • HEIDELBERG • LONDON
NEW YORK • OXFORD • PARIS • SAN DIEGO
SAN FRANCISCO • SINGAPORE • SYDNEY • TOKYO
Academic Press is an imprint of Elsevier

Cover photo credit:
Hoch, K., Volk, D.E.
Structures of Thymosin Proteins
Vitamins and Hormones (2016) **102**, pp. 1–24.

Academic Press is an imprint of Elsevier
50 Hampshire Street, 5th Floor, Cambridge, MA 02139, United States
525 B Street, Suite 1800, San Diego, CA 92101-4495, United States
The Boulevard, Langford Lane, Kidlington, Oxford OX5 1GB, United Kingdom
125 London Wall, London, EC2Y 5AS, United Kingdom

First edition 2016

ISBN: 978-0-12-804818-4
ISSN: 0083-6729

Publisher: Zoe Kruze
Acquisition Editor: Alex White
Editorial Project Manager: Helene Kabes
Production Project Manager: Vignesh Tamil
Cover Designer: Greg Harris

Typeset by SPi Global, India

CONTENTS

CONTRIBUTORS

G. Covelo
Facultad de Biología, Universidade de Santiago de Compostela, Santiago de Compostela, Spain

C. Díaz-Jullien
Facultad de Biología, Universidade de Santiago de Compostela, Santiago de Compostela, Spain

M. Freire
Facultad de Biología, Universidade de Santiago de Compostela, Santiago de Compostela, Spain

E. Garaci
University of Rome "Tor Vergata"; San Raffaele Pisana Scientific Institute for Research, Hospitalization and Health Care, Rome, Italy

K. Hoch
Texas Children's Microbiome Center, TCH Pathology, Houston, TX, United States

K. Ioannou
Faculty of Biology, National and Kapodistrian University of Athens, Panepistimiopolis, Athens, Greece; King's College London, Rayne Institute, London, United Kingdom

Y. Jung
Pusan National University, Pusan, Republic of Korea

J. Kim
Pusan National University, Pusan, Republic of Korea

R. King
SciClone Pharmaceuticals, Inc., Foster City, CA, United States

H.K. Kleinman
George Washington University, Washington, DC; NIDCR/NIH, Bethesda, MD, United States

A. Kumar
Nanomedicine Research Laboratory, University of Delaware, Newark, DE, United States

W. Mandaliti
University of Rome "Tor Vergata", Rome, Italy

E.D. Marks
Nanomedicine Research Laboratory, University of Delaware, Newark, DE, United States

R. Nepravishta
University of Rome "Tor Vergata", Rome, Italy; Faculty of Pharmacy Catholic University "Our Lady of Good Counsel", Tirane, Albania

G. Ousler
Kresge Eye Institute, Wayne State University School of Medicine, Detroit, MI; ORA Inc., Andover, MA, United States

M. Paci
University of Rome "Tor Vergata", Rome, Italy

F. Pica
University of Rome "Tor Vergata", Rome, Italy

G.T. Pipes
Cardiovascular Drug Discovery, Discovery Biology Research & Development, Bristol-Myers Squibb, Pennington, NJ, United States

L. Renault
Institute for Integrative Biology of the Cell (I2BC), CEA, CNRS, Univ. Paris-Sud, Université Paris-Saclay, Gif-sur-Yvette, France

D. Rimmer
Kresge Eye Institute, Wayne State University School of Medicine, Detroit, MI; ORA Inc., Andover, MA, United States

R.C. Robinson
Institute of Molecular and Cell Biology, A*STAR (Agency for Science, Technology and Research), Biopolis; Yong Loo Lin School of Medicine, National University of Singapore; NTU Institute of Structural Biology; School of Biological Sciences, Nanyang Technological University; Lee Kong Chan School of Medicine, Singapore

P. Samara
Faculty of Biology, National and Kapodistrian University of Athens, Panepistimiopolis, Athens, Greece

C.S. Sarandeses
Facultad de Biología, Universidade de Santiago de Compostela, Santiago de Compostela, Spain

G. Sosne
Kresge Eye Institute, Wayne State University School of Medicine, Detroit, MI, United States

O.E. Tsitsilonis
Faculty of Biology, National and Kapodistrian University of Athens, Panepistimiopolis, Athens, Greece

C. Tuthill
SciClone Pharmaceuticals, Inc., Foster City, CA, United States

P.S. Vallebona
University of Rome "Tor Vergata", Rome, Italy

D.E. Volk
Center for Proteomics and Systems Biology, Brown Foundation Institute of Molecular Medicine for the Prevention of Human Diseases; Department of Nanomedicine and Biomedical Engineering, The University of Texas Health Science Center at Houston, Houston, TX, United States

B. Xue
Institute of Molecular and Cell Biology, A*STAR (Agency for Science, Technology and Research), Biopolis, Singapore

J. Yang
Cardiovascular Drug Discovery, Discovery Biology Research & Development, Bristol-Myers Squibb, Pennington, NJ, United States

PREFACE

Thymosin peptides originally were isolated from the thymus gland and were found to stimulate the development of T cells; consequently, they were initially considered to be thymic hormones. Later research revealed several related peptides and thymosins were found in many tissues. In humans, there exist three families of beta thymosins, beta4, beta10, and beta15; of these, beta4 has been most studied. Beta4 and beta15 families each are encoded by two separate genes and beta10 is encoded by a single gene. There appears to be a single human gene responsible for the prothymosin alpha gene family (generating thymosin alpha 1), deriving from the thymus gland.

There are monomeric and polymeric beta thymosins. The beta thymosins are involved in actin assembly, cytoskeletal remodeling, cell regulation, the cardiovascular system, and dermal healing; in addition, they have therapeutic effects on diseases of the ocular surface. Alpha thymosin appears to have modulatory functions in the immune system.

This volume concentrates on the structure and functions of thymosins and their activities. Also reviewed are the biological and clinical conditions in which thymosins are active and potentially therapeutic.

Initially, there is concentration on structure and biological function. The second part of the book describes biological and clinical conditions involving the thymosins.

To begin, K. Hoch and D.E. Volk describe "Structures of thymosin proteins." Chapter 2 by L. Renault reports on "Intrinsic, functional, and structural properties of β-thymosins and β-thymosin/WH2 domains in the regulation and coordination of actin self-assembly dynamics and cytoskeletal remodeling." B. Xue and R.C. Robinson further review "Actin-induced structure in the beta-thymosin family of intrinsically disordered proteins." Then, M. Freire, C.S. Sarandeses, G. Covelo, and C. Díaz-Jullien focus on "Phosphorylation of prothymosin α. An approach to its biological significance." Following, R. Nepravishta, W. Mandaliti, P.S. Vallebona, F. Pica, E. Garaci, and M. Paci report on the "Mechanism of action of thymosinα1: Does it interact with membrane by recognition of exposed phosphatidylserine on cell surface? A structural approach." In Chapter 6, J. Kim and Y. Jung discuss "Thymosin beta 4 as a potential regulator of hepatic stellate cells."

With respect to the role of thymosin alpha 1 on the immune system, R. King and C. Tuthill author "Immune modulation with thymosin alpha 1 treatment" and then P. Samaria, K. Ioannou, and C.E. Tsitsilonis review "Prothymosin alpha and immune responses: Are we close to potential clinical applications?"

The remainder of the volume concentrates on thymosin beta 4. "Cardioprotection by thymosin beta 4" is described by G.T. Pipes and J. Yang. E.D. Marks and A. Kumar report on "Thymosin β4: Roles in development, repair, and engineering of the cardiovascular system." H.K. Kleinman and G. Sosne describe "Thymosin β4 promotes dermal healing." The concluding chapter is "Thymosin beta 4: A potential novel therapy for neurotrophic keratotherapy, dry eye, and ocular surface diseases" by G. Sosne, D. Rimmer, H.K. Kleinman, and G. Ousler.

The illustration on the cover is Fig. 7, reproduced from Chapter 1:"Structures of thymosin proteins" by K. Hoch and D.E. Volk. The legend is: Crystal structure of alpha actin bound to gelsolin–thymosin beta-4 C-terminal helix chimera. The helix is the minus-end capping helix and it competes with DNAse I binding at this site. Atomic coordinates were obtained from the Protein Data Bank (PDB ID 1T44; Irobi et al., 2004).

As usual, Helene Kabes of Elsevier, Oxford, UK, played a major role in the development of the published work in tandem with the Reed Elsevier group in Chennai, India.

Gerald Litwack
Toluca Lake, North Hollywood, CA
April 2, 2016

CHAPTER ONE

Structures of Thymosin Proteins

K. Hoch*, D.E. Volk†,‡,1

*Texas Children's Microbiome Center, TCH Pathology, Houston, TX, United States
†Center for Proteomics and Systems Biology, Brown Foundation Institute of Molecular Medicine for the Prevention of Human Diseases, The University of Texas Health Science Center at Houston, Houston, TX, United States
‡Department of Nanomedicine and Biomedical Engineering, The University of Texas Health Science Center at Houston, Houston, TX, United States
[1]Corresponding author: e-mail address: david.volk@uth.tmc.edu

Contents

Abstract

The thymosin proteins are all short, highly charged, intrinsically unstructured proteins under natural conditions. However, structure can be induced in many of the thymosin proteins by providing charge neutralization at low pH or by the addition of Zn^{2+} ions, organic reagents such as trifluoroethanol, hexafluoropropanol, or *n*-dodecyltrimethylammonium bromide, or interactions with their natural binding partner proteins. The differing structures of thymosin alpha and thymosin beta proteins have been studied by circular dichroism, nuclear magnetic resonance, and crystallographic methods in order to better understand

Vitamins and Hormones, Volume 102
ISSN 0083-6729
http://dx.doi.org/10.1016/bs.vh.2016.04.009

the role of these proteins. In this structural biology review the structures of prothymosin, parathymosin, thymosin alpha-1, and several beta thymosin proteins, in both native states and under secondary structure-inducing conditions are discussed.

1. INTRODUCTION

Although the thymosin proteins were originally discovered from fractionations of calf thymus tissue, and thus so named, they are genetically unrelated while being distributed widely throughout most tissues and play important, yet very different, roles in cells. They are highly charged proteins with no or few aromatic amino acids and thus lack stable tertiary structure unless induced by interactions with partnering proteins or unnatural solvent conditions. The active peptides are typically short. The beta thymosins are each about 43 amino acids long, while thymosin α1 (Tα1) is only 28 amino acids long, although its precursor, prothymosin, is nearly 100 bases long.

Tα1 and prothymosin α have been used to treat a variety of viral infections, including HIV (Mosoian et al., 2006, 2010), chronic hepatitis B (Iino et al., 2005; You et al., 2006), chronic hepatitis C (Andreone et al., 2001; Kullavanuaya et al., 2001), cytomegalovirus (Bozza et al., 2007), and invasive aspergillosis (Segal & Walsh, 2006), due to their immunological effects (Markova et al., 2003; Romani et al., 2006). The thymosin beta proteins are major sequestering agents of monomeric actin protein, thus allowing cells to have a high concentration of G-actin at the ready for quick use. As such, these proteins are clinically important. The structures of the thymosin proteins are explored under a variety of conditions in this review.

2. STRUCTURES OF PROTHYMOSIN α AND PARATHYMOSIN

2.1 Native Structure of Prothymosin α

Prothymosin α (ProTα), a 110-amino acid protein first discovered by Haritos, Tsolas, and Horecker (1984), has a very unusual amino acid sequence containing no aromatic (Tyr, Trp, Phe, His) or sulfur-containing residues (Met, Cys), while being very acidic, containing 35 glutamates and 19 aspartates (Goodall, Dominguez, & Horecker, 1986). As the name suggests, prothymosin α is a precursor to the 28 amino acid protein Tα1 discovered by Goldstein et al. (1977). Using traditional biophysical methods, such as X-ray scattering, dynamic light scattering, CD, NMR, mass spectrometry, and gel-filtration, early studies showed that ProTα has a random coil structure at physiological pH (Cordero, Sarandeses, Lopez, & Nogueira, 1992; Gast et al., 1995; Watts, Cary, Sautiera, & Crane-Robinson, 1990).

2.2 pH-Induced Structures of Prothymosin α

Structure can be induced in ProTα by a number of methods (Table 1). First, it was shown by CD and NMR that lowering the pH induced a small amount of structure to prothymosin (Watts et al., 1990). Later it was shown that the presence of about 50% trifluoroethanol (TFE) at pH 2.4 induced approximately 69% helical structure in ProTα, as measured by CD spectra (Gast et al., 1995), but significantly less helical structure was observed without TFE: ~0% at pH 7.4, 8% at pH 4.6, and 13% at pH 2.4. The presence of no secondary structure at neutral pH and a little secondary structure at low pH was first observed by NMR and CD (Watts et al., 1990). Subsequently, it was shown that low pH alone was enough to cause a partially folded collapsed structure, presumably due to neutralization of the acidic residues (Uversky et al., 1999). Interestingly, far-UV CD data suggested no structural changes occur between pH 5.5 and 9.5, but that a dynamic, partially collapsed structure(s) forms between pH 3.5 and 5.5. This structure was described as a compact denatured structure, and no long-range NOE signals were detected in NMR spectra. At neutral pH, ProTα could have a charge as large as −44 based on its primary amino acid sequence. Therefore, neutralization of this acidic protein at low pH can lead to structural collapse near the amino acid neutralization sites. Secondary structure was also induced by changes in temperature and the addition of *n*-dodecyltrimethylammonium bromide (Pomco et al., 2001).

Table 1 Helical Structure of Prothymosin α as a Function of pH

pH	TFE	Helical Content (%)
2.4	Yes	69[a]
2.0	No	20[b]
2.4	No	13[a]
2.5	No	~15[c]
4.6	No	8[a]
7.0–7.5	No	0[a,b,c]

[a]Values reported by Gast et al. (1995).
[b]Values reported by Watts et al. (1990).
[c]Values reported by Uversky et al. (1999).

2.3 Structure of the Prothymosin α Carboxy-Terminal Peptide

During apoptosis, ProTα undergoes cleavage by caspases, loses its C-terminal tail (residues 100–109), and sequesters cytochrome *c* (Markova et al., 2003). Due to the apparent immunomodulating role of the ProTα C-terminal tail elucidated by Tsitsilonis and coworkers (Skopeliti et al., 2006), they sought to investigate the structure of the ProTα carboxy-terminus, residues 100–109 (TKKQKTDEDD) which includes the nuclear localization signal (NLS) TKKQKT (Manrow, Sburlati, Hanover, & Berger, 1991). By comparing attenuated total reflectance Fourier-transform infrared (ATF FT-IR) spectra of the C-terminus, truncated versions of it, including the NLS, and a scrambled version of it, they (Skopeliti et al., 2009) observed wave number ranges for the amide I, amide II, and amide III regions of the ATF FT-IR spectra that were most consistent with the formation of antiparallel β-pleated sheets. Their model structures based on similar hairpin peptides deposited in the Protein Data Bank (Berman et al., 2000) were also consistent with these structures (Fig. 1).

2.4 Structure of Prothymosin α in a Complex with the Keap1 Kelch Domain

ProTα and Neh2 have nearly identical Kelch-binding domains which have been investigated by both crystallographic and NMR methods.

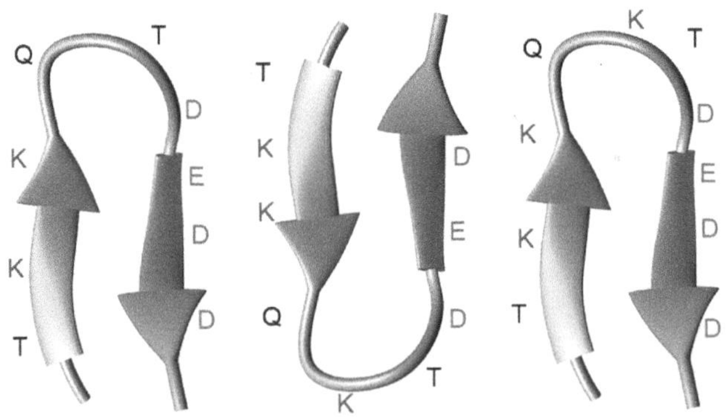

Fig. 1 One of the models proposed for the nuclear localization signal after proteolitic cleavage from thymosin α1 (Skopeliti et al., 2006) to account for CD measurements most consistent with an antiparallel β-pleated sheet. An alternating pattern of linear extended beta structures would also neutralize the charges. Images were created using Chimera (Petterson et al., 2004). (See the color plate.)

The Kelch-binding domain of ProTα (NEENGE) is nearly identical to that of Neh2, DEETGE. Yokoyama and coworkers (Padmanabhan, Nakamura, & Yokoyama, 2008; PDB ID 2Z32) cocrystallized a ProTα peptide, covering residues 39–54, with the Kelch domain of Keap1 and found that it binds to the bottom of the highly basic β-propeller domain, as shown in Fig. 2A. In this complex the ProTα peptide forms a hairpin conformation with a tight β-turn spanning residues E42, E43, N44, and G45, as shown in Fig. 2A and C. The peptide sequence EETGE (residues 78–82) in Nrf2 is equivalent to ProTα residues EENGE (residues 42–46), and the structure of this ProTα peptide is nearly identical to that observed for the ETGE (Padmanabhan et al., 2006; PDB ID 1X2R) and DLG (Tong et al., 2007) motifs of Nrf2 bound to Keap1-DC. Although the crystallized peptide included ProTα amino acids A39–D54 (AQNEENGEQEADNEVD), electron densities were not observed for amino acids 49–54 or the side chains of residues Q47 and E48.

The ProTα hairpin structure is stabilized by both intermolecular and intramolecular interactions (Padmanabhan et al., 2008). As shown in Fig. 2B, the structure is stabilized by a series of charge–charge and hydrogen bonding interactions with the Kelch domain. A side chain carboxyl atom of E42 interacts with the side chain amide group of Q530, and similar interactions between E43 and R415 and R483, and between E46 and N382 and R380, also occur. E43 and E46 side chain oxygen atoms also hydrogen bond with the hydroxyl groups of S508 and S363, respectively. The side chain amide group of Q40 interacts with the hydroxyl group of Y572. As shown in Fig. 2C, the ProTα hairpin is further stabilized internally by hydrogen bonding interactions between the N41 carbonyl oxygen and the G45 amide proton, and between the E46 carbonyl oxygen and the N41 amide group.

Choy and coworkers investigated the structure of ProTα alone and in the complex formed between ProTα and Kelch domain of the Kelch-like ECH-associated protein 1 (Keap1) using NMR methods. First using molecular mechanics simulations (Cino, Wong-ekkabut, Karttunen, & Choy, 2011), they showed that both ProTα and Neh2 have a tendency to form bound-state-like β-turn structures which suggested that preformed structural elements play an important role in a concerted binding and folding mechanism. The simulations also suggest that a short antiparallel beta sheet is formed by residues on either side of this β-turn.

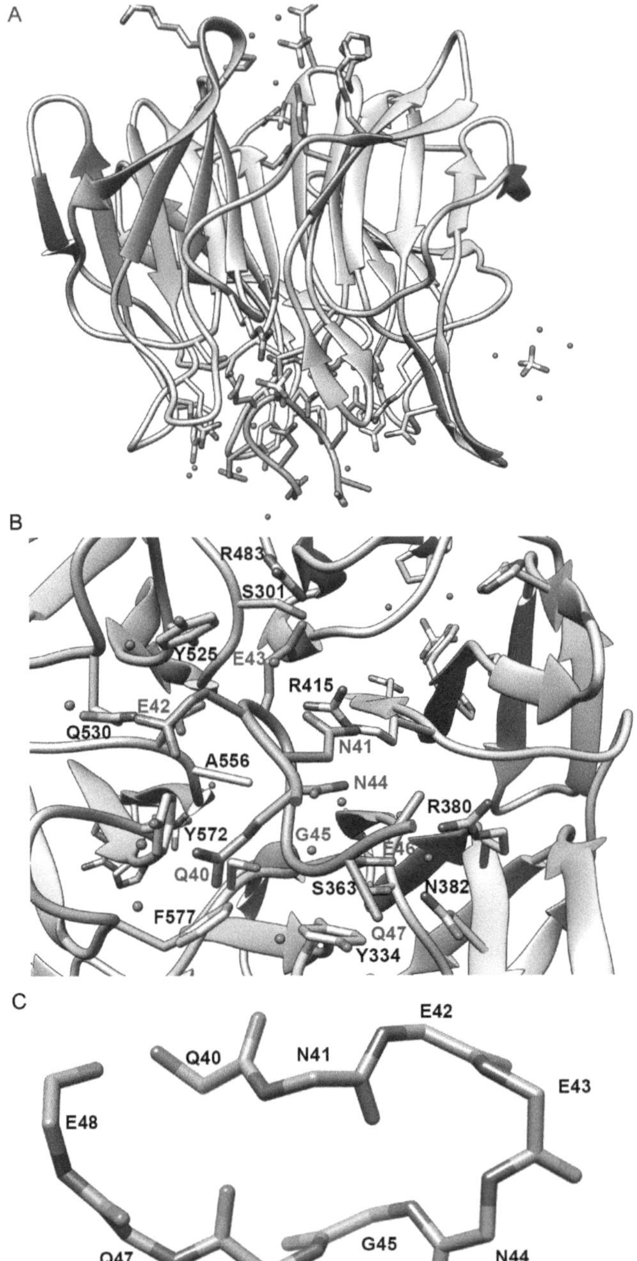

Fig. 2 (A) Crystal structure of ProTα bound to the Kelch domain of Keap1. (B) Structure of ProTα when bound to the Kelch domain. (C) Structure of the ProTα hairpin that interacts with the Kelch domain. Atomic coordinates were obtained from the Protein Data Bank (Berman et al., 2000; PDB ID 2Z32; Padmanabhan et al., 2008). (See the color plate.)

The structure of ProTα in complex with Kelch was also followed by two-dimensional $^{1}H,^{15}N$-HSQC (heteronuclear single quantum coherence) NMR in a series of titration experiments (Khan et al., 2013). Following the changing amide correlations as a function of Kelch:ProTα ratio, amino acids involved in the binding interaction could be followed (Khan et al., 2013). In the absence of Kelch the HSQC spectrum displayed very little dispersion of signals, which is typical of proteins with little native structure. However, upon interaction with Kelch, significant changes to intensity or chemical shifts were noted for the $^{1}H,^{15}N$-HSQC signals of ProTα residues Asn36, Gly37, Asn38, Ala39, Asn40, Asn43, Gly44, Asp49, and Asn50. However, the lack of significant changes in both the HSQC signal dispersion and the secondary structure propensity scores (Marsh, Singh, Jia, & Foreman-Kay, 2006) for each amino acid suggests the presence of only transient structures. The formation of only transient structure, even while binding to a Kelch domain (in solution phase), was further confirmed by relatively small changes to NMR relation rates R_1 and R_2 and heteronuclear $^{1}H,^{15}N$-NOE signals. The largest changes in R_2, though still small, were noted for residues in the region 30–53, consistent with these residues being part of the interaction site.

2.5 Structure of Prothymosin α in the Presence of Zn^{2+}

Uversky et al. (2000) further characterized ProTα in the presence of different divalent counterions. Because the ProTα sequence is nearly 50% acidic residues, neutralization by counterions could counteract charge–charge repulsion between these acid groups and lead to the formation of some secondary structure. However, high concentrations, 250 mM, of Ca^{2+}, Mg^{2+}, Mn^{2+}, Cu^{2+}, and Ni^{2+}, all at pH 7.5, had no measurable effect on the secondary structure of ProTα, as measured by CD or by fluorescence intensity of ANS dye. Each of these metals provided less than 1% helical structure. On the other hand, they found that the presence of a small amount of Zn^{2+}, 12.5 mM, induced a structure with 12.0% helical structure, which is comparable to the helical structure, 10.7%, induced by dropping the pH to 2.0.

2.6 Structure of Parathymosin α

In the course of isolating prothymosin α from rat thymus, a structurally similar protein, parathymosin α (also called Zn^{2+}-binding protein or

Table 2 Comparison of Primary Structures in Selected Regions of Parathymosin α and Prothymosin α

Peptide Regions	Primary Sequence
Prothymosin α (14–25)	KDLKEKKEVVEE
Parathymosin α (14–25)	KDLKEKKDKVEE
Prothymosin α (19–29)	KKEVVEEAENG
Parathymosin α (33–43)	KKDKVEEEENG
Prothymosin α (90–94)	KRVAE
Parathymosin α (83–87)	KRTAE

Bold letters indicate conserved amino acids. *Underlined residues* indicate amino acids repeated in data rows 1 and 2 of this table.
Adapted from Komiyama et al. (1986).

macromolecular translocation inhibitor II), was discovered in Horecker's lab (Haritos, Salvin, Blacher, Stein, & Horecker, 1985). While it was initially shown that the primary structures of parathymosin α and prothymosin α residues 14–20 and 23–25 shared sequence identity, they later described two additional areas of sequence homology between the two proteins, as shown in Table 2.

Ten of the twelve amino acids within parathymosin positions K14–E25 are identical to those present in prothymosin α (and also thymosin α). Interestingly, the 11-amino acid parathymosin sequence from K33 to G43 is also nearly identical to prothymosin residues K19–G29, with 8 conserved amino acids shared between them. Although both proteins are natively unfolded at physiological conditions, later studies showed that secondary structures can be induced in thymosin proteins by the introduction of solvents, micells, or interacting proteins.

Across species, parathymosin α shares about 45% identity with prothymosin α proteins (Table 3) and where bases are not identical they are highly homologous. Despite this, the proteins have very different tissue distributions (Clinton et al., 1989; Haritos et al., 1985). Parathymosin was also shown to bind, through its highly acidic central domain, to the glucocorticoid receptor protein, another intrinsically unstructured protein (Okamoto & Isohashi, 2000, 2005).

Table 3 Comparison of Primary Structures of Parathymosin and Prothymosin α Proteins

Protein	Primary Sequence
Bovine ParaT[a]	-SEKSVEAAAELSAKDLKEKKDKVEEKAGRKERKKEVV-EEEENGAEEEEEE
Rat PataT[a]	-SEKSVEAAAELSAKDLKEKKDKVEEKAGRKERKKEVV-EEEENGAEEEEEE
Human ParaT[a]	MSEKSVEAAAELSAKDLKEKKEKVEEKASRKERKKEVV-EEEENGAEEEEGG
Human ProTα[b]	MSDAAVDTSSEITTKDLKEKKEVVEEAENGRDAPANGNAENEENGEQEADNE
Murine ProTα[c]	MSDAAVDTSSEITTKDLKEKKEVVEEAENGRDAPANGNAQNEENGEQEADNE
Rat ProTα[d]	MSDAAVDTSSEITTKDLKEKKEVVEEAENGRDAPANGNAQNEENGEQEADNE
Bovine ParaT[a]	--------TAE-DGEDDDDGDDEDEEEEEEEDEGPVRVRTAE-- EEDEADPKRQK-TENGASA
Rat ParaT[a]	--------TAE-DGEDDDEGDEEDEEEEEEEDEGPVRKRTAE-- EEDEADPKRQK-TENGASA
Human ParaT[a]	--EEEEEETAE-DGEEEDEGEEEDEEEEEEEDDEGPALKRAAE-- EEDEADPKRQK-TENGASA
Human ProTα[b]	VDEEEEEEEEEGDGEEED-G-DEDEEAESATG----- KRAAEDDEDDDVDTKKQK-TDEDD
Murine ProTα[c]	VDEEEEGGEEEEEEEEGD-GEEEDGDEDEEA-EAPTGKRVAEDDEDDDVD- KKQK-TEEDD
Rat ProTα[d]	VDEEEEEGGEEEEEEEEGDGEEEDGDEDEEA- EAPTGKRVAEDDEDDDVETKKQKKTDEDD

[a]Sequence data from Clinton, Frangou-Lazaridis, Panneerselvam, and Horecker (1989).
[b]Sequence data from Eschenfeldt and Berger, 1986.
[c]Sequence data from Schmidt and Werner (1991).
[d]Sequence data from Frangou-Lazaridis, Clinton, Goodall, and Horecker (1988).

3. STRUCTURES OF Tα1

3.1 Structure of Tα1 in Water

Tα1 was first isolated by Allen Goldstein (Goldstein et al., 1977) and it represents the N-terminal acetylated first 28 amino acids of prothymosin α. More detailed structural information is available for Tα1 alone (Elizondo-Riojas, Chamow, Tuthill, Gorenstein, & Volk, 2011; Grottesi et al., 1998; Nepravishta et al., 2015; Volk, Tuthill, Elizondo-Riojas, & Gorenstein, 2012) than for its precursor ProTα. Like ProTα, as well as thymosins β4 (Czisch, Schleicher, Hörger, Voelter, & Holak, 1993) and β9 (Stoll, Voelter, & Holak, 1997), Tα1 (Grottesi et al., 1998) is an intrinsically disordered protein at neutral pH and body temperature in water. As first shown by Grottesi et al., a lack of sequential NOE signals in NMR spectra or appropriate CD bands suggested that Tα1 lacked any structure. However, they also showed that structure could be induced by using mixed solvents or interactions with membrane-like environments.

3.2 Structure of Tα1 in Mixed Solvents

In the presence of 40% TFE Tα1 adopts a structure containing two structured regions separated by a flexible linker region, as shown in Fig. 3. Grottesi et al. first used NMR data collected on a 400 MHz magnet, together with CD data, to determine that a C-terminal helix was present from residues K17 to E24, even though the structural model they presented appears to be helical from residues 14 to 16 also. It was later found (Elizondo-Riojas et al., 2011) that the C-terminal helix extended from residues T12 to N28. The smaller helix first reported by Grottesi et al. was a result of missing NMR assignments for residues on the C-terminal end, namely, residues E25, E27, and N28, as well as incorrect resonance assignments for the alpha protons of residues near the N-terminal end of this helix, namely, residues I11, T12, T13, and L16. The difference in data between the two studies is due simply to the difference in NMR spectrometers available (400 vs 800 MHz) at the time of each study. The Grottesi study, while not expressly enumerating the number of sequential and medium-range NOEs, appears to have used somewhere between 50 and

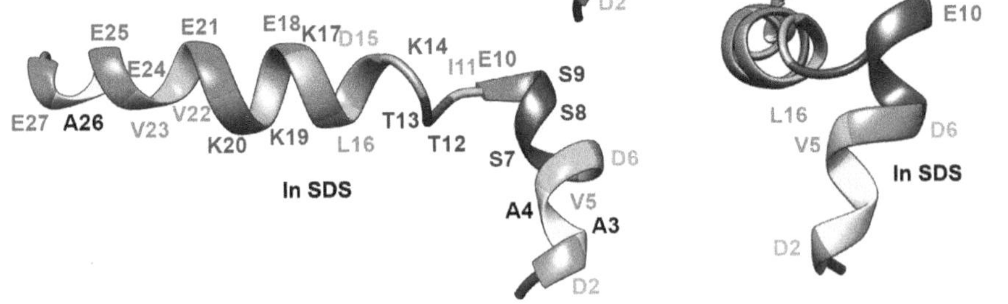

Fig. 3 (A) NMR-derived structures of thymosin α1 protein in 40% TFE solution and in a membrane-like environment with SDS micelles. (B) Same structures rotated 90 degrees. Atomic coordinates were obtained from the Protein Data Bank for thymosin α1 in 40%TFE (PDB ID 2L9I; Elizondo-Riojas et al., 2011) and in SDS (PDB ID 2MNQ; Nepravishta et al., 2015). Images were created using Chimera (Petterson et al., 2004). (See the color plate.)

80 NOE interactions according to their NOE signal plot. The Elizondo-Riojas study provided a robust set NMR assignments (BioMagResBank accession number 17458) of 415 NOE distance restraints, including 106 medium range interactions ($1 < |i-j| < 5$), 139 sequential interactions ($|i-j| = 1$), as well as 170 intraresidue interactions, where *i* and *j* refer to residue (amino acid) number. This structure was also calculated in a box of explicit water and explicit TFE molecules.

As noted previously (Volk et al., 2012), this C-terminal helix of Tα1 has a distribution of charges unlike an amphiphilic helix, which would typically contain both a hydrophilic side and a hydrophobic side. For Tα1, as shown in Fig. 4A, the C-terminal helix contains on one face seven charged residues

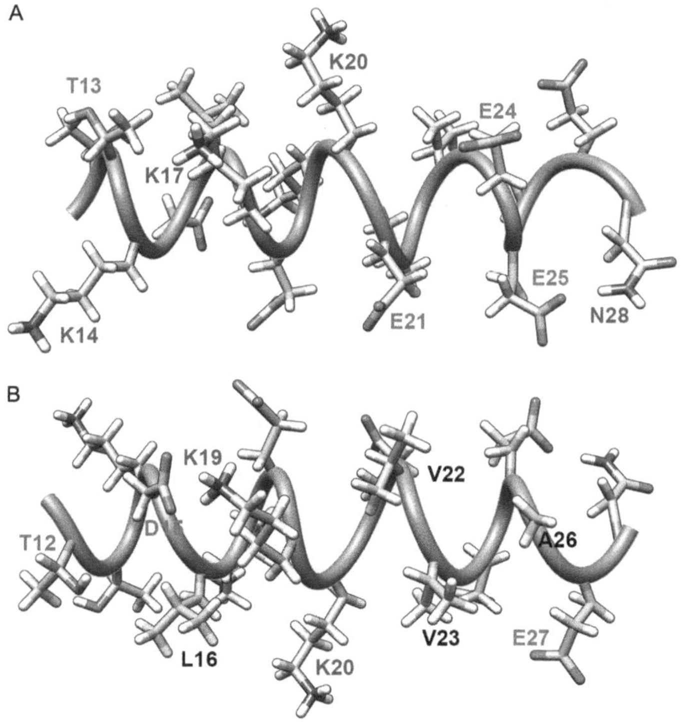

Fig. 4 *Ribbon* and *stick* diagrams of the C-terminal helix of thymosin α1 spanning residues T12–N28. Aliphatic residues are indicated by *black* labels, while positive and negative side chains are indicated by *blue* or *red* labels, respectively. (A) Looking down on the hydrophilic side of the helix. (B) Looking down at the opposite side of the helix. Images were created using Chimera (Petterson et al., 2004), and coordinates were obtained from the Protein Data Bank (PDB ID 2L9I; Elizondo-Riojas et al., 2011). (See the color plate.)

(K14, K17, E18, K20, E21, E24, and E25), residue T13, and the C-terminal amino acid N28. This face represents a typical hydrophilic side of a helix. In contrast, the opposing face of the helix, shown in Fig. 4B, is neither a hydrophobic face nor a hydrophilic face. Although the second side of the helix does contain four hydrophobic amino acids (L16, V22, V23, and A26), it also contains at least three charged amino acids (D15, K19, E27) and T12. K20 could arguably be said to be on this second face.

Like many helices, an asymmetrical distribution of charges is present running along the C-terminal helix axis, such that the N-terminus side of this helix (left halves of Fig. 4A and B) is positively charged, while the C-terminus of this helix (right halves of Fig. 4A and B) maintains a negative charge. Note that all of the positive lysine residues (K14, K17, K19, and K20) occur in the first half of the helix, being offset only by a single negatively charged residue, D15. In contrast, the N-terminal half of the helix contains the remaining 80% of the negative residues (E21, E24, E25, and E27) in the helix. Thus, a dipole charge is formed along the helical axis.

It is probably fair to say the secondary structure at the N-terminus of Tα1 is somewhat in flux, even in 40% TFE. Grottesi et al. first reported that NOE signals arising from residues 5 to 8 indicated the presence of some structure in this region, which they described as a β-turn between residues 5 and 8. In the later study (Elizondo-Riojas et al., 2011), it was found that the first 10 or so residues, in the presence of 40% TFE, form what looks like a distorted α-helix, due to the presence of two double beta-turns. The first double turn, of type (I,I + 1), consists of back-to-back type-I turns spanning residues D2 through V5 and A3 through D6. The second double turn, of type (I,I + 2), consists of turns covering T7–E10 and S9–T12. The split between these two elements, centered about residues D6 and T7, almost looks like the beta turn indicated in the earlier work of Grottesi et al., and if not for this disturbance in structure, this portion of the TFE structure would be helical. It is likely the dynamic motion of the peptide's N-terminus under these conditions is confounding the observed NOE signal strengths and that the structure may be alternating between an alpha helix and less structured orientations.

3.3 Structure of Tα1 in Membrane-Like Environments

In addition to investigating the structure of Tα1 in 40% TFE, Paci and coworkers also investigated the structure of Tα1 in membrane-like environments (Grottesi et al., 1998; Nepravishta et al., 2015). Using both CD and

NMR methods, they investigated the structure of Tα1 in the presence of negatively charged molecules to mimic cell membranes. In the presence of dimyristoylphosphatidylcholine (DMPC) alone, in molar ratios of 3:1 down to 1:3, no structural changes in Tα1 were observed by CD specta (Grottesi et al., 1998). However, increased helical structure was detected when Tα1 was subjected to unilamellar vesicles containing both DMPC and dimyristoylphosphatidic acid (DMPA) in either a 10:1 or 9:1 ratio of DMPC to DMPA. For the DMPC:DMPA experiments, relative concentrations of Tα1:DMPC of 3:1 or 1:1 were found to induce the most structural change. They reported, but did not illustrate, that similar results were found upon addition of sodium dodecylsulfate (SDS) to Tα1 and that higher concentrations of SDS up to 1 M did not induce further structuring.

Very recently, Paci and coworkers (Nepravishta et al., 2015) reported the full proton, carbon, and nitrogen NMR resonance assignments for Tα1 (BioMagResBank accession number 19901), as well as the induced structure (PDB ID 2MNQ) of Tα1, in the presence of 80 mM deuterated SDS at pH 6.5 in H_2O/D_2O (9/1). Both CD and NMR methods were used to characterize these structural changes. As illustrated in the bottom sections of panels A and B of Fig. 3, they found a helical structure for Tα1 that is disrupted in the region between residues S9 and K19. The deposited structures of the C-terminal helix from residues D15 to E25 determined in SDS (Nepravishta et al., 2015) and 40% TFE (Elizondo-Riojas et al., 2011) are very similar to each other, as shown by Fig. 3A and B. Whereas the C-terminal helix begins to disrupt at residue K14 in SDS, it is not disrupted in the 40% TFE structure until residue T12 or so, at which point the helix begins to be stretch out, but is still twisting correctly until S8.

In contrast to the C-terminal helix, the N-terminal region of the Tα1 is quite different between the two structures, as shown in Fig. 3. In SDS, the protein forms a helix from residues D2 to E10, as shown in Fig. 3, bottom sections, while in 40% TFE, two double beta-turns are present, with a disruptive twist around residues V5, D6, T7, and T8 (see Fig. 5 for a detailed view), that destroys the potential alpha helix from residues D2 to E10 (possibly to I11). While these two structural models might in fact reflect a physical difference that exists under these two conditions, it is just as likely to be a result of either the differing number of NOEs used in each structure calculation, or the confounding effects of motion and concomitant population averaging of NOE signals on the detected NOE signal intensities that were used to derive NMR-based distance restraints.

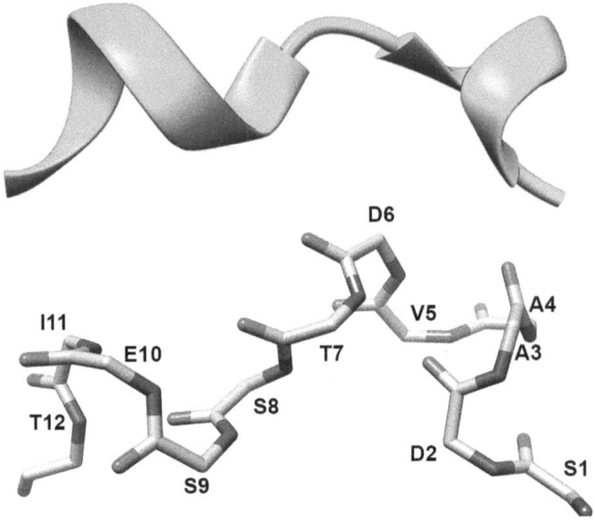

Fig. 5 The N-terminal domain of thymosin α1 in 40% TFE. The N-terminal domain of Tα1 in TFE contains two double beta-turns and nearly forms a complete helix. However, an unusual kink along residues 6 and 7 disrupts the helical structure. Atomic coordinates were obtained from the Protein Data Bank (PDB ID 2L9I; Elizondo-Riojas et al., 2011). (See the color plate.)

4. STRUCTURES OF BETA THYMOSIN PROTEINS

4.1 Structure of Thymosin β9

Like most thymosin proteins, calf thymosin β9, first isolated by Horecker's group (Hannappel, Davoust, & Horecker, 1982), lacks structure under normal conditions, and a human homolog, termed thymosin β10, is known (Erickson-Viitanen, Ruggieri, Natalini, & Horeker, 1983). It contains 41 amino acids of which 32 are identical to those of thymosin β4, as shown in Table 4. While no structure was evident in both CD and NMR spectra acquired on samples of thymosin β9 in water, in the presence of 40% deuterated 1,1,1,3,3,3-hexafluoropropanol (HFP) in 2.5 mM sodium phosphate at pH 5.1, a helical structure was found (Stoll et al., 1997; PDB ID 1HJ0), as illustrated in Fig. 6. The structure contains two alpha helices, one spanning residues P4 through T27 and a second one spanning residues E32–K41, separated by a bend containing residues L28, P29, T30, and K31. In contrast to Tα1, the helices in thymosin β9 are both more acidic on the N-terminal ends and more basic on the C-terminal ends.

Table 4 Comparison of Primary Structures of Beta Thymosin Proteins

Protein	**Primary Sequence**
Thymosin β9[a]	ADKPDLGEINSFDKA**KLKKTET**QEKNTLPTEETIEQEKQAK
Thymosin β10[b]	ADKPDMGEIASFDKA**KLKKTET**QEKNTLPTKETIEQEKRSEIS
Thymosin β4[c]	SDKPDMAEIEKFDKS**KLKKTET**QEKNPLPSKETIEQEKQAGES
Thymosin β11[d]	SDKPDLQEVASFDKT**KLKKTET**QEKNPLPTKETIEQEKQA--S
Thymosin β12[e]	SDKPDISEVTSFDKT**KLKKTET**QEKNPLPSKETIEQEKAAATS
Thymosin β15[f]	SDKPDLSEVEKFDRS**KLKK**-NTEEKNTLPSKETIQQEKECVQTS

[a]Thymosin β9 sequence obtained from the Protein Data Bank (PDB ID 1HJ0; Stoll et al., 1997).
[b]Human thymosin β10 sequence obtained from McCreary, Kartha, Bell, and Toback (1988).
[c]Human thymosin β4 sequence obtained from Low, Hu, and Goldstein (1981).
[d]Sequence from Erickson-Viitanen et al. (1983).
[e]Thymosin β12 Sequence from Low, Liu, and Jou (1992).
[f]Human thymosin β15 sequence obtained from Banyard, Hutchinson, and Zetter (2007).

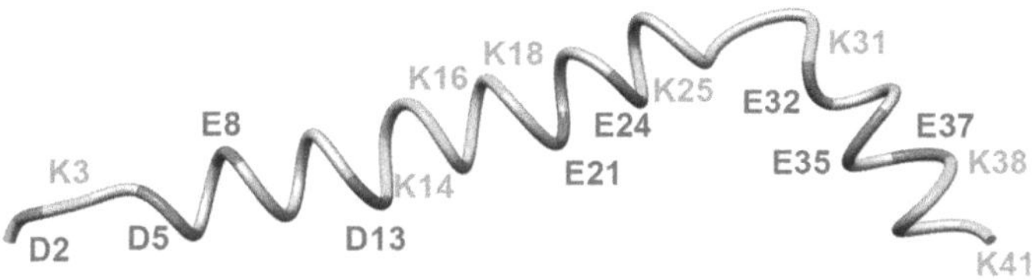

Fig. 6 NMR-derived structure of thymosin β9 in hexafluoropropanol. Atomic coordinates were obtained from the Protein Data Bank (PDB ID 1HJ0; Stoll et al., 1997). (See the color plate.)

4.2 Structure of Human Thymosin β10

Human thymosin β10 is an analog of calf thymosin β9 that shares 87.8% sequence identity with it (Erickson-Viitanen et al., 1983; Rho et al., 2005), as shown in Table 4. Like thymosin β4, it binds to G-actin monomers to prevent formation of F-actin (Hannappel & Wartenberg, 1993; Heintz, Reichert, Mihelic, Voelter, & Faulstich, 1993; Huff, Müller, & Hannappel, 1997; Huff, Zerzawy, & Hannappel, 1995; Jean et al., 1994; Yu, Lin, Morrison-Bogorad, Atkinson, & Yin, 1993). Although there are no crystallographic or NMR-derived structures for thymosin β10, due to its high identity with thymosin β9, molecular mechanics calculations were performed using homology modeling for thymosin β10 and for several mutants (Rho et al., 2005). During 200 ps of molecular dynamics simulations the thymosin β10 protein and mutants all converted to structures with two turns between the two helices. It could be argued that this extra bending is due to a lack of fluoroethanol in the calculations; however, the thymosin β9 protein maintained its structure over similar calculation conditions, suggesting that the double bend between the helices is a real feature of thymosin β10.

4.3 Structure of Thymosin β4

The initial studies of the 43 amino acid thymosin β4 peptide used NMR-based methods and CD to determine if any native structure was present (Czisch et al., 1993). Although salts had little effect on the structure of thymosin β4, temperature played a major role. NMR signals indicative of helical structure, including strong HN–HN NOEs and medium-range interactions, suggested the presence of helices from residues 5 to 19 and from residues 30 to 37 at low temperatures (1°C and 4°C) but that the C-terminal helix was melted out by warming to 14°C. CD measurements suggested that 15% helix was present at 4°C but that only 7% helix was present at 20°C.

Subsequently, the structure of thymosin β4 in mixed solvents was investigated (Zarbock et al., 1990). By taking advantage of slightly different chemical shifts in two different solvent systems, namely, 50% HFP or 60% TFE, Zarbock et al. was able to derive nearly complete NMR resonance assignments (BioMagResBank accession # 1065) for all protons of thymosin β4. Despite the slightly differing chemical shifts, essentially identical NOE patterns collected from samples in the two solvents suggested a common structure. Based on the sequential HN–NH NOEs and several medium-range NOEs, such as $\alpha N(i,i+3)$, $\alpha N(i,i+4)$, and $\alpha\beta(i,i+3)$, they found helical structure extending from residues P4 to K16 and from residues S30 to A40. Due to the lack of long-range NOE signals, which is generally true for all NMR-based studies of thymosin proteins, the relative orientation of the two helices could not be determined. Like thymosin β9, the first helix is more acidic on the N-terminal side while being more basic on the C-terminal end.

4.4 Solution Phase Structure of Thymosin β4 Interacting with Actin

Thymosin β4 contains an evolutionarily conserved actin monomer-binding motif called WASP homology 2, or WH2 (Paunola, Mattila, & Lappalainen, 2002), and it is the most abundant actin-sequestering protein found in platelets (Safer, Elzinga, & Nachmias, 1991; Weber, Nachmias, Pennise, Pring, & Safer, 1992) and neutrophils (Cassimeris, Safer, Nachmias, & Zigmond, 1992). Due to the lack of any crystallographic data for actin–thymosin β4 (or thymosin β9)-binding interactions, the structure of thymosin β4 in complex with monomeric G-actin was first investigated using solution phase NMR methods (Domanski et al., 2004). Using a combination of uniformly 15N-labeled thymosin β4 and selectively 15N-labeled thymosin β4, they observed NOE interactions at 2°C suggestive of partial helical structure from residues D5 to K16, as had been previously reported (Czisch et al., 1993), as well as chemical shifts sufficiently different from random coil values to suggest a weak helical tendency from residues K31 to E37. Upon addition of the monomeric actin to thymosin β4, it was found (Domanski et al., 2004) that thymosin β4 adopted a structure with helices from residues D5 to L17 and from K31 to A40, with an extended conformation between them from residues K18 to N26. The solution phase studies at cold temperatures, in mixed solvents, or bound to monomeric actin represented very similar results: a C-terminal helix from about residues P4 or D5 to K16 or L17 and a second helix from residue S30 or K31 to K37 or A40. Note that the critical actin-binding motif LKKTET (residues 17–22) is situated between the two helices.

4.5 Crystallographic Structures of Thymosin β4 Chimeras Interacting with Actin

Due to difficulties in crystallizing thymosin β4 alone or in complex with actin, the first crystallographic data for thymosin β4 came in the form of a complex formed between actin alpha and a chimeric protein consisting of the gelsolin domain I and the C-terminal portion of mouse thymosin b4 (Irobi et al., 2004; PDB ID 1T44) and a complex formed between actin and ciboulot (PDB ID 1SQK; Hertzog et al., 2004). As shown in Fig. 7, gelsolin (green) and the C-terminal portion of mouse thymosin β4 (goldenrod) form a complex in which a portion of mouse thymosin β4 (corresponding to residues E21–N26 of human thymosin β4) forms an extended structure. The first three bases of this region contain part of the critical actin-binding sequence. The remaining residues of thymosin β4, corresponding to human thymosin β4 residues P27–Q39, forms a helix. Importantly, this structure showed that thymosin β4 sequesters actin monomers by capping both ends of actin, and that profilin–thymosin β4 exchange is mediated by binding sites that overlap slightly. Robinson's group (Aguda, Xue, Irobi, Preat, & Robinson, 2006) followed up this seminal work by also providing structures of actin bound to gelsolin-N-WASP domain 2 (PDB ID 2FF3) and to gelsolin–ciboulot (PDB ID 2FF6) and showed that the C-terminal ends of thymosin β4 and WH2 motifs play different structural roles when binding to actin.

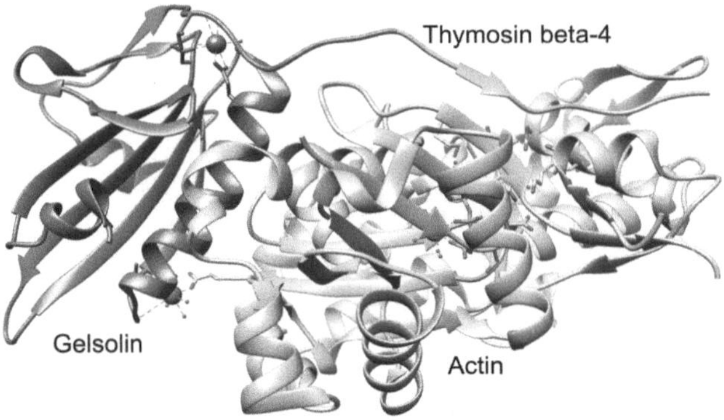

Fig. 7 Crystal structure of alpha actin bound to a gelsolin–thymosin beta-4 C-terminal helix chimera. The helix is the minus-end capping helix, and it competes with DNAse I binding at this site. Atomic coordinates were obtained from the Protein Data Bank (PDB ID 1T44; Irobi et al., 2004). (See the color plate.)

While monomeric thymosin β4 sequesters actin in the monomeric form tandem repeats of thymosin β4 promotes actin elongation (Fig. 8). As shown in Fig. 8A, the complex formed between actin monomers and thymosin β4 (here crystallized as an actin–thymosin-β4 hybrid) exhibits two terminal helices on thymosin β4 that cap both the barbed and the pointed faces of actin, thereby interfering with multimerization and fiber elongation (PDB ID 4PL7; Xue, Leyrat, Grimes, & Robinson, 2014). However, as shown in Fig. 8B depicting the complex formed (PDB ID 4PL8; Xue, Leyrat, Grimes, & Robinson, 2014) when a peptide with tandem thymosin β4 units binds to actin, peptides with tandem thymosin β4 units promote the elongation of monomeric actin into longer fibers, in this case crystallized as a dimeric form. The nucleotide-binding pocket is more open, and the C-terminal helix of thymosin β4 is disrupted (Fig. 8B, far right side). In this manner, proteins with multiple WH2/thymosin β4 domains promote the formation of actin fibers rather than sequestering monomeric actin.

A series of related structures may be of interest to the reader, including actin bound to: Cobl segments (PDB ID 3TU5; Durer et al., 2012), RPEL domains (PDB ID 4B1X, 4B1Y, 2V51; Mouilleron, Guettler, Langer, Treisman, & McDonald, 2008; Mouilleron, Wiezlak, O'Rielly, Treisman, & McDonald, 2012), drosophila ciboulot (PDB ID 3U9Z; Didry et al., 2012), or drosophila ciboulot–thymosin β4 constructs (PDB ID 3U9D and 3U8X; Didry et al., 2012).

The tandem thymosin β4 structure depicted in Fig. 9, which is a closer look at part of thymosin β4 shown in Fig. 8B, details many of the

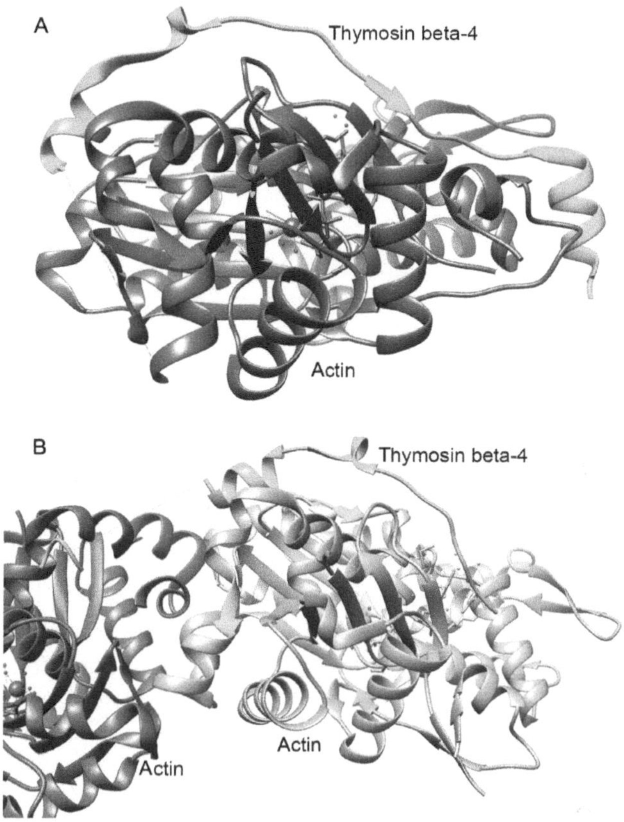

Fig. 8 (A) Crystal structure of an actin thymosin beta-4 hybrid (PDB ID 4PL7). (B) Crystal structure (PDB ID 4PL8) of actin bound to a peptide construct containing tandem thymosin β4 units. The two actin proteins are shown in *purple* or *goldenrod*, and the tandem thymosin β4 peptide is colored *cyan*. While much of the second thymosin β4 unit (D5–E24) was detected binding to the actin unit on the *right*, the final C-terminal helix is disrupted. (See the color plate.)

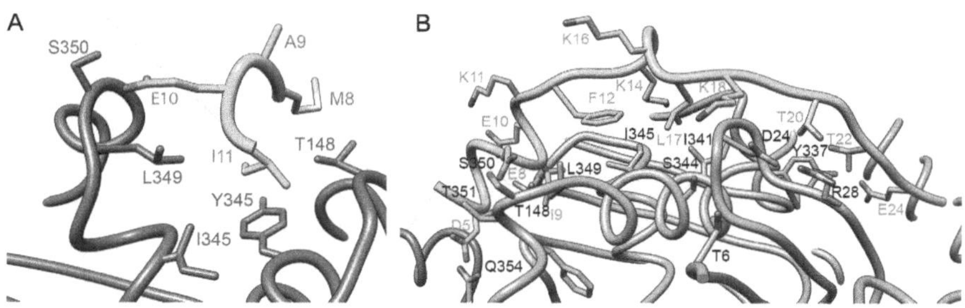

Fig. 9 Expanded views of the actin dimer bound to a peptide containing tandem thymosin β4 units. (A) The interactions of the first thymosin β4 unit. (B) The interactions of the second thymosin β4 unit. The peptides have individually been renumbered to match the thymosin β4 sequence for convenience. Atomic coordinates were obtained from the Protein Data Bank (PDB ID 4PL8; Xue et al., 2014). (See the color plate.)

intermolecular interactions which stabilize the N-terminal portion of thymosin β4. Thymosin β4 glutamate E8 interacts with the hydroxyl groups of both S350 and T351 in actin, glutamate E10 interacts with the T148 hydroxyl group of actin, and glutamate E24 interacts with the hydroxyl group of Y337. The positive side chain of lysine K14 also interacts with actin glutamate E334. In addition to these interactions, hydrophobic packing of some thymosin side chains, for example, I9, F12, and L17, with actin side chains, including I341, I345, and Y337, help to stabilize the N-terminal portion of thymosin β4 in these structures. It has been noted that the FxKxK motif of thymosin β4 (note F12 and K14 mentioned earlier) is crucial for actin binding (Didry et al., 2012).

5. CONCLUSIONS

The thymosin proteins are all short, highly charged, intrinsically unstructured proteins under natural conditions. However, structure can be induced in many of the thymosin proteins by providing charge neutralization at low pH or by the addition of Zn^{2+} ions, organic reagents such as TFE, HFP, or *n*-dodecyltrimethylammonium bromide, or interactions with their natural binding partner proteins. The differing structures of thymosin proteins have been studied by circular dichroism, nuclear magnetic resonance, and crystallographic methods in order to better understand the role of these proteins. This review covered much of the major findings resulting from such structural biology studies.

ACKNOWLEDGMENTS

This work was supported, in part, by Grants from the National Institutes of Health (U54CA1516, 1RO1CA16027, UL1 TR000371, and 1RO1DE024392-01).

REFERENCES

Aguda, A. H., Xue, B., Irobi, E., Preat, T., & Robinson, R. C. (2006). The structural basis of actin interaction with multiple WH2/beta-thymosin motif-containing proteins. *Structure, 14*, 469–476.

Andreone, P., Cursaro, C., Gramenzi, A., Margotti, M., Ferri, E., Talarico, S., et al. (2001). In vitro effect of thymosin-α1 and interferon-α on Th1 and Th2 cytokine synthesis in patients with chronic hepatitis C. *Journal of Viral Hepatitis, 8*, 194–201.

Banyard, J., Hutchinson, L. M., & Zetter, B. R. (2007). Thymosin beta-NB is the human isoform of rat thymosin beta15. *Annals of the New York Academy of Sciences, 1112*, 286–296.

Berman, H. M., Westbrook, J., Feng, Z., Gilliland, G., Bhat, T. N., Weissig, H., et al. (2000). The Protein Data Bank. *Nucleic Acids Research, 28*, 235–242.

Bozza, S., Gaziano, R., Bonifazi, P., Zelante, T., Pitzurra, L., Montagnoli, C., et al. (2007). Thymosin alpha1 activates the TLR9/MyD88/IRF7-dependent murine cytomegalovirus

sensing for induction of anti-viral responses in vivo. *International Immunology, 19*, 1261–1270.
Cassimeris, L., Safer, D., Nachmias, V. T., & Zigmond, S. H. (1992). Thymosin beta 4 sequesters the majority of G-actin in resting human polymorphonuclear leukocytes. *Journal of Cell Biology, 119*, 1261–1270.
Cino, E. A., Wong-ekkabut, J., Karttunen, M., & Choy, W.-Y. (2011). Microsecond molecular dynamics simulations of intrinsically disordered protein involved in the oxidative stress response. *PLoS One, 6*, e27371.
Clinton, M., Frangou-Lazaridis, M., Panneerselvam, C., & Horecker, B. L. (1989). The sequence of human parathymosin deduced from a cloned human kidney cDNA. *Biochemical and Biophysical Research Communications, 158*, 855–862.
Cordero, O. J., Sarandeses, C. S., Lopez, J. L., & Nogueira, M. (1992). On the anomalous behavior on gel-filtration and SDS-electrophoresis of prothymosin-alpha. *Biochemistry International, 28*, 1117–1124.
Czisch, M., Schleicher, M., Hörger, S., Voelter, W., & Holak, T. A. (1993). Conformation of thymosin beta 4 in water determined by NMR spectroscopy. *European Journal of Biochemistry, 218*, 335–344.
Didry, D., Cantrelle, F. X., Husson, C., Roblin, P., Moorthy, A. M., Perez, J., et al. (2012). How a single residue in individual beta-thymosin/WH2 domains controls their functions in actin assembly. *EMBO Journal, 31*, 1000–1013.
Domanski, M., Hertzog, M., Coutant, J., Gutsche-Perlroizen, R., Bontems, F., Carlier, M.-F., et al. (2004). Coupling of folding and binding of thymosin β4 upon interaction with monomeric actin monitored by nuclear magnetic resonance. *Journal of Biological Chemistry, 279*, 23637–23645.
Durer, Z. A., Kudryashov, D. S., Sawaya, M. R., Altenbach, C., Hubbel, W., & Reisler, E. (2012). Structural states and dynamics of the D-loop in actin. *Biophysical Journal, 103*, 930–939.
Elizondo-Riojas, M.-A., Chamow, S. M., Tuthill, C. W., Gorenstein, D. G., & Volk, D. E. (2011). NMR structure of human thymosin alpha-1. *Biochemical and Biophysical Research Communications, 416*, 356–361.
Erickson-Viitanen, S., Ruggieri, S., Natalini, P., & Horeker, B. L. (1983). Thymosin β10, a new analogue of thymosin β4 in mammalian tissues. *Archives of Biochemistry and Biophysics, 225*, 407–413.
Eschenfeldt, W. H., & Berger, S. L. (1986). The human prothymosin α gene is polymeric and induced upon growth stimulation: Evidence using a clode cDNA. *Proceedings of the National Academy of Sciences of the United States of America, 83*, 9403–9407.
Frangou-Lazaridis, M., Clinton, M., Goodall, G. J., & Horecker, B. L. (1988). Prothymosin alpha and parathymosin: Amino acid sequences deduced from the cloned rat spleen cDNAs. *Archives of Biochemistry and Biophysics, 263*, 305–310.
Gast, K., Damaschun, H., Eckert, K., Schultz-Forster, K., Maurer, H. R., Müller-Frohne, M., et al. (1995). Prothymosin alpha: A biologically active protein with random coil conformation. *Biochemistry, 34*, 13211–13218.
Goldstein, A. L., Low, T. L., McAdoo, M., McClure, J., Thurman, G. B., Rossio, J. L., et al. (1977). Thymosin α1: Isolation and sequence analysis of an immunologically active thymosin polypeptide. *Proceedings of the National Academy of Sciences of the United States of America, 74*, 725–729.
Goodall, G. J., Dominguez, F., & Horecker, B. L. (1986). Molecular cloning of cDNA for human prothymosin alpha. *Proceedings of the National Academy of Sciences of the United States of America, 83*, 8926–8928.
Grottesi, A., Sette, M., Palamara, A. T., Rotilio, G., Garaci, E., & Paci, M. (1998). The conformation of peptide thymosin α1 in solution and in a membrane-like environment by circular dichroism and NMR spectroscopy. A possible model for its interaction with the lymphocyte membrane. *Peptides, 19*, 1731–1738.

Hannappel, E., Davoust, S., & Horecker, B. L. (1982). Thymosins β8 and β9: Two new peptides isolated from calf thymus homologous to thymosin β4. *Proceedings of the National Academy of Sciences of the United States of America*, *79*, 1708–1711.

Hannappel, E., & Wartenberg, F. (1993). Actin-sequestering ability of thymosin β4, thymosin β4 fragments, and thymosin β4-like peptides as assessed by the DNAse I inhibition assay. *Biological Chemistry Hoppe-Seyler*, *374*, 117–122.

Haritos, A. A., Salvin, B. B., Blacher, R., Stein, S., & Horecker, B. L. (1985). Parathymosin alpha: A peptide from rat tissues with structural homology to prothymosin alpha. *Proceedings of the National Academy of Sciences of the United States of America*, *82*, 1050–1053.

Haritos, A. A., Tsolas, O., & Horecker, B. L. (1984). Distribution of prothymosin alpha in rat tissues. *Proceedings of the National Academy of Sciences of the United States of America*, *81*, 1391–1393.

Heintz, D., Reichert, A., Mihelic, M., Voelter, W., & Faulstich, H. (1993). Use of bimanyl actin derivative (TMB-actin) for studying complexation of beta-thymosins. Inhibition of actin polymerization by thymosin beta 9. *FEBS Letters*, *329*, 9–12.

Hertzog, M., van Heijenoort, C., Didry, D., Gaudier, M., Coutant, J., Gigant, B., et al. (2004). The beta-thymosin/WH2 domain; structural basis for the switch from inhibition to promotion of actin assembly. *Cell*, *117*, 611–623.

Huff, T., Müller, C. S., & Hannappel, E. (1997). C-terminal truncation of thymosin β10 by an intracellular protease and its influence on the interaction with G-actin studied by ultrafiltration. *FEBS Letters*, *414*, 39–44.

Huff, T., Zerzawy, D., & Hannappel, E. (1995). Interactions of β-thymosins, thymosin β4-sulfoxide, and N-terminally truncated thymosin β4 with actin studied by equilibrium centrifugation, chemical cross-linking and viscometry. *European Journal of Biochemistry*, *230*, 650–657.

Iino, S., Toyota, J., Kumada, H., Kiyosawa, K., Kakumu, S., Sata, M., et al. (2005). The efficacy and safety of thymosin alpha-1 in Japanese patients with chronic hepatitis B; results from a randomized clinical trial. *Journal of Viral Hepatitis*, *12*, 300–306.

Irobi, E., Aguda, A. H., Larsson, M., Guerin, C., Yin, H. L., Burtnick, L. D., et al. (2004). Structural basis of actin sequestration by thymosin-B4: Implications for arp2/3 activation. *EMBO Journal*, *23*, 3599–3608.

Jean, C., Rieger, K., Blanchoin, L., Carlier, M. F., Lenfant, M., & Pantaloni, D. (1994). Interaction of G-actin with thymosin beta 4 and its variants thymosin beta 9 and thymosin beta met9. *Journal of Muscle Research and Cell Motility*, *15*, 278–286.

Khan, H., Cino, E. A., Brickenden, A., Fan, J., Yang, J., & Choy, W. Y. (2013). Fuzzy complex formation between the intrinsically disordered prothymosin α and the Kelch domain of Keap1 involved in the oxidative stress response. *Journal of Molecular Biology*, *425*, 1011–1027.

Komiyama, T., Pan, L.-X., Haritos, A. A., Wideman, J. W., Pan, Y.-C. E., Chang, M., et al. (1986). The primary structure of rat parathymosin. *Proceedings of the National Academy of Sciences of the United States of America*, *83*, 1242–1245.

Kullavanuaya, P., Treeprasertsuk, S., Thong-Ngam, D., Chaermthai, K., Gonlachanvit, S., & Suwanagool, P. (2001). The combined treatment of interferon alpha-2a and thymosin alpha 1 for chronic hepatitis C: The 48 weeks end of treatment results. *Journal of the Medical Association of Thailand*, *84*, 462–468.

Low, T. L. K., Hu, S.-K., & Goldstein, A. L. (1981). Complete amino acid sequence of bovine thymosin beta 4: A thymic hormone that induces terminal deoxynucleotidyl transferase activity in thymocyte populations. *Proceedings of the National Academy of Sciences of the United States of America*, *78*, 1162–1166.

Low, T. L. K., Liu, D. T.-W., & Jou, J.-H. (1992). Primary structure of thymosin β12, a new member of the β-thymosin family isolated from perch liver. *Archives of Biochemistry and Biophysics*, *293*, 32–39.

Manrow, R. E., Sburlati, A. R., Hanover, J. A., & Berger, S. L. (1991). Nuclear targeting of prothymosin alpha. *Journal of Biological Chemistry, 266*, 3916–3924.

Markova, O. V., Evstafieva, A. G., Mansurova, S. E., Moussine, S. S., Palamarchuk, L. A., Pereverzev, M. O., et al. (2003). Cytochrome c is transformed from anti- to pro-oxidant when interacting with truncated oncoprotein prothymosin alpha. *Biochimica et Biophysica Acta, 1557*, 109–117.

Marsh, J. A., Singh, V. K., Jia, Z., & Foreman-Kay, J. D. (2006). Sensitivity of secondary structure propensities to sequence differences between alpha- and gamma-synuclein: Implications for fibrillation. *Protein Science, 15*, 2795–2804.

McCreary, V., Kartha, S., Bell, G. I., & Toback, F. G. (1988). Sequence of a human kidney cDNA clone encoding thymosin beta 10. *Biochemical and Biophysical Research Communications, 152*, 862–866.

Mosoian, A., Teixeira, A., Burns, C. S., Sander, L. E., Gusella, G. L., He, C., et al. (2010). Prothymosin-a inhibits HIV-1 via toll-like receptor 4-mediated type I interferon induction. *Proceedings of the National Academy of Sciences of the United States of America, 107*, 10178–10183.

Mosoian, A., Teixeira, A., High, A. A., Christian, R. E., Hunt, D. F., Shabanowitz, J., et al. (2006). Novel function of prothymosin alpha as a potent inhibitor of human immunodeficiency virus type 1 gene expression in primary macrophages. *Journal of Virology, 80*, 9200–9206.

Mouilleron, S., Guettler, S., Langer, C. A., Treisman, R., & McDonald, N. Q. (2008). Molecular basis for G-actin binding to RPEL motifs from the serum response factor coactivator MAL. *EMBO Journal, 27*, 3198–3208.

Mouilleron, S., Wiezlak, M., O'Rielly, N., Treisman, R., & McDonald, N. Q. (2012). Structures of the Phactr1 RPEL domain and RPEL motif complexes with G-actin reveal the molecular basis for actin binding cooperativity. *Structure, 20*, 1960–1970.

Nepravishta, R., Mandaliti, W., Eliseo, T., Vallebona, P. S., Pica, F., Garaci, E., et al. (2015). Thymosin α1 inserts N terminus into model membranes assuming a helical conformation. *Expert Opinion on Biological Therapy, 15*(Suppl. 1), S71–S81.

Okamoto, K., & Isohashi, F. (2000). Purification and primary structure of a macromolecular-translocation inhibitor II of glucocorticoid-receptor binding to nuclei from rat liver. Inhibitor II is the 11.5-kDa $Zn2^{+}$-binding protein (parathymosin). *European Journal of Biochemistry, 267*, 155–162.

Okamoto, K., & Isohashi, F. (2005). Macromolecular translocation inhibitor II ($Zn2^{+}$-binding protein, parathymosin) interacts with the glucocorticoid receptor and enhanced transcription in vivo. *Journal of Biological Chemistry, 280*, 36986–36993.

Padmanabhan, B., Nakamura, Y., & Yokoyama, S. (2008). Structural analysis of the complex of Keap1 with a prothymosin alpha peptide. *Acta Crystallographica Section F, Structural Biology and Crystallization Communications, 64*, 233–238.

Padmanabhan, B., Tong, K. I., Ohta, T., Nakamura, Y., Scharlock, M., Ohtsuji, M., et al. (2006). Structural basis for defects of Keap1 activity provoked by its point mutations in lung cancer. *Molecular Cell, 21*, 689–700.

Paunola, E., Mattila, P. K., & Lappalainen, P. (2002). WH2 domain: A small, versatile adapter for actin monomers. *FEBS Letters, 513*, 92–97.

Petterson, E. F., Goddard, T. D., Huang, C. C., Couch, G. S., Greenblatt, D. M., Meng, E. C., et al. (2004). UCSF Chimera—A visualization system for exploratory research and analysis. *Journal of Computational Chemistry, 25*, 1605–1612.

Pomco, C., Suarez, M. J., Czarnecki, J., Ruso, J. M., Sarmiento, F., & Prieto, G. (2001). Secondary structure of prothymosin alpha evidence for conformational transitions induced by changes in temperature and concentration of n-dodecyltrimethylammonium bromide. *European Biophysics Journal, 30*, 242–249.

Rho, S. B., Lee, K. W., Chun, T., Lee, S.-H., Park, K., & Lee, J.-H. (2005). The identification of apoptosis-related residues in the human thymosin β-10 by mutational analysis and computational modeling. *Journal of Biological Chemistry*, *280*, 34003–34007.

Romani, L., Bistoni, F., Perruccio, K., Montagnoli, C., Gaziano, R., Bozza, S., et al. (2006). Thymosin α1 activates dendritic cell tryptophan catabolism and established a regulatory environment for balance of inflammation and tolerance. *Blood*, *108*, 2265–2274.

Safer, D., Elzinga, M., & Nachmias, V. T. (1991). Thymosin beta 4 and Fx, an actin-sequestering peptide, are indistinguishable. *Journal of Biological Chemistry*, *266*, 4029–4032.

Schmidt, G., & Werner, D. (1991). Nucleotide synthesis of the murine prothymosin α cDNA and its deduced primary and secondary sequence. *Biochimica et Biophysica Acta*, *1088*, 442–444.

Segal, B. H., & Walsh, T. J. (2006). Current approaches to diagnosis and treatment of invasive Aspergillosis. *American Journal of Respiratory and Critical Care Medicine*, *173*, 707–717.

Skopeliti, M., Iconomidou, V. A., Derhovanessian, E., Pawelec, G., Voelter, W., Kalbacher, H., et al. (2009). Prothymosin a immunoactive carboxy-terminal peptide TKKQKTDEDD stimulates lymphocyte reactions, induces dendritic cell maturation and adopts a β-sheet conformation in a sequence-specific manner. *Molecular Immunology*, *46*, 784–792.

Skopeliti, M., Voutsas, I. F., Klimentzou, P., Tsiatas, M. L., Beck, A., Bamias, A., et al. (2006). The immunologically active site of prothymosin alpha is located at the carboxy-terminus of the polypeptide. Evaluation of its in vitro effects in cancer patients. *Cancer Immunology, Immunotherapy*, *55*, 1247–1257.

Stoll, R., Voelter, W., & Holak, T. A. (1997). Conformation of thymosin beta 9 in water/fluoroalcohol solution determined by NMR spectroscopy. *Biopolymers*, *41*, 623–634.

Tong, K. I., Padmanbhan, B., Kobayashi, A., Shang, C., Hirotsu, Y., Yokoyama, S., et al. (2007). Different electrostatic potentials define ETGE and DLG Motifs as Hinge and Latch in oxidative stress response. *Molecular and Cellular Biology*, *27*, 7511–7521.

Uversky, V. N., Gillespie, J. R., Millet, I. S., Khodyakova, A. V., Vasilenko, R. N., Vasiliev, A. M., et al. (2000). Zn^{2+}-mediated structure formation and compaction of the "Natively Unfolded" human prothymosin α. *Biochemical and Biophysical Research Communications*, *267*, 663–668.

Uversky, V. N., Gillespie, J. R., Millet, I. S., Khodyakova, A. V., Vasiliev, A. M., Chernovskaya, T. V., et al. (1999). Natively unfolded human prothymosin alpha adopts partially folded collapsed conformation at acidic pH. *Biochemistry*, *38*, 15009–15016.

Volk, D. E., Tuthill, C. W., Elizondo-Riojas, M.-A., & Gorenstein, D. G. (2012). NMR structural studies of thymosin α1 and β-thymosins. *Annals of the New York Academy of Sciences*, *1270*, 73–78.

Watts, J. D., Cary, P. D., Sautiera, P., & Crane-Robinson, C. (1990). Thymosins: Both nuclear and cytoplasmic. *European Journal of Biochemistry*, *192*, 561–643.

Weber, A., Nachmias, V. T., Pennise, C. R., Pring, M., & Safer, D. (1992). Interaction of thymosin beta 4 with muscle and platelet actin: Implications for actin sequestration in resting platelets. *Biochemistry*, *31*, 6179–6185.

Xue, B., Leyrat, C., Grimes, J. M., & Robinson, R. C. (2014). Structural basis of thymosin-beta 4/profilin exchange. *Proceedings of the National Academy of Sciences of the United States of America*, *111*, E4596–E4605.

You, J., Zhuang, L., Cheng, H.-Y., Yan, S.-M., Lan, Y., Huang, J.-H., et al. (2006). Efficacy of thymosin alpha-1 and interferon alpha in treatment of chronic viral hepatitis B: A randomized controlled study. *World Journal of Gastroenterology*, *12*, 715–721.

Yu, F. X., Lin, C. S., Morrison-Bogorad, M., Atkinson, M. A., & Yin, H. L. (1993). Thymosin beta 10 and thymosin beta 4 are both actin monomer sequestering proteins. *Journal of Biological Chemistry*, *268*, 502–509.

Zarbock, J., Oschkinat, H., Hannapel, E., Kalbacher, H., Voelter, W., & Holak, T. A. (1990). The solution conformation of thymosin β4: A nuclear magnetic resonance and simulated annealing study. *Biochemistry*, *29*, 7814–7821.

CHAPTER TWO

Intrinsic, Functional, and Structural Properties of β-Thymosins and β-Thymosin/WH2 Domains in the Regulation and Coordination of Actin Self-Assembly Dynamics and Cytoskeleton Remodeling

L. Renault[1]
Institute for Integrative Biology of the Cell (I2BC), CEA, CNRS, Univ. Paris-Sud, Université Paris-Saclay, Gif-sur-Yvette, France
[1]Corresponding author: e-mail address: louis.renault@i2bc.paris-saclay.fr

Contents

Abstract

β-Thymosins are a family of heat-stable multifunctional polypeptides that are expressed as small proteins of about 5 kDa (~45 amino acids) almost exclusively in multicellular

Vitamins and Hormones, Volume 102
ISSN 0083-6729
http://dx.doi.org/10.1016/bs.vh.2016.04.006

animals. They were first isolated from the thymus. As full-length or truncated polypeptides, they appear to stimulate a broad range of extracellular activities in various signaling pathways, including tissue repair and regeneration, inflammation, cell migration, and immune defense. However, their cell surface receptors and structural mechanisms of regulations in these multiple pathways remain still poorly understood. Besides their extracellular activities, they belong to a larger family of small, intrinsically disordered actin-binding domains called WH2/β-thymosin domains that have been identified in more than 1800 multidomain proteins found in different taxonomic domains of life and involved in various actin-based motile processes including cell morphogenesis, motility, adhesions, tissue development, intracellular trafficking, or pathogen infections. This review briefly surveys the main recent findings to understand how these small, intrinsically disordered but functional domains can interact with many unrelated partners and can thus integrate and coordinate various intracellular activities in actin self-assembly dynamics and cell signaling pathways linked to their cytoskeleton remodeling.

1. INTRODUCTION

β-Thymosins are a family of heat-stable multifunctional polypeptides that are expressed as small and apparently monomodular proteins of about 5 kDa (~45 amino acids) almost exclusively in metazoa. They were first discovered during the mid-1960s when establishing the function and hormone-like mechanism of the thymus in the immune system of higher multicellular animals. The interest for βT emerged when thymus extracts containing these and dozens of other unrelated polypeptides, named collectively "thymosins," were found to restore many of the immunological deficiencies observed in animals lacking thymus gland (Goldstein, 2007). The thymus is now recognized as a primary lymphatic organ of the immune system that participates in the maturation of the immune T lymphocytes (T stands for thymus). Full length or truncated variants of βT have thus been found in several extracellular human fluids including blood serum and plasma (Mannherz & Hannappel, 2009) and shown to possess hormonal or paracrine activities modulating the immune response and the cascade of tissue healing and regeneration following injuries. Various paracrine effects have been reported for βT, for example, in angiogenesis, cardiac protection, stimulation of dermal and corneal wound healing, cell migration, or reduction of inflammation (Huff, Muller, Otto, Netzker, & Hannappel, 2001; Mannherz & Hannappel, 2009; Sosne, Qiu, Goldstein, & Wheater, 2010). βT represent therefore promising hints for the development of novel therapeutic peptides and polypeptides in human and animal health and are studied in a number of clinical trials involving tissue repair and regeneration (Crockford, Turjman, Allan, & Angel, 2010; Goldstein & Kleinman, 2015). However, the multiple interacting partners or cell surface receptors, structural regulatory mechanisms and full molecular signal transduction pathways which support all their activities as extracellular modulators in tissue repair, cell migration, and immune defense remain elusive. Besides their various above paracrine effects, βT are expressed as well at very high concentrations within most cells of metazoa where their main direct binding partners appear to be the cytoskeletal ATP-binding protein actin. In the last years, various important structural and functional properties of these small actin-binding domains have been underlined to understand their versatile intracellular regulations in the dynamics of actin cytoskeleton remodeling.

Actin is an ATP-binding protein of 42 kDa specific to eukaryotic cells and represents one of the most conserved and abundant proteins in eukaryotes. Monomeric/globular actin (G-actin) can spontaneously self-assemble into double-stranded helical polymers called actin filaments (F-actin). This reversible transition between G- and F-actin, or higher order F-actin structures upon interaction with various actin-binding proteins (ABPs), builds one of the three main cytoskeletons specific to eukaryotic cells. Actin filaments are polarized with different rate constants for monomer association and dissociation at the two filament ends. The fast-growing end of actin filaments is called the barbed end (or (+) end), and the slow-growing end the pointed end (or (−) end). After monomer association to filaments, actin subunits undergo additionally conformational changes that accelerate the hydrolysis of the nucleotide ATP bound to actin. Since the rate constants for actin-ATP and actin-ADP association and dissociation differ significantly, this ATP→ADP hydrolysis in polymerized actin subunits further increase the polarity of filaments. At a steady state, actin monomers are constantly added to barbed ends and depolymerized from pointed ends. This phenomenon is called treadmilling. The actin cytoskeleton is constantly remodeled in eukaryotic cells to control and coordinate many actin-based motile processes, including cell polarity establishment and maintenance, polarized cell migration, cell adhesion, cytokinesis, intracellular transport, phagocytosis, or intracellular pathogen infections. The actin cytoskeleton is regulated by more than 100 functionally different ABPs (dos Remedios et al., 2003). Each ABP regulates elementary functions with G- or F-actin, and works often in concert with other ABPs to regulate the high turnover of assembly and disassembly of F-actin. In addition, many ABPs use complex multidomain architectures to integrate and coordinate multiple signals and interactions with the dynamic remodeling of actin cytoskeleton. Within metazoan cells, β-thymosins represent essential regulators of the pool of G-actin in combination with another important G-actin binding protein called profilin (Huff et al., 2001; Mannherz & Hannappel, 2009; Xue & Robinson, 2013). With their small size of 5 kDa, β-thymosins are the shortest functional ABPs known. They belong to a larger family of small, ubiquitous, intrinsically disordered actin-binding domains called WH2/β-thymosin domains (βT domains) that are otherwise present in a large number of multidomain ABPs. Based on their intrinsically disordered nature, small sequence variations, specific arrangements in repeats, or associations with other unrelated actin-binding domains, they have emerged in the last 10 years as versatile regulators of the actin self-assembly dynamics by interacting with G- and/or F-actin, and sometimes additional proteins or neighboring domains of ABPs (Carlier, Husson, Renault, & Didry, 2011; Dominguez, 2009; Renault, Deville, & van Heijenoort, 2013).

This review briefly surveys the main recent findings to understand some of the main intrinsic properties, molecular regulatory mechanisms, and bases for the functional versatility of intrinsically disordered β-thymosins and WH2/β-thymosin domains in the regulation of actin self-assembly dynamics and intracellular signaling pathways linked to their cytoskeleton remodeling. These fundamental elements have the potential to be valuable to understand and control how these small, intrinsically disordered but functional domains can interact with many unrelated partners and regulate various intracellular and extracellular activities.

2. THE SEQUENCES OF β-THYMOSINS AND β-THYMOSIN/WH2 DOMAINS

Like many functional intrinsically disordered polypeptides/regions, βT/WH2 domains display highly variable sequences and a weak sequence signature, which makes them difficult to identify or delimit properly (Figs. 1 and 2). They appear to be widely distributed, being predicted in more than 1800 proteins, mostly multidomain proteins, of different taxonomic domains of life in the SMART (Letunic, Doerks, & Bork, 2015) or Pfam (Finn et al., 2016) databases.

2.1 Isolated β-Thymosins Expressed as Small Proteins of ~5 kDa

About 15 highly similar and homologous βT variants of 40–44 amino acids have been found in many invertebrate and vertebrate cells. They are

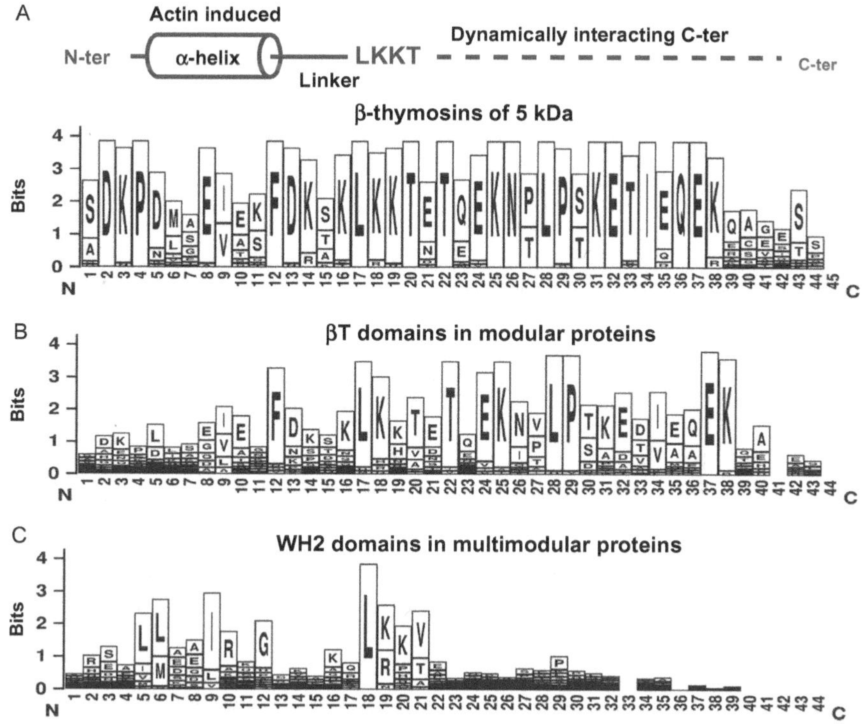

Fig. 1 Outside 5 kDa β-thymosins, βT/WH2 domains display poor sequence homology. The homology of 30 sequences corresponding to (A) 5 kDa β-thymosins, (B) βT, or (C) WH2 domains found in proteins from different species is displayed as a sequence logo for their respective sequence pattern (Crooks, Hon, Chandonia, & Brenner, 2004). The height of each amino acid letter is proportional to their relative frequency at that position in the sequence. Above are indicated the main sequence elements and how they interact with G-actin.

expressed in a species- and cell-specific manner with most mammalian cells expressing two variants (Huff et al., 2001). In human, there are five β-thymosin genes: thymosin β10, β15A, and β15B, and two thymosin β4 on the X- and Y-chromosomes. Several of these genes have been observed to be misregulated in different types of cancers (Sribenja, Wongkham, Wongkham, Yao, & Chen, 2013), which suggests a complex balance between all the multiple intracellular and extracellular activities of full-length or truncated variants of β-thymosins, or their possible cell-type functional specificities.

Thymosin β4 (Tβ4) is the most abundant member of the family and expressed in most mammalian cell types. Besides being present at substantial extracellular concentrations of about 5–80 nM in blood plasma (Huff et al., 2001; Mannherz & Hannappel, 2009), Tβ4 is present in very high

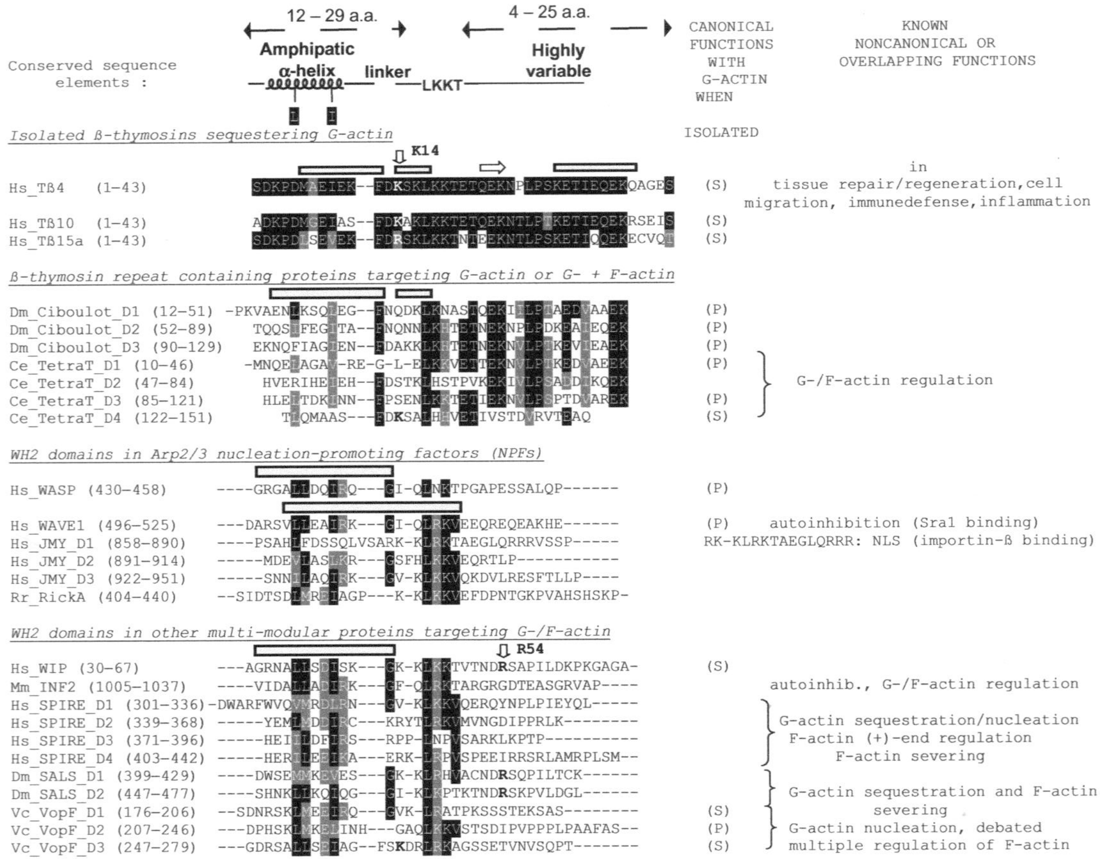

Fig. 2 Sequence alignment of βT/WH2 domains found in proteins from different species, with their canonical function as isolated domains with G-actin-ATP and other known noncanonical or overlapping functions in actin assembly regulation and signaling as isolated/truncated domains or when incorporated in larger multimodular organizations as explained in the text. (S) and (P) stand for sequestering and profilin-like functions, respectively. The *gray boxes* and *horizontal arrow* above the sequences indicate the residues that form α-helix or β-sheet, respectively, upon G-actin binding or Sra1 binding (autoinhibited state of WAVE1 complex).

concentrations in highly motile cells like white blood cells or macrophages (in the range of 300–500 μM) or cells that rapidly respond to external signals to remodel profoundly their shape like platelets (between 140 and 290 μM in various mammalian platelets). The very high intracellular concentrations of Tβ4 are connected with the very high concentrations of its main binding partner within metazoan cells, namely actin monomers. In the lamellipodial membrane protrusions of migrating cells, the concentrations of G-actin and F-actin are estimated to be 150 and 500 μM, respectively (Koestler et al., 2009). Inside these cells, β-thymosins act as major G-actin sequesterers and contribute to maintain a large pool of monomeric actin for next incoming polymerizations. Like actin Tβ4 is localized both in the cytoplasm and the nucleus of cells (Mannherz & Hannappel, 2009). Tβ4 is liberated from platelets during blood coagulation and partially cross-linked to fibrin by the factor XIIIa transglutaminase. Highly reactive sites in Tβ4 for the reaction are reported in Fig. 3. This covalent cross-linking to the fibrin clot may

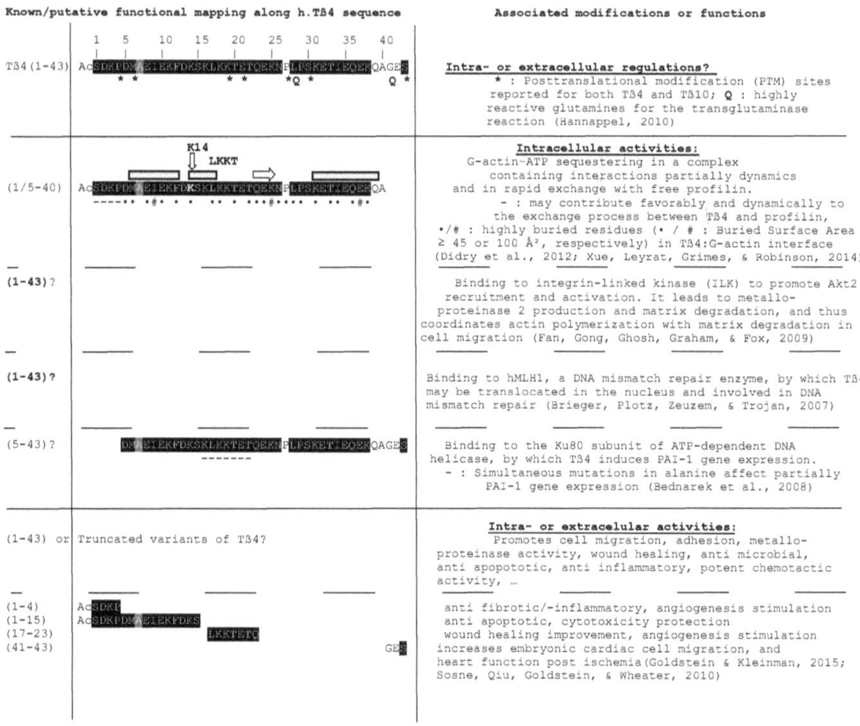

Fig. 3 Functional mapping of human Tβ4 sequence. Known important or putative functional residues/motifs are reported.

guarantee a high local concentration of the polypeptide at the site of injury and help subsequent wound healing processes with elusive molecular mechanisms (Mannherz & Hannappel, 2009).

2.2 β-Thymosin Domains Expressed as Repeats and in Modular Proteins

Besides being expressed as small isolated, functional polypeptides/proteins of 5 kDa in metazoans (referred hereafter as β-thymosins), other polypeptides with high sequence similarity to β-thymosins (referred hereafter more broadly as β-thymosin/βT domains) have been found as individual or repeated domains in many modular proteins of metazoans (Renault et al., 2013; Van Troys, Dhaese, Vandekerckhove, & Ampe, 2007; Xue & Robinson, 2013). Modular proteins with βT domains appear to be linked to a variety of motile and morphogenetic processes in which actin is involved (Carlier et al., 2011; Van Troys et al., 2007). More than 350 protein sequences of metazoans are thus predicted in the SMART or Pfam databases to contain βT domains, with a taxonomic distribution mostly in arthropoda (20%) and chordata (almost 70%) where proteins contain from 1 to 4 βT domains (Renault et al., 2013). The phylogenetic distributions of isolated β-thymosins or proteins made of βT domain repeats are often mutually exclusive (Van Troys et al., 2007). For example, the fruit fly *Drosophila melanogaster* or the worm *Caenorhabditis elegans* expresses the protein Ciboulot or tetrathymosinβ, respectively, which are proteins build exclusively by βT domain repeats: Ciboulot isoforms contain 2–3 consecutive βT domains and tetrathymosinβ4. Exceptions however exist such as mouse that expresses short Tβ4, Tβ10, Tβ15 isoforms and additionally a two-repeat βT resulting from read-through transcription and alternative splicing of two tandem duplicated mouse Tβ15 genes (Dhaese et al., 2009).

Cumulative studies in the last 10 years have demonstrated that βT domains are also structurally and functionally related to actin-binding WH2 domains.

2.3 WH2 Domains in Multimodular Proteins

WH2 (Wiskott-Aldrich syndrome protein (WASP) homology domain 2) domains were initially found in actin nucleation-promoting factors (NPFs) of the WASP/Scar/WAVE family that activate the Arp2/3 complex to generate branching networks of filaments in many physiological or pathological actin-based motile processes involving membrane deformations (lamellipodia, membrane ruffling, phagocytosis, etc.) and intracellular transport (Campellone & Welch, 2010; Dominguez, 2009). NPFs promote de novo filament branching via their conserved and intrinsically disordered, C-terminal constitutively active moiety, conventionally called V-C-A. The latter is composed of 1–3 actin-binding WH2 domains (referred as WH2 or V), followed by a connector/central domain (C) and an acidic short extension (A) that bind and activate the Arp2/3 complex of 220 kDa in the presence of the side of a preexisting filament and the V/WH2 region.

WH2 domains have been since identified or predicted as single or multiple units in a large number of modular signaling/regulatory proteins involved in various actin-based motile processes such as cell motility, adhesion, morphogenetic processes, or pathogen infection (Carlier et al., 2011; Dominguez, 2009; Renault et al., 2013). In contrast to isolated β-thymosins or multirepeats of βT domains, they are not found as isolated single domains nor building entirely proteins with consecutive repeats but are often found in modular signaling/regulatory proteins that contain other actin-binding domains or other domains interacting with ABPs. They are much more widely distributed in the major taxonomic domains of life than βT domains, suggesting βT domains may derive from ancient WH2 modules. More than 1700 sequences of multimodular proteins from eukaryotes, including metazoan or fungi, and few dozen from prokaryotes or viruses are predicted to contain WH2 domains in the SMART or Pfam databases. Most of them include a single WH2 in their sequences, but sequences with WH2 repeats are also found and include more often 2–4 consecutive WH2 modules (Renault et al., 2013). More than 50% of the eukaryotic proteins with WH2 domains are predicted or known to be involved in adherens junction processes, about 1–2% in embryonic developmental processes, the rest of proteins being more broadly classified as participating to the actin cytoskeleton regulation.

2.4 General Features of the Sequences of βT/WH2 Domains

The sequences of βT/WH2 domains vary from 25 to 55 amino acids (Carlier et al., 2011; Chereau et al., 2005). It is mainly only identified by two common features: (i) a central consensual but minor actin-binding motif LKKT/V flanked by variable N- and C-terminal extensions containing 12–29 and 4–25 amino acids, respectively, and (ii) an N-terminal segment before the motif LKKT/V that is prone or predicted to fold into an amphipathic α-helix (Figs. 1 and 2). βT share a more extended and conserved C-terminal half than WH2 (Fig. 1). Sequence identity within the separated βT and WH2 domain families is ~40% and ~20%, respectively, and drops to less than 15% when comparing sequences across the families. Structural studies using NMR spectroscopy or circular dichroism have demonstrated early that Tβ4 (Czisch, Schleicher, Horger, Voelter, & Holak, 1993; Domanski et al., 2004; Safer, Sosnick, & Elzinga, 1997), Tβ9 (Stoll, Voelter, & Holak, 1997), or Scar1 isolated WH2 domain (Kelly, Kranitz, Dotsch, & Mullins, 2006) are largely disordered when they are isolated in their native state and physiological buffers, and β-thymosins may contain two helices at the N- and C-termini, secondary structures inducible by either solvent or the presence of actin.

3. GENERAL PROPERTIES OF INTRINSICALLY DISORDERED PROTEINS/DOMAINS

Intrinsically disordered regions/domains or proteins (IDRs or IDPs) are far from anecdotal in prokaryotic or eukaryotic genomes. About 30% of eukaryotic proteins and 44% of the proteins of the human genome are predicted to contain disordered segments of more than 30 amino acids (van der Lee et al., 2014). The higher presence of IDRs/IDPs in eukaryotes than prokaryotes is correlated with their frequent involvement in signal transduction or transcriptional regulations (van der Lee et al., 2014), including intracellular signaling pathways that mediate and coordinate the remodeling of cytoskeletons specific to eukaryotes (Guharoy, Szabo, Martos, Kosol, & Tompa, 2013; Renault et al., 2013). Many further investigations are required to understand the main functions, interfacial properties, and regulatory mechanisms of this new class of proteins that defy the structure–function paradigm and are frequently involved in human diseases, such as amyloidoses and neurodegenerative diseases, but also cancer, cardiovascular disease, and diabetes (Uversky, 2010).

3.1 Intrinsic and Protein–Protein Interaction Properties of IDPs/IDRs

IDPs or IDRs are characterized by low proportions of hydrophobic, order-promoting residues and high proportions of polar or charged residues, glycines and prolines, low sequence complexity, and repetitiveness (Tompa, 2003). They are unable to fold into a unique and stable tertiary structure in solution and/or in the crowded environment of cells.

Many IDPs operate through transient or permanent protein–protein interactions. IDPs' interactions with their partner are generally characterized by moderate affinities in the μM range but high specificity, large interaction surfaces, conformational plasticity, and adaptability. Most of their interactions and/or affinities is often mediated by short molecular recognition features (MoRFs) or elements (MoREs) (Oldfield et al., 2005; van der Lee et al., 2014). Binding of these MoREs/MoRFs is often accompanied by disorder to order transitions leading to the formation of secondary or tertiary structures upon binding to partners. However, numerous IDPs appear also to be functional in complexes through retaining various degrees of dynamics and disorder, or adopting multiple conformations in protein–protein interfaces, leading to the introduction of fuzzy complexes (Fuxreiter, 2012; Fuxreiter & Tompa, 2012).

Particular interesting features of protein–protein interactions involving IDPs/IDRs include the following specificities that also support the interactions of β-thymosins or βT/WH2 domains with G-actin: (i) They often interact in extended conformations with their partners and thus provide more binding interface than folded proteins for the same number of amino acids. (ii) Their specificity is decoupled from binding affinity (Oldfield et al., 2005). Interaction surfaces are high because of the extended nature of bound IDPs leading to important specificity but the entropic cost for binding is important thus decreasing the binding affinity. It allows highly specific reversible interactions particularly relevant in regulatory or/and cytoskeletal processes involving multiple dynamic and coordinated interactions. (iii) Their interactions or adaptability to different partners are frequently regulated by posttranslational modifications (PTMs) or alternative splicing. In addition, the intrinsic flexibility of IDPs may facilitate their passive or controlled transport across intact cell membranes to be either externalized or internalized. In this regard the mechanisms of release of intracellular Tβ4 to initiate its extracellular signaling or of its uptake from external fluids remain enigmatic and will require additional studies. Its presence in most human body fluids studied (saliva, tears, blood, plasma, wound fluid) has been so far speculated to result from damaged cells. In the extracellular activities of Tβ4 for cardiac injury protection, Tβ4 has been described as an effective cell penetrating peptide for treating mammalian myocardial infarction (Sopko et al., 2011).

3.2 Main Functions of IDPs/IDRs

Functions of IDPs or IDRs have been classified into six general categories belonging to three overall classes (Tompa, Szasz, & Buday, 2005; van der

Lee et al., 2014). The intracellular functions of β-thymosins or βT/WH2 domains with actin may be currently linked more specifically to three of these functional categories.

The first class of IDP/IDRs is composed of entropic chains including flexible linkers and entropic springs (van der Lee et al., 2014). In that case, the function is directly related to the physical properties of the peptide chain and does not require partner binding or order gain. The second class of IDP/IDR functions is linked to transient binding and includes two main regulations. First intrinsic disorder plays an important role in chaperones which assist protein or RNA folding by transient binding. IDRs may second serve as display sites for PTMs. IDRs are generally enriched in sites for PTMs including phosphorylation, methylation, acetylation, etc. Their inherent flexibility first enables transient but specific interaction with different catalytic sites of modifying enzymes. It also provides high accessibility and recognition of the PTMs within the IDR by effector proteins that will mediate different downstream outcomes upon alternative binding with IDRs modulated by alternative PTMs (van der Lee et al., 2014). Tβ4 or Tβ10 is also subject to PTMs whose role needs to be further investigated (Hannappel, 2010) (Fig. 3). Phosphorylations in the four consecutive βT repeats of the protein Csp24 from *Hermissenda crassicornis* appear to modulate its association with actin (Redell, Xue-Bian, Bubb, & Crow, 2007).

The third class of IDP/IDR functions is linked to permanent binding and includes three main regulations (van der Lee et al., 2014). First some IDPs/IDRs may act as scavengers which bind, store, and neutralize small ligands. It remains to be seen if such properties are shared by the truncated polypeptide 1–15 of Tβ4 that protects cells from the cytotoxic effects of various reagents (Sosne et al., 2010) (Fig. 3). IDPs/IDRs may second act as effectors and interact with other proteins to modulate their activity. Interaction may be accompanied by partial or global folding of the IDP. This corresponds to the interaction and different regulations of β-thymosins and individual βT/WH2 domains with G-actin (Didry et al., 2012) as described below, or possibly of their multifunctional repeats with G- and F-actin. Finally, some IDPs/IDRs may act as disordered assemblers which recruit different partners and enable scaffolding of multiprotein complexes (Fuxreiter et al., 2014). They may contain several overlapping binding interfaces within a relatively small number of amino acids. Several determinant-binding sites such as MoREs and short linear peptide motifs (SLiMs) are distributed along the sequence in such a way that these disordered assemblers form interaction hubs to recruit multiple different partners. This too corresponds to the interaction and versatile regulations of WH2 domains in multimodular proteins (Renault et al., 2013).

4. MAIN INTRACELLULAR FUNCTION OF β-THYMOSINS AND FUNCTIONAL VERSATILITY OF βT/WH2 DOMAINS IN ACTIN ASSEMBLY

4.1 β-Thymosins as Genuine G-Actin Sequestering Proteins

In cells, actin is assembled in filaments that coexist with a stationary amount of actin monomers. In the absence of any regulators of actin assembly, the filament treadmilling is two orders of magnitude slower than in vivo and more than 99% of cellular actin would be F-actin because the critical concentration (Cc) above which polymerization occurs is only 0.1 μM. To reach a ratio of F-actin/G-actin in cell lamellipodia of about 500/150 μM (Koestler et al., 2009), most actin filaments are capped at their barbed ends by capping proteins that prevent both monomer association and dissociation and thus nonproductive polymerization. Elongating filaments are created through controlled uncapping, severing, or de novo nucleation mechanisms. Rapid formation of new filaments depends on high intracellular concentration of actin monomers. The availability and preservation of the pool of G-actins that are competent or not to polymerize is tightly regulated through the binding of two main G-actin binding proteins: β-thymosins and profilin.

The short isolated polypeptides β-thymosins that are present at high intracellular concentrations in cells (between 100 and 500 μM) (dos Remedios et al., 2003) discriminate between the ATP- and ADP-bound states of free G-actin, whose conformations remain elusive, and bind preferentially G-actin-ATP over G-actin-ADP. Their binding to G-actin inhibits the nucleotide exchange in G-actin. Tβ4 interacts only weakly with filaments (Carlier et al., 1996). Their binding affinity for G-actin-ATP can be very sensitive to the ionic strength, and be artificially increased at low, nonphysiological ionic strength by 10- to 30-fold for Tβ4 up to at least 500-fold for chimera of Tβ4 and Ciboulot βT domains compared to their affinities measured near physiological ionic strengths (De La Cruz et al., 2000; Didry et al., 2012). This makes sometimes difficult to infer the behaviors of all β-thymosin variants in physiological conditions because many affinity measurements have been obtained at low ionic strength in order to prevent actin polymerization, or at various ionic strengths between different biochemical analyses. Near physiological ionic strength, Tβ4 binds G-actin-ATP with an equilibrium dissociation constant (K_d) between 1.8–22 μM and G-actin-ADP with a K_d at least 30- to 50-fold higher (Carlier, Jean, Rieger, Lenfant, & Pantaloni, 1993; De La Cruz et al., 2000; Didry et al., 2012). Different mammalian variants of β-thymosins appear to bind to G-actin-ATP with only subtle differences in their K_d that differ only by two- to threefold at low ionic strength (Huff, Muller, & Hannappel, 2007). Upon binding G-actin-ATP, Tβ4 or Tβ10 completely sequesters actin monomers, preventing them to form nuclei that would initiate the formation of new filaments or to join either end of a preexisting filament. The sequestration and release of G-actin-ATP from β-thymosins is thought to be passively regulated, β-thymosins acting merely as actin-buffering proteins. The buffering of G-actin-ATP by β-thymosins is regulated on one hand by their high intracellular concentrations and moderate affinities for their main target (G-actin-ATP) and on the other hand by the competition with the other main G-actin-binding protein profilin.

Profilin isoforms are small ubiquitous eukaryotic proteins of 13–19 kDa that are present at relatively high concentrations in cells, ie, between 10 and 100 μM (dos Remedios et al., 2003). In contrast to β-thymosins, they enhance the exchange of ADP for ATP and bind equally well to G-actin-ATP and G-actin-ADP and with better affinities than β-thymosins. Near physiological ionic strength, their K_d for G-actin-ATP and G-actin-ADP binding is 0.1 and 0.17 μM, respectively (Kinosian, Selden, Gershman, & Estes, 2000). Their complex with G-actin prevents both monomers to form nuclei promoting spontaneous polymerization and to participate in pointed-end filament elongation. The profilin:G-actin complex can only associate with the most dynamic barbed-end of F-actin. Profilin dissociates from the actin protomer as the latter is incorporated into the filament, and thus participates actively in the unidirectional/polarized (+)-end elongation of filaments. During polymerization phases, the profilin:G-actin complex gives up its actin to the elongating barbed ends. The liberated free profilin competes with β-thymosins for its bound G-actin-ATP, effectively restoring the pool of polymerization-competent actin. In addition, profilin, Tβ4, and G-actin-ATP have been reported to be able to form a ternary complex displaying a moderate K_d of 2–8 μM for the binding of Tβ4 to the profilin:G-actin complex (Yarmola, Parikh, & Bubb, 2001). Biochemical analyses suggest that the ternary complex is more efficient in lowering the combined concentration of polymerization competent actin that is free or profilin-bound (Xue, Leyrat, Grimes, & Robinson, 2014; Yarmola et al., 2001).

In the context of exchange between profilin and Tβ4 for binding G-actin, the concentration of free profilin is tightly regulated in cells (Mannherz & Hannappel, 2009) and may locally increase following filament uncapping and barbed-end elongation from profilin:G-actin. The dissociation of Tβ4 from G-actin stimulated by high local concentrations of free profilin has been proposed to couple actin polymerization to matrix degradation during endothelial cell migration. Profilin-dependent dissociation of G-actin:Tβ4 complexes simultaneously stimulates filament barbed-end assembly from profilin:G-actin and facilitates Tβ4 binding to integrin-linked kinase (ILK) in the lamellipodia. Tβ4:ILK complexes then recruit and activate Akt2, resulting in metalloproteinase-2 production and matrix degradation (Fan, Gong, Ghosh, Graham, & Fox, 2009). Two other intracellular partners of free Tβ4 have been identified in connection with its role in cancer. Tβ4 was first found to interact with a key enzyme of DNA mismatch repair, hMLH1 that is frequently mutated in human cancers. Since hMLH1 contains a canonical nuclear localization sequence (NLS), it was proposed to translocate Tβ4 into the nucleus where the latter may be involved in DNA mismatch repair and the regulation of cell mortality (Brieger, Plotz, Zeuzem, & Trojan, 2007). Tβ4 was also found to stimulate plasminogen activator inhibitor type 1 (PAI-1) gene expression in endothelial cells by interacting with the Ku80 subunit of ATP-dependent DNA helicase II (Bednarek et al., 2008). The observed interaction did not depend on the N-terminal tetrapeptide AcSDKP. Further investigations are required to understand the regions of Tβ4 and the molecular mechanisms that are at the basis of these different regulations (Fig. 3).

The other regulators of actin assembly that control an efficient F-actin/G-actin ratio and high turnover in cells include barbed-end tracking nucleators/elongators and ADF/cofilin depolymerization factors (Dominguez, 2009; Xue & Robinson, 2013). Barbed-end tracking nucleators/elongators are for example formin or ENA/VASP multimodular ABPs which allow fast-processive barbed assembly rate from profilin:G-actin-ATP complexes. They compete with and may displace capping proteins from barbed ends. ADF/cofilin proteins bind, sever preferentially F-actin-ADP, and enhance pointed-end depolymerization. They consequently increase the amount of profilin:G-actin and enhance the barbed-end assembly rate of elongating filaments. On filament dissociation, G-actin-ADP is indeed preferentially captured by profilin that has a higher affinity for G-actin-ADP than β-thymosins have. Profilin stimulates the nucleotide exchange to regenerate G-actin-ATP. The latter is subject to competition between profilin and β-thymosins, which restores the pool of G-actin-ATP sequestered by β-thymosins.

4.2 Extension of the Functions of βT/WH2 Domains in Actin Assembly

In the past years, the functions of βT and WH2 domains in actin assembly have been significantly extended. They now appear as versatile regulators of actin cytoskeleton dynamics (Campellone & Welch, 2010; Carlier et al., 2011; Dominguez, 2009; Renault et al., 2013).

4.2.1 Functional Versatility of βT and WH2 as Individual Domains

When they are studied in actin self-assembly as single domains isolated from their larger proteins, most other WH2/βT studied do not regulate the sequestering function of β-thymosins. Instead, they form 1:1 complexes with G-actin-ATP in which actin is prevented to nucleate or to bind pointed ends but can associate to filament barbed ends like profilin:G-actin complexes. Regarding the ATP bound to actin, they inhibit nucleotide exchange in their complexes with G-actin similarly as Tβ4 (unlike profilin). They are therefore not fully homologous to profilin regarding F-actin elongation. The isolated WH2 domains of WASP/Scar1/WAVE1 proteins (Higgs & Pollard, 1999), the three βT repeats *of Drosophila* Ciboulot, or its isolated first βT domain (CibD1) (Hertzog, Yarmola, Didry, Bubb, & Carlier, 2002) enhance actin-based motility by directing polarized barbed-end assembly with different rate constants for association of the actin:WH2/βT to barbed ends. The barbed-end growth rates (k_+) they control in their complexes with G-actin-ATP are typically slower than the k_+ of profilin:G-actin-ATP or free G-actin-ATP. The isolated first βT domain of *Drosophila* Ciboulot or the full protein containing its three βT repeats displays a similar k_+ of 3 $\mu M^{-1}\,s^{-1}$ (Didry et al., 2012), while the k_+ of profilin:G-actin-ATP and free G-actin-ATP-Mg in similar buffer conditions are 7 and 11 $\mu M^{-1}\,s^{-1}$, respectively (Gutsche-Perelroizen, Lepault, Ott, & Carlier, 1999). Only a very small number of WH2 domains have been shown to control G-actin-ATP sequestration as isolated single domains. Illustrative examples include the WH2 domains of human WASP-interacting protein (WIP) (Didry et al., 2012) and the first or third WH2 domains of the bacterial effector protein VopL from the pathogen *Vibrio parahaemolyticus* (Pernier et al., 2013), whose sequences are only weakly distantly related to Tβ4 sequence (Fig. 2). However, VopF WH2 domains do not sequester G-actin-ATP in actin self-assembly when they are associated in repeats with their adjacent C-terminal dimerization actin-binding domain (Pernier et al., 2013). Understanding their functions in multimodular organizations is thus complex and requires understanding what are the main functions of each individual domains and how intermediate and full combinations of domains work in synergy and in presence of all their respective partners.

The small intrinsically disordered WH2/βT domains can act as signal integrators in multimodular ABPs by integrating and overlapping several binding interfaces within their short functional sequence. This is notably the case of single or repeated WH2 domains found in NPFs of the WASP/WAVE family (Campellone & Welch, 2010; Dominguez, 2009). They thus may be distantly related to β-thymosins and βT domains because of their actin-binding sequence code for additional interactions specific to their multimodular structural organization. In contrast to β-thymosins, the activity of these multimodular ABPs is actively regulated in cells. Their WH2 domains in company of other neighboring IDRs control dynamic interactions between the inactive and active states of NPFs (Renault et al., 2013). NPFs are autoinhibited in their inactive state by head/middle-to-tail intramolecular interactions that involved functionally distinct IDRs of NPFs (Renault et al., 2013). The structural plasticity and dynamics of NPF IDRs, including WH2 domains, drive variable and specific interactions and conformations with different partners. Their bindings of moderate affinity allow controlling multiple conformational equilibria that can be easily modulated or displaced by small modification or interactions like PTMs or by different partners. The structural study of WAVE1 WH2 domain in the autoinhibited interface of WAVE1 complex (Chen et al., 2010) or in its activated state bound to G-actin-ATP (Chereau et al., 2005) shows that it is the same N-terminal portion of the WH2, ie, its N-terminal amphipathic helix and its central consensus LKKT/V motif, that provides the most buried contacts of the two different protein:protein interfaces (Renault et al., 2013). The WH2 domain adopts different conformations between the two different protein:protein interfaces with its N-terminal portion folding into an α-helix of variable length (Fig. 2). It forms interactions that are mutually exclusive between its two partners: Sra1 in the autoinhibited WAVE1 complex and G-actin in its active state. All these structural elements suggest that the MoREs of WH2/βT domains with actin and ABPs include their N-terminal amphipathic helix and central consensus LKKT/V motif.

Other examples that illustrate the overlapping and/or synergistic functions of WH2 sequences are found in the mammalian multimodular JMY NPF and INF2 (inverted formin 2) formin. The junction-mediating and regulatory protein (JMY) is an apoptosis and transcriptional activator of p53 and a multifunctional actin NPF containing in its C-terminal V/WH2-C-A regions three consecutive WH2 repeats. The function of its three consecutive WH2 repeats is further extended by integrating an overlapping importin-β-binding nuclear-localizing sequence (NLS) (Zuchero, Belin, & Mullins, 2012). The NLS is overlapping with the central region of the first WH2 including its consensus LKKT/V motif (Fig. 2). The mutually exclusive interactions between G-actin and importin-α/β bindings to the repeats regulate JMY cytoplasmic or nuclear localization, and consequently, the nuclear transcription induced by the protein according to G-actin cytoplasmic concentration and binding. INF2 formin contains in its C-terminus a single WH2 and an additional IDR called DAD. The WH2-DAD IDRs are located after its formin homology 1 (FH1) and 2 (FH2) domains that are the basic catalytic units of all formins in actin assembly. These two functional IDRs regulate first via their weak interactions the inactive conformation of INF2. They are involved in interdomain head-to-tail autoinhibition with an unrelated N-terminal DID DAD (Chhabra, Ramabhadran, Gerber, & Higgs, 2009). In the active state of INF2 ABP, they contribute at the regulation of two distinct activities. They first modulate the usual activity of the FH2 in actin assembly and act in synergy with their FH2 domain to accelerate actin polymerization from monomers. Second, their presence extends the usual functions of an FH2 domain with F-actin and leads to the severing of filaments containing actin-ADP subunits (Chhabra & Higgs, 2006). This atypical formin via the addition of its C-terminal WH2-DAD IDRs therefore accelerates both actin polymerization and depolymerization in mitochondrial fission regulation and other cytoplasmic actin-based processes.

4.2.2 Functional Versatility as Repeats or in Association with Other Domains/Proteins

The repeats of WH2/βT domains in proteins or their association with other actin-binding domains or domains that bind ABPs expand significantly the panoply of their functions in actin assembly (Carlier et al., 2011; Dominguez, 2009). Repeats are frequently found in intrinsically disordered sequences and may represent an efficient way to expand IDP functions in evolution (Tompa, 2003). They can be either functionally equivalent or nonredundant, introduce novel functions when the repeat number exceeds a threshold level, or favor the dynamics and turnover of multiple interactions. In many modular organizations containing WH2/βT repeats, consecutive WH2/βT are functionally nonequivalent. They exhibit variable affinity for binding G-actin-ATP and lead in repeats to complexes in which each of their consecutive WH2/βT domains does not always steadily bind an actin monomer (Dhaese et al., 2009; Hertzog et al., 2002; Husson, Renault, Didry, Pantaloni, & Carlier, 2011; Pernier et al., 2013; Van Troys et al., 2004; Xue & Robinson, 2013), which suggest additional roles than G-actin capture and regulation. In contrast to individual WH2/βT domains, WH2/βT repeats associated with other actin-binding domains can target more significantly the side and/or barbed end of filaments to regulate conjointly the dynamics of F-actin assembly or disassembly.

The specificities and synergy of multiple βT domains were first highlighted in *C. elegans* tetrathymosinβ that contains four consecutive βT domains (Fig. 2) and binds multiple G-actins in vitro (Van Troys et al., 2004). Some repeats were shown to exhibit specificity for G-actin (repeats 1 and 4), F-actin (repeat 3) binding, or both (repeat 2). Together the four repeats cooperate to display the full biochemical activity of tetrathymosinβ in G- and F-actin assembly regulation, including its desequestering ability in presence of Tβ4 (Van Troys et al., 2004). Spire (Quinlan, Heuser, Kerkhoff, & Mullins, 2005), Cordon-bleu (Ahuja et al., 2007), JMY (Zuchero, Coutts, Quinlan, Thangue, & Mullins, 2009), or the pathogen VopF/VopL effectors (Liverman et al., 2007; Tam, Serruto, Dziejman, Brieher, & Mekalanos, 2007) contain different modular organizations with 3–4 WH2/βT repeats (Fig. 2) that accelerate de novo filament nucleation from pure G-actin. They have been then proposed to represent a new subfamily of actin nucleators (Ahuja et al., 2007; Quinlan et al., 2005). However, compared to other known actin nucleators like formin, WH2 repeats of Spire or Cobl display a relatively weak nucleating efficiency in vitro in view of their affinity for G-actin, and their nucleating activity is abolished in vitro when G-actin is bound to profilin (Bosch et al., 2007; Husson et al., 2011). Nucleation may thus not represent the primary in vivo function of the repeats in these multimodular ABPs involved in different developmental processes (Carlier et al., 2011). The 4 WH2 of Spire bind to filament barbed ends with a good affinity corresponding to a K_d of approximately 0.01 μM, while they bind cooperatively 4 G-actin with a lower K_d of $\sim$0.15 μM^4 in identical physiological ionic strength (Bosch et al., 2007). The 4 WH2 repeats of Spire, or the first two WH2 domains of Cobl associated with their N-terminal lysine-rich motif, modulate both the dynamics of assembly and disassembly of filaments. They regulate filament severing with different efficiencies, and Spire regulates also barbed-end capping in the presence of profilin (Bosch et al., 2007; Carlier et al., 2011;

Husson et al., 2011). Similarly, the presence of two consecutive WH2 in the V regions of N-WASP introduce a capacity to interact with filament barbed end (Gaucher et al., 2012), and a higher branching efficiency in N-WASP V-C-A than in WASP or WAVE V-C-A (Yamaguchi et al., 2000). More recently, the tandem of WH2 domains from *D. melanogaster* sarcomere length short protein (Fig. 2) has been shown to regulate both monomeric and filamentous actin, by sequestering actin monomers into nonpolymerizable complexes and enhancing actin filament disassembly by severing (Toth et al., 2016).

In conclusion, understanding the functions of WH2/βT domains requires studying the regulation of dynamic interactions of weak affinity, which requires the combination of multiple techniques of analyses of protein:protein interactions (Husson et al., 2010). Deciphering the molecular bases of their versatility or multifunctionality in different metabolic pathways requires scrutinizing acutely their conformational adaptability and dynamics in their complexes. Their specific regulation may require local disorder and fuzzy complexes as it has been observed at the level of their elementary functional unit with a single βT bound to one G-actin (Didry et al., 2012).

5. INDIVIDUAL WH2/βT IDRs AND G-ACTIN FORM FUZZY COMPLEXES TO REGULATE EITHER SEQUESTERING OR PROFILIN-LIKE FUNCTIONS IN ACTIN ASSEMBLY

Upon binding to G-actin-ATP, Tβ4 was shown by NMR spectroscopy to fold completely and to display a central extended region flanked by two N- and C-terminal helices (Domanski et al., 2004; Safer et al., 1997). To understand the different structural regulations in Tβ4:G-actin and other βT/WH2:G-actin complexes, many crystal structures of various WH2/βT domains bound to G-actin have been obtained using different polymerization inhibiting methods (Chereau et al., 2005; Didry et al., 2012; Irobi et al., 2004; Xue et al., 2014). These crystal structures show that highly variable WH2/βT domains display a similar overall extended fold and binding path on G-actin with their N-terminal half but elusive interactions after their central LKKT/V motif. Their N-terminal amphipathic helix always caps the barbed face of G-actin by binding into the hydrophobic cleft between its subdomains 1 and 3. This N-terminal helical folding and capping is essential because it prevents the actin association with the slow-growing pointed ends of filaments (Fig. 4). Considerable efforts have been required to understand how their C-terminal half prevents or allows barbed-end self-assembly

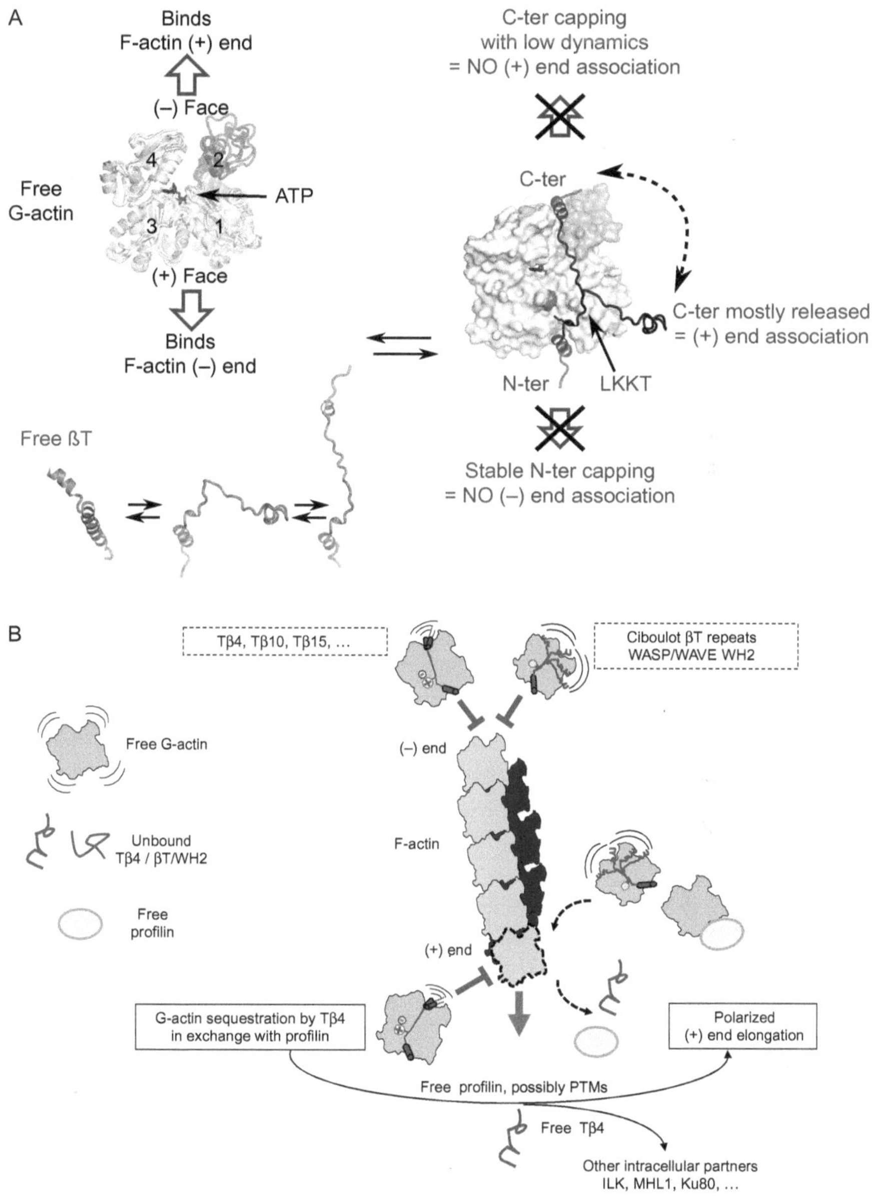

Fig. 4 Structural basis for the regulations of G-actin by Tβ4 and other βT/WH2 domains in actin assembly. (A) Conformational variability of free G-actin, and free and actin-bound βT. Actin subdomains 1 and 3 (G-actin barbed face), and 2 and 4 (G-actin pointed face) are indicated on free G-actin. Its intrinsic conformational flexibility is shown by four overlaid G-actin conformations (PDB: 1HLU, 2VCP, 2PDB or SAXS model (Didry et al., 2012)), and the flexibility of free βT by the solution structure of free Tβ9, which adopts

(Continued)

by interacting or not with G-actin pointed face or by controlling different dynamics following different C-terminal sequence variations (Aguda, Xue, Irobi, Preat, & Robinson, 2006; Chereau et al., 2005; Didry et al., 2012; Domanski et al., 2004; Eadie et al., 2000; Hertzog et al., 2004; Irobi et al., 2004; Xue et al., 2014).

It appears now that the intrinsic dynamics of both the WH2/βT domains and the four subdomains of G-actin, which move between each other, play an important role in WH2/βT:G-actin complexes to define their different functions and regulations in actin assembly. Protein engineering was necessary to stabilize and visualize the N- or C-terminal binding of Tβ4 with G-actin by protein crystallography. A more stable but partial complex was first obtained using a chimera of a C-terminal segment of Tβ4 fused to gelsolin segment 1 in complex with G-actin (Irobi et al., 2004). In this hybrid complex, Tβ4 C-terminal residues (amino acid 30–39) are seen to fold as well into an α-helix that caps the pointed face of G-actin by binding between its subdomains 2 and 4 (Fig. 4A). The C-terminal helical folding and capping appears therefore as another MoRE of Tβ4 sequence for its sequestering function with G-actin-ATP. WH2/βT domains with profilin-like function were proposed in contrast to display no interactions after their central LKKT/V motif in order to conserve G-actin pointed face free in complexes for association with filament barbed ends (Aguda et al., 2006; Chereau et al., 2005; Irobi et al., 2004). This structural model of regulation is not correct to understand all the functional specificities and

Fig. 4—Cont'd a mostly helical folding in 40% 1,1,1,3,3,3-hexafluoro-2-propanol (HFP) (Stoll et al., 1997), and different SAXS conformations of bound CibD1 βT, shown here without G-actin (Didry et al., 2012). The right panel overlay two full SAXS models of G-actin bound to CibD1 βT obtained either at low (sequestering function like Tβ4) or near physiological (profilin-like function) ionic strength (Didry et al., 2012). Tβ4 adopts the same sequestering conformation with G-actin (Didry et al., 2012; Xue et al., 2014). (B) βT/WH2 domains interact with actin monomers to either sequester G-actin, or promote specifically F-actin barbed-end elongation like profilin. They display always a stable N-terminal binding and capping over G-actin barbed face that prevents G-actin association at (−) ends. In contrast, their C-terminal interactions over G-actin pointed face display different dynamics between the different sequences of βT/WH2 domains: with low dynamics like in Tβ4 they sequester, with high dynamics in most other βT/WH2 they allow association at (+) ends and thus may control different (+) end growth rates (Didry et al., 2012). The sequestration of G-actin by Tβ4 is in rapid exchange with free profilin. Other intracellular interactions and regulations may be thus stimulated by massive and spatially restricted profilin-dependent dissociation of G-actin:Tβ4, or possibly by PTMs that may also switch or weaken the interactions with G-actin. (See the color plate.)

modulations of individual WH2/βT in actin assembly. Several recent findings have amended the regulatory model and the functional specificities of β-thymosins like Tβ4 compared to other individual WH2/βT.

The structure–function relationships of individual WH2/βT were first revised by three main findings (Didry et al., 2012): (i) the first βT of Ciboulot (CibD1) that regulates barbed-end elongation upon binding to G-actin can reverse its function and sequester G-actin at low ionic strength, suggesting that it can also efficiently interact with G-pointed face (Fig. 4A); (ii) a single mutation before the LKKT/V motif of Tβ4 or the first βT of Ciboulot (CibD1) is sufficient to switch their respective functions in actin assembly; and (iii) a large variety of sequences can efficiently sequester G-actin-ATP (at physiological ionic strength) (Didry et al., 2012). To understand how βT/WH2 domains could either inhibit or promote unidirectional actin assembly upon G-actin binding, it was necessary to combine mutational analysis of the domains with actin-binding and polymerization assays and several structural approaches, including protein crystallography, SAXS, and NMR (Didry et al., 2012). These techniques were complementary to analyze all static/stable interactions at high resolution in actin:βT complexes by protein crystallography, and the conformational dynamics of βT IDRs in functionally different complexes by SAXS and NMR (Didry et al., 2012; Husson et al., 2010). Altogether the structural analyses demonstrate that the entire sequence of all βT can interact with G-actin but with different dynamics along their short polypeptide chain. Single βT domains with different functions do not need to target alternative actin-binding sites via specific sequence variations in their central and C-terminal regions as previously thought, but to exhibit alternative dynamics of their C-terminal half interactions with G-actin pointed face. Their N-terminal half interactions are in contrast much more stable and static in their complex with G-actin (Fig. 4). It confirms that sequestration requires βT/WH2 with an extended C-terminus to reach and cap G-actin pointed end, and explains how sequestration may accommodate a large variety of dissimilar C-terminal sequences. At physiological ionic strength, the local interaction dynamics of βT domains with G-actin-ATP are primarily controlled by strong electrostatic interactions of a single residue along their sequence. In Tβ4, this key residue is Lys14 that is located in the linker region between the N-terminal amphipathic helix and the central consensus LKKT/V motif. It provides a salt bridge with the highly conserved Glu334 of actin subdomain 3. Thus, a single salt bridge with G-actin located just upstream (Lys14 in Tβ4 or Tβ10/15) or downstream (Arg54 in the sequestering WIP WH2) of their

LKKT/V motif can induce a sequestering function in actin assembly with different and distantly related βT or WH2 sequences. The removal of this strong electrostatic interaction between WH2/βT and G-actin by a point mutation in Tβ4 βT or WIP WH2 on their key residue (K14Q/A in Tβ4, R54N in WIP) is sufficient to reverse their sequestering function into a profilin-like function. Similarly, CibD1 βT can be switched into an efficient sequestering βT following a single point mutation that introduces a salt bridge with G-actin next to the LKKT/V motif like in Tβ4 βT. Thus, all β-thymosin variants that display an intact N-terminal amphipathic helix, the Lys14 in their linker region, and an extended C-terminus can be predicted to sequester G-actin despite their small sequence variations (Didry et al., 2012).

The full interactions of Tβ4 with G-actin-ATP was recently stabilized and trapped by protein crystallography using another chimeric protein in which Tβ4 was fused C-terminally to *Pichia pastoris* actin through a flexible linker (Xue et al., 2014). Confirming previous studies (Didry et al., 2012; Irobi et al., 2004), the full Tβ4 was shown to adopt an extended conformation on the surface of G-actin. Its N- and C-terminal helices cap the barbed and pointed faces of actin monomers, respectively, to prevent G-actin to associate with either ends of filaments (Figs. 3 and 4). The crystal structure underlines additionally that the shorter N-terminal amphipathic helix of Tβ4 is an additional important structural and functional features of β-thymosins that are not found in most other βT/WH2 (Fig. 2). It favors the formation of the ternary complex between Tβ4, G-actin-ATP, and profilin. The combined analyses of crystal structures of two different hybrid Tβ4:G-actin-ATP complexes and molecular dynamics simulations suggest that the intrinsic dynamics of actin monomers may be affected by Tβ4 binding. Tβ4 C-terminal α-helix binding with G-actin pointed face may favor a more closed actin nucleotide binding cleft between subdomains 2 and 4 than in free or profilin-bound G-actin-ATP (Xue et al., 2014). A challenge in the future will be to investigate further the dynamics of free and bound actin in different cytoskeletal molecular assemblies to understand the coordinated and dynamic regulations of actin in its different monomeric and polymerized states.

6. CONCLUSIONS

In summary, the following intrinsic, functional, and structural properties of βT/WH2 domains may be drawn from their intrinsically disordered nature and multifunctionality in actin self-assembly regulation. Because they are intrinsically disordered in their native state, they are very flexible and can adapt their folding and interactions to various conformations/surfaces of proteins. They may bind in extended conformations with their partners, such as with actin (Didry et al., 2012; Safer et al., 1997; Xue et al., 2014), and thus may offer a large surface of interaction despite their small sequence compared to folded proteins. The extended protein:protein interface they cover is therefore specific and may modulate efficiently the interactions or dynamics of their partners (Didry et al., 2012; Xue et al., 2014). The full or partial folding/refolding of intrinsically disordered βT/WH2 on their partner leads to an entropic cost for binding (van der Lee et al., 2014). The functional polypeptides hence interact with weak affinity, which makes their interactions and regulations difficult to identify and characterize. In addition, the structural analysis of WH2/βT:G-actin complexes has revealed that they can build fuzzy complexes in which the inherent disorder of WH2/βT is partially conserved and functional in actin assembly (Fig. 4). Their intrinsic flexibility appears to provide three functional benefits with actin: (i) it may be adapted and tuned to the intrinsic dynamics of actin subdomains while allowing to cover an extended binding interface with their partner (covering both the functionally important barbed- and pointed faces of G-actin); (ii) it leads to a large functional variety of 1:1 WH2/βT:G-actin complexes in which single/small sequence variations in WH2/βT can control different dynamics of their C-terminal interactions with G-actin pointed face and allow to control either G-actin sequestration, or unidirectional barbed-end assembly with different rates of elongation; and (iii) it allows rapid exchange with other G-/F-actin-binding regulators such as profilin (Fig. 4B).

Since small sequence variations are sufficient to change their functional specificity (Didry et al., 2012), PTMs may regulate different conformations states and functions of WH2/βT domains as previously observed (Redell et al., 2007). β-Thymosins have been reported to undergo PTMs whose biological relevance is mostly unknown (Hannappel, 2010). In a human proteomic study, a fraction of Tβ4 and Tβ10 was found to be acetylated on the ε-amino group of Lys14 (Hannappel, 2010) (Fig. 3). By abolishing the salt bridge with Glu334, acetylation of Lys14 can be predicted to decrease significantly the affinity for G-actin (Didry et al., 2012) and to switch the sequestering factors Tβ4/10 into factors promoting unidirectional actin assembly at filament barbed ends. Other conserved PTMs in Tβ4/10 are targeting other highly buried residues in Tβ4:G-actin interface

(Fig. 3). They may switch or weaken the interaction with G-actin and promote interactions with other partners (Fig. 4B). Future experiments are required to understand better how PTMs may regulate in vivo the functions of β-thymosins or other WH2/βT domains.

Their full sequence interacts with G-actin, either in a stable or dynamic manner, and contributes to their functions with actin. The different MoREs that are important for the regulation of Tβ4 interactions with G-actin are widely distributed along its sequence and include a short N-terminal amphipathic α-helix which exists transiently in the free state and stably in the bound state, a basic residue at position 14 in their linker region that provides a strong electrostatic interaction with an acidic residue of actin, the central consensus LKKT/V motif, and a short C-terminal α-helix more specific to Tβ4 sequence (Fig. 3). It remains to be seen if these regions are also important for other intracellular or extracellular regulations of 5 kDa β-thymosins. Finally, their specific regulations can also result from the addition of their small individual effects that can compensate their low affinity. Their activities or additional functions may then specifically emerge only beyond a certain threshold of binding sites, such as for WH2 repeats in actin assembly. Their low interaction can also be compensated by a synergy with another protein/domains which is required to reveal their interactions and regulations, which is otherwise nonexistent or insignificant in their isolated state. They can then modulate the known functions of the other partner(s) with their small sequence.

Understanding the structure–function relationships of βT/WH2 domains will require combining further integrative structural and functional approaches to study their subtle and coordinated regulations in many pathways with multiple partners. This will help to control better the pleiotropic activities of β-thymosins in cell survival and repair, and hopefully the engineering of variants of βT with enhanced therapeutic potency in various diseases.

REFERENCES

Aguda, A. H., Xue, B., Irobi, E., Preat, T., & Robinson, R. C. (2006). The structural basis of actin interaction with multiple WH2/beta-thymosin motif-containing proteins. *Structure*, *14*(3), 469–476.

Ahuja, R., Pinyol, R., Reichenbach, N., Custer, L., Klingensmith, J., Kessels, M. M., et al. (2007). Cordon-bleu is an actin nucleation factor and controls neuronal morphology. *Cell*, *131*(2), 337–350.

Bednarek, R., Boncela, J., Smolarczyk, K., Cierniewska-Cieslak, A., Wyroba, E., & Cierniewski, C. S. (2008). Ku80 as a novel receptor for thymosin beta4 that mediates

its intracellular activity different from G-actin sequestering. *The Journal of Biological Chemistry, 283*(3), 1534–1544.
Bosch, M., Le, K. H., Bugyi, B., Correia, J. J., Renault, L., & Carlier, M. F. (2007). Analysis of the function of Spire in actin assembly and its synergy with formin and profilin. *Molecular Cell, 28*(4), 555–568.
Brieger, A., Plotz, G., Zeuzem, S., & Trojan, J. (2007). Thymosin beta 4 expression and nuclear transport are regulated by hMLH1. *Biochemical and Biophysical Research Communications, 364*(4), 731–736.
Campellone, K. G., & Welch, M. D. (2010). A nucleator arms race: Cellular control of actin assembly. *Nature Reviews. Molecular Cell Biology, 11*(4), 237–251.
Carlier, M. F., Didry, D., Erk, I., Lepault, J., Van Troys, M. L., Vandekerckhove, J., et al. (1996). Tbeta 4 is not a simple G-actin sequestering protein and interacts with F-actin at high concentration. *The Journal of Biological Chemistry, 271*(16), 9231–9239.
Carlier, M. F., Husson, C., Renault, L., & Didry, D. (2011). Control of actin assembly by the WH2 domains and their multifunctional tandem repeats in Spire and Cordon-Bleu. *International Review of Cell and Molecular Biology, 290*, 55–85.
Carlier, M. F., Jean, C., Rieger, K. J., Lenfant, M., & Pantaloni, D. (1993). Modulation of the interaction between G-actin and thymosin beta 4 by the ATP/ADP ratio: Possible implication in the regulation of actin dynamics. *Proceedings of the National Academy of Sciences of the United States of America, 90*(11), 5034–5038.
Chen, Z., Borek, D., Padrick, S. B., Gomez, T. S., Metlagel, Z., Ismail, A. M., et al. (2010). Structure and control of the actin regulatory WAVE complex. *Nature, 468*(7323), 533–538.
Chereau, D., Kerff, F., Graceffa, P., Grabarek, Z., Langsetmo, K., & Dominguez, R. (2005). Actin-bound structures of Wiskott-Aldrich syndrome protein (WASP)-homology domain 2 and the implications for filament assembly. *Proceedings of the National Academy of Sciences of the United States of America, 102*(46), 16644–16649.
Chhabra, E. S., & Higgs, H. N. (2006). INF2 Is a WASP homology 2 motif-containing formin that severs actin filaments and accelerates both polymerization and depolymerization. *The Journal of Biological Chemistry, 281*(36), 26754–26767.
Chhabra, E. S., Ramabhadran, V., Gerber, S. A., & Higgs, H. N. (2009). INF2 is an endoplasmic reticulum-associated formin protein. *Journal of Cell Science, 122*(Pt. 9), 1430–1440.
Crockford, D., Turjman, N., Allan, C., & Angel, J. (2010). Thymosin beta4: Structure, function, and biological properties supporting current and future clinical applications. *Annals of the New York Academy of Sciences, 1194*, 179–189.
Crooks, G. E., Hon, G., Chandonia, J. M., & Brenner, S. E. (2004). WebLogo: A sequence logo generator. *Genome Research, 14*(6), 1188–1190.
Czisch, M., Schleicher, M., Horger, S., Voelter, W., & Holak, T. A. (1993). Conformation of thymosin beta 4 in water determined by NMR spectroscopy. *European Journal of Biochemistry, 218*(2), 335–344.
De La Cruz, E. M., Ostap, E. M., Brundage, R. A., Reddy, K. S., Sweeney, H. L., & Safer, D. (2000). Thymosin-beta(4) changes the conformation and dynamics of actin monomers. *Biophysical Journal, 78*(5), 2516–2527.
Dhaese, S., Vandepoele, K., Waterschoot, D., Vanloo, B., Vandekerckhove, J., Ampe, C., et al. (2009). The mouse thymosin beta15 gene family displays unique complexity and encodes a functional thymosin repeat. *Journal of Molecular Biology, 387*(4), 809–825.
Didry, D., Cantrelle, F. X., Husson, C., Roblin, P., Moorthy, A. M., Perez, J., et al. (2012). How a single residue in individual beta-thymosin/WH2 domains controls their functions in actin assembly. *The EMBO Journal, 31*(4), 1000–1013.
Domanski, M., Hertzog, M., Coutant, J., Gutsche-Perelroizen, I., Bontems, F., Carlier, M. F., et al. (2004). Coupling of folding and binding of thymosin beta4 upon

interaction with monomeric actin monitored by nuclear magnetic resonance. *The Journal of Biological Chemistry, 279*(22), 23637–23645.

Dominguez, R. (2009). Actin filament nucleation and elongation factors—Structure-function relationships. *Critical Reviews in Biochemistry and Molecular Biology, 44*(6), 351–366.

dos Remedios, C. G., Chhabra, D., Kekic, M., Dedova, I. V., Tsubakihara, M., Berry, D. A., et al. (2003). Actin binding proteins: Regulation of cytoskeletal microfilaments. *Physiological Reviews, 83*(2), 433–473.

Eadie, J. S., Kim, S. W., Allen, P. G., Hutchinson, L. M., Kantor, J. D., & Zetter, B. R. (2000). C-terminal variations in beta-thymosin family members specify functional differences in actin-binding properties. *Journal of Cellular Biochemistry, 77*(2), 277–287.

Fan, Y., Gong, Y., Ghosh, P. K., Graham, L. M., & Fox, P. L. (2009). Spatial coordination of actin polymerization and ILK-Akt2 activity during endothelial cell migration. *Developmental Cell, 16*(5), 661–674.

Finn, R. D., Coggill, P., Eberhardt, R. Y., Eddy, S. R., Mistry, J., Mitchell, A. L., et al. (2016). The Pfam protein families database: Towards a more sustainable future. *Nucleic Acids Research, 44*(D1), D279–D285.

Fuxreiter, M. (2012). Fuzziness: Linking regulation to protein dynamics. *Molecular BioSystems, 8*(1), 168–177.

Fuxreiter, M., & Tompa, P. (2012). Fuzzy complexes: A more stochastic view of protein function. *Advances in Experimental Medicine and Biology, 725*, 1–14.

Fuxreiter, M., Toth-Petroczy, A., Kraut, D. A., Matouschek, A., Lim, R. Y., Xue, B., et al. (2014). Disordered proteinaceous machines. *Chemical Reviews, 114*(13), 6806–6843.

Gaucher, J. F., Mauge, C., Didry, D., Guichard, B., Renault, L., & Carlier, M. F. (2012). Interactions of isolated C-terminal fragments of neural Wiskott-Aldrich syndrome protein (N-WASP) with actin and Arp2/3 complex. *The Journal of Biological Chemistry, 287*(41), 34646–34659.

Goldstein, A. L. (2007). History of the discovery of the thymosins. *Annals of the New York Academy of Sciences, 1112*, 1–13.

Goldstein, A. L., & Kleinman, H. K. (2015). Advances in the basic and clinical applications of thymosin beta4. *Expert Opinion on Biological Therapy, 15*(Suppl. 1), S139–S145.

Guharoy, M., Szabo, B., Martos, S. C., Kosol, S., & Tompa, P. (2013). Intrinsic structural disorder in cytoskeletal proteins. *Cytoskeleton (Hoboken, N.J.), 70*, 550–571.

Gutsche-Perelroizen, I., Lepault, J., Ott, A., & Carlier, M. F. (1999). Filament assembly from profilin-actin. *The Journal of Biological Chemistry, 274*(10), 6234–6243.

Hannappel, E. (2010). Thymosin beta4 and its posttranslational modifications. *Annals of the New York Academy of Sciences, 1194*, 27–35.

Hertzog, M., van Heijenoort, C., Didry, D., Gaudier, M., Coutant, J., Gigant, B., et al. (2004). The beta-thymosin/WH2 domain; structural basis for the switch from inhibition to promotion of actin assembly. *Cell, 117*(5), 611–623.

Hertzog, M., Yarmola, E. G., Didry, D., Bubb, M. R., & Carlier, M. F. (2002). Control of actin dynamics by proteins made of beta-thymosin repeats: The actobindin family. *The Journal of Biological Chemistry, 277*(17), 14786–14792.

Higgs, H. N., & Pollard, T. D. (1999). Regulation of actin polymerization by Arp2/3 complex and WASp/Scar proteins. *The Journal of Biological Chemistry, 274*(46), 32531–32534.

Huff, T., Muller, C. S., & Hannappel, E. (2007). Thymosin beta4 is not always the main beta-thymosin in mammalian platelets. *Annals of the New York Academy of Sciences, 1112*, 451–457.

Huff, T., Muller, C. S., Otto, A. M., Netzker, R., & Hannappel, E. (2001). beta-Thymosins, small acidic peptides with multiple functions. *The International Journal of Biochemistry & Cell Biology, 33*(3), 205–220.

Husson, C., Cantrelle, F. X., Roblin, P., Didry, D., Le, K. H., Perez, J., et al. (2010). Multifunctionality of the beta-thymosin/WH2 module: G-actin sequestration, actin filament growth, nucleation, and severing. *Annals of the New York Academy of Sciences, 1194*, 44–52.

Husson, C., Renault, L., Didry, D., Pantaloni, D., & Carlier, M. F. (2011). Cordon-Bleu uses WH2 domains as multifunctional dynamizers of actin filament assembly. *Molecular Cell, 43*(3), 464–477.

Irobi, E., Aguda, A. H., Larsson, M., Guerin, C., Yin, H. L., Burtnick, L. D., et al. (2004). Structural basis of actin sequestration by thymosin-beta4: Implications for WH2 proteins. *The EMBO Journal, 23*(18), 3599–3608.

Kelly, A. E., Kranitz, H., Dotsch, V., & Mullins, R. D. (2006). Actin binding to the central domain of WASP/Scar proteins plays a critical role in the activation of the Arp2/3 complex. *The Journal of Biological Chemistry, 281*(15), 10589–10597.

Kinosian, H. J., Selden, L. A., Gershman, L. C., & Estes, J. E. (2000). Interdependence of profilin, cation, and nucleotide binding to vertebrate non-muscle actin. *Biochemistry, 39*(43), 13176–13188.

Koestler, S. A., Rottner, K., Lai, F., Block, J., Vinzenz, M., & Small, J. V. (2009). F- and G-actin concentrations in lamellipodia of moving cells. *PLoS One, 4*(3), e4810.

Letunic, I., Doerks, T., & Bork, P. (2015). SMART: Recent updates, new developments and status in 2015. *Nucleic Acids Research, 43*(Database issue), D257–D260.

Liverman, A. D., Cheng, H. C., Trosky, J. E., Leung, D. W., Yarbrough, M. L., Burdette, D. L., et al. (2007). Arp2/3-independent assembly of actin by Vibrio type III effector VopL. *Proceedings of the National Academy of Sciences of the United States of America, 104*(43), 17117–17122.

Mannherz, H. G., & Hannappel, E. (2009). The beta-thymosins: Intracellular and extracellular activities of a versatile actin binding protein family. *Cell Motility and the Cytoskeleton, 66*(10), 839–851.

Oldfield, C. J., Cheng, Y., Cortese, M. S., Romero, P., Uversky, V. N., & Dunker, A. K. (2005). Coupled folding and binding with alpha-helix-forming molecular recognition elements. *Biochemistry, 44*(37), 12454–12470.

Pernier, J., Orban, J., Avvaru, B. S., Jegou, A., Romet-Lemonne, G., Guichard, B., et al. (2013). Dimeric WH2 domains in Vibrio VopF promote actin filament barbed-end uncapping and assisted elongation. *Nature Structural & Molecular Biology, 20*(9), 1069–1076.

Quinlan, M. E., Heuser, J. E., Kerkhoff, E., & Mullins, R. D. (2005). Drosophila Spire is an actin nucleation factor. *Nature, 433*(7024), 382–388.

Redell, J. B., Xue-Bian, J. J., Bubb, M. R., & Crow, T. (2007). One-trial in vitro conditioning regulates an association between the beta-thymosin repeat protein Csp24 and actin. *Neuroscience, 148*(2), 413–420.

Renault, L., Deville, C., & van Heijenoort, C. (2013). Structural features and interfacial properties of WH2, beta-thymosin domains and other intrinsically disordered domains in the regulation of actin cytoskeleton dynamics. *Cytoskeleton (Hoboken, N.J.), 70*(11), 686–705.

Safer, D., Sosnick, T. R., & Elzinga, M. (1997). Thymosin beta 4 binds actin in an extended conformation and contacts both the barbed and pointed ends. *Biochemistry, 36*(19), 5806–5816.

Sopko, N., Qin, Y., Finan, A., Dadabayev, A., Chigurupati, S., Qin, J., et al. (2011). Significance of thymosin beta4 and implication of PINCH-1-ILK-alpha-parvin (PIP) complex in human dilated cardiomyopathy. *PLoS One, 6*(5), e20184.

Sosne, G., Qiu, P., Goldstein, A. L., & Wheater, M. (2010). Biological activities of thymosin beta4 defined by active sites in short peptide sequences. *FASEB Journal, 24*(7), 2144–2151.

Sribenja, S., Wongkham, S., Wongkham, C., Yao, Q., & Chen, C. (2013). Roles and mechanisms of beta-thymosins in cell migration and cancer metastasis: An update. *Cancer Investigation*, *31*(2), 103–110.

Stoll, R., Voelter, W., & Holak, T. A. (1997). Conformation of thymosin beta 9 in water/fluoroalcohol solution determined by NMR spectroscopy. *Biopolymers*, *41*(6), 623–634.

Tam, V. C., Serruto, D., Dziejman, M., Brieher, W., & Mekalanos, J. J. (2007). A type III secretion system in Vibrio cholerae translocates a formin/spire hybrid-like actin nucleator to promote intestinal colonization. *Cell Host & Microbe*, *1*(2), 95–107.

Tompa, P. (2003). Intrinsically unstructured proteins evolve by repeat expansion. *Bioessays*, *25*(9), 847–855.

Tompa, P., Szasz, C., & Buday, L. (2005). Structural disorder throws new light on moonlighting. *Trends in Biochemical Sciences*, *30*(9), 484–489.

Toth, M. A., Majoros, A. K., Vig, A. T., Migh, E., Nyitrai, M., Mihaly, J., et al. (2016). Biochemical activities of the Wiskott-Aldrich syndrome homology region 2 domains of sarcomere length short (SALS) protein. *The Journal of Biological Chemistry*, *291*(2), 667–680.

Uversky, V. N. (2010). Targeting intrinsically disordered proteins in neurodegenerative and protein dysfunction diseases: Another illustration of the D(2) concept. *Expert Review of Proteomics*, 7(4), 543–564.

van der Lee, R., Buljan, M., Lang, B., Weatheritt, R. J., Daughdrill, G. W., Dunker, A. K., et al. (2014). Classification of intrinsically disordered regions and proteins. *Chemical Reviews*, *114*(13), 6589–6631.

Van Troys, M., Dhaese, S., Vandekerckhove, J., & Ampe, C. (2007). Multirepeat beta-thymosins. In P. Lappalainen (Ed.), *Actin-monomer-binding proteins*, New York: Springer Science + Business Media, p. 71.

Van Troys, M., Ono, K., Dewitte, D., Jonckheere, V., De Ruyck, N., Vandekerckhove, J., et al. (2004). Tetrathymosinbeta is required for actin dynamics in Caenorhabditis elegans and acts via functionally different actin-binding repeats. *Molecular Biology of the Cell*, *15*(10), 4735–4748.

Xue, B., Leyrat, C., Grimes, J. M., & Robinson, R. C. (2014). Structural basis of thymosin-beta4/profilin exchange leading to actin filament polymerization. *Proceedings of the National Academy of Sciences of the United States of America*, *111*(43), E4596–E4605.

Xue, B., & Robinson, R. C. (2013). Guardians of the actin monomer. *European Journal of Cell Biology*, *92*(10–11), 316–332.

Yamaguchi, H., Miki, H., Suetsugu, S., Ma, L., Kirschner, M. W., & Takenawa, T. (2000). Two tandem verprolin homology domains are necessary for a strong activation of Arp2/3 complex-induced actin polymerization and induction of microspike formation by N-WASP. *Proceedings of the National Academy of Sciences of the United States of America*, *97*(23), 12631–12636.

Yarmola, E. G., Parikh, S., & Bubb, M. R. (2001). Formation and implications of a ternary complex of profilin, thymosin beta 4, and actin. *The Journal of Biological Chemistry*, *276*(49), 45555–45563.

Zuchero, J. B., Belin, B., & Mullins, R. D. (2012). Actin binding to WH2 domains regulates nuclear import of the multifunctional actin regulator JMY. *Molecular Biology of the Cell*, *23*(5), 853–863.

Zuchero, J. B., Coutts, A. S., Quinlan, M. E., Thangue, N. B., & Mullins, R. D. (2009). p53-cofactor JMY is a multifunctional actin nucleation factor. *Nature Cell Biology*, *11*(4), 451–459.

CHAPTER THREE

Actin-Induced Structure in the Beta-Thymosin Family of Intrinsically Disordered Proteins

B. Xue*[,1], R.C. Robinson*[,†,‡,§,¶]

*Institute of Molecular and Cell Biology, A*STAR (Agency for Science, Technology and Research), Biopolis, Singapore
†Yong Loo Lin School of Medicine, National University of Singapore, Singapore
‡NTU Institute of Structural Biology, Nanyang Technological University, Singapore
§School of Biological Sciences, Nanyang Technological University, Singapore
¶Lee Kong Chan School of Medicine, Singapore
[1]Corresponding author: e-mail address: bxue@imcb.a-star.edu.sg

Contents

Abstract

Thymosin β4 (Tβ4) is a 43-amino acid signature motif peptide that defines the beta-thymosin (βT) family of proteins. βTs are intrinsically unstructured in their free states and undergo disorder-to-order transitions in carrying out their biological functions. This property poses challenges in determining their 3D structures, mainly favoring structural studies on the complexes formed between βTs and their interaction partners. One of the βTs' primary binding partners is monomeric actin, a major component of the cytoskeleton in eukaryotic cells. Tβ4's role in this system is to maintain the highly concentrated pool of monomeric actin that can be accessed through profilin by actin filament nucleating machineries. Here, we give an account of the structures of βTs that have been illuminated by nuclear magnetic resonance (NMR) and X-ray crystallography.

Vitamins and Hormones, Volume 102
ISSN 0083-6729
http://dx.doi.org/10.1016/bs.vh.2016.04.007

NMR has been the method of choice for probing regions that have intrinsic conformational preference within the largely disordered βTs in their native states in solution. X-ray crystallography has demonstrated at atomic detail how βTs interact with actin. Detailed analysis of these structures highlights the disorder-to-order transition of Tβ4 in binding to actin and its isoform specificity.

1. INTRODUCTION

Beta-thymosins (βTs) are a family of proteins that are comprised of one or more of a ~43 amino acid signature motif (βT repeat) that was first defined for the prototypical single motif protein, thymosin β4 (Tβ4). Tβ4 was discovered in the 1970s in an effort to extract hormones from thymus tissue (Goldstein, 2007). In the early 1990s, Tβ4 was found to bind and sequester monomeric actin, a major component of the cytoskeleton in eukaryotic cells (Safer, Elzinga, & Nachmias, 1991, Safer, Golla, & Nachmias, 1990). Around that time, the first nuclear magnetic resonance (NMR) structural studies on isolated Tβ4 began, and later on the Tβ4:actin complex (Czisch, Schleicher, Horger, Voelter, & Holak, 1993; Domanski et al., 2004; Zarbock et al., 1990). X-ray crystallography studies were only possible once strategies were developed to stabilize the intrinsically disordered Tβ4 in its complex with actin (Aguda, Xue, Irobi, Preat, & Robinson, 2006; Hertzog, van Heijenoort, Didry, et al., 2004; Irobi et al., 2004; Xue, Leyrat, Grimes, & Robinson, 2014). The accumulation of these structural data, together with the discovery of homologs through protein and nucleotide sequencing, allowed for a precise definition of the features of the βT repeat: it has a consensus LKKT motif at the center of the sequence, which is flanked by two moderately conserved segments that tend to weakly form amphipathic α-helices in isolation, yet become stabilized when bound to actin. Since similar LKKT motifs, N-terminal α-helices and actin-binding properties can also be found in the Wiskott–Aldrich syndrome protein (WASP) homology domain 2 (WH2)-containing proteins, βTs and WH2s are often grouped in the same functional superfamily (Paunola, Mattila, & Lappalainen, 2002), whether or not they arose from a common ancestor (Edwards, 2004).

βT repeats appear within proteins either singly or as multiple repeats. Tβ4 and close homologs, such as Tβ10 and Tβ15, contain one βT repeat, whereas other members contain from 2 to 27 such repeats (Xue & Robinson, 2013). One widely studied multi-βT repeat protein is the three βT repeat *Drosophila* ciboulot, which was the first protein to yield an actin-bound βT repeat structure by X-ray crystallography (Hertzog et al., 2004). Generally, single- and multi-repeat βTs appear to be mutually exclusive in their phylogenetic distributions (Van Troys, Dhaese, Vandekerckhove, & Ampe, 2007), with few exceptions (Dehal, Satou, Campbell, et al., 2002; Dhaese et al., 2009). Single- and multirepeat βTs also show differences in their actin-binding functionality: all of the studied single-repeat βTs act like Tβ4 in inhibiting their bound actin from forming or joining a filament, whereas multirepeat βTs, such as ciboulot and the four-repeat TetraThymosinβ, often allow their bound actin to join the fast-growing, barbed end of a preexisting actin filament, a profilin-like activity (Hertzog et al., 2004; Van Troys et al., 2004). Recent studies have attributed the functional difference to the stability of the C-terminal helices within the βT repeats (Didry, Cantrelle, Husson, et al., 2012; Xue et al., 2014).

From a structural point of view, βTs belong to a class of proteins termed intrinsically disordered/unstructured proteins (IDPs/IUPs) (Andreeva, Howorth, Chothia, Kulesha, & Murzin, 2014; Oldfield & Dunker, 2014; van der Lee, Buljan, Lang, et al., 2014). Members of this class of proteins contain regions, or in the case of βTs, the entire protein, that lack ordered structure in their native states, but undergo disorder-to-order transitions in carrying out their biological functions (Uversky, 2002). A large collection of IDPs/IUPs can be found in DisProt, the curated database of disordered proteins (Sickmeier, Hamilton, LeGall, et al., 2007), among which are Tβ4 and other actin-binding proteins, such as WASP, tropomodulin, and troponin. Despite being generally unstructured, IDPs/IUPs do have regions with intrinsic conformational preferences, which are further classified into two motifs by SCOP2 (Andreeva et al., 2014), flexible disordered helical regions and extended binding regions. We give a brief historical account of the structural studies on βTs and present a detailed analysis of the interaction between Tβ4 and actin, which reveals that βTs contain both of the IDPs/IUPs motifs.

2. BETA-THYMOSINS IN SOLUTION

2.1 Tβ4 in Mixed Organic-Aqueous Solvent

Due to the small size and unstructured nature of βTs, solution NMR has been the method of choice to study their isolated structures. The first βT NMR structure determined was that of bovine Tβ4, which is identical in amino acid sequence to human Tβ4 (Zarbock et al., 1990). The experiments were conducted in solutions of 60% (v/v) trifluoroethanol-d_3 or 50% (v/v) hexafluoroisopropyl-d_2 in water, because no obvious structure could be detected by circular dichroism and ^{1}H NMR spectra for Tβ4 in aqueous solutions. The presence of alcohols in the solution promoted the formation of ordered helical structures in Tβ4 at residues 4–16 and 30–40. Due to insufficient nuclear Overhauser enhancement (NOE) signals for residues 17 and 18, the relative orientation of these two helices could not be determined. In addition to the helices, there was also a loop region between residues 24–29 of Tβ4.

2.2 Tβ4 in Aqueous Solution

The NMR study was subsequently continued to probe the influences of temperature and buffer pH on the structure of Tβ4 in aqueous solutions (Czisch et al., 1993). The measurements were performed at 1, 4, 14, and 29°C for samples at pH 3.0, and at 1 and 14°C for samples at pH 6.5, which were the conditions favorable for minimizing amide exchange. Temperature was found to have a significant effect on the secondary structure Tβ4. Residues 5–16 had a tendency to form an α-helix, which was stable at up to 14°C and became destabilized at higher temperatures. Residues 31–37 also showed a propensity to adopt a helical conformation at 1 and 4°C, albeit less well defined than residues 5–16; however, residues 31–37 became completely random at 14°C. In agreement, the helical content percentages determined from the circular dichroism spectra were 15%, 9%, and 7% for 4°C, 9°C, and 20°C, respectively. The first four residues at the N-terminus of Tβ4 were always disordered irrespective of the temperature.

2.3 Tβ10 in Mixed Organic-Aqueous Solvent

Following these studies on bovine Tβ4, the NMR study was extended to bovine Tβ10 which is a close homolog of Tβ4 and was originally called Tβ9 (Stoll, Voelter, & Holak, 1997). Two helical regions at residues 4–27 and 32–41 of Tβ10 were detected in a solution of 40% (v/v) 1,1,1,3,3,3-hexafluoro-2-propanol-d2 in water, which were separated by a poorly defined loop region between residues 28–31 (Fig. 1A). The first helix of Tβ10 was, however, twice the length of its counterpart in Tβ4. Similar to Tβ4, the lack of medium or long-range NOEs for the loop region impeded the determination of the relative orientation of the two helices in Tβ10. Moreover, the N-terminus of Tβ10 also showed random-coil structure.

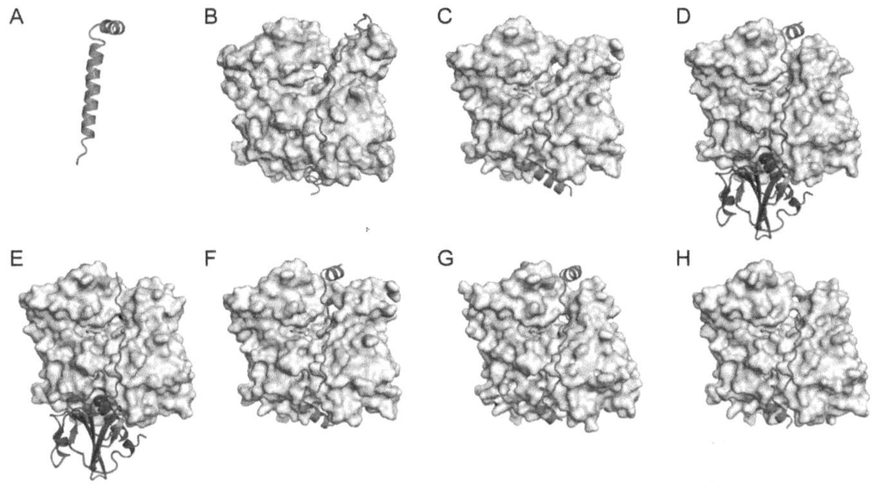

Fig. 1 Structures of beta-thymosins. The structures were solved by X-ray crystallography unless stated otherwise. Beta-thymosins are shown as *red* cartoons; actin is shown as a grey surface, with its bound nucleotide shown as *green* sticks; gelsolin domain 1 (G1) in two of the structures is shown as a *blue* cartoon. PDB IDs are given following the abbreviations of the structures, whose meanings can be found in the main text. (A) Solution NMR structure of Tβ10 (1HJ0). (B) NMR-based model of Tβ4:actin (1UY5). (C) Cib1N:actin (1SQK). (D) G1-Tβ4:actin (1T44). (E) G1-Cib23:actin (2FF6). (F) Structure-based model of Tβ4:actin. (G) Tβ4FL:actin (4PL7). (H) Tβ4N:actin (4PL8). This and the subsequent structural figures were prepared with the PyMOL Molecular Graphics System (DeLano Scientific LLC). (See the color plate.)

2.4 Tβ4 Mutants in Mixed Organic-Aqueous Solvent

To further correlate Tβ4's structure with its actin-binding activity, NMR spectra were recorded in 60% (v/v) trifluoroethanol for wild-type Tβ4 and its mutants (Simenel, Van Troys, Vandekerckhove, Ampe, & Delepierre, 2000). Consistent with the previously reported Tβ4 structures in alcohols, the wild-type conformation under these conditions consisted of two α-helices: residues 5–16, of which Lys11 to Lys16 form a more stable core; and the second helix from residues 31–39. Several point mutations were made in the first helix and in the LKKT motif (residues 17–22), the two main regions known to be responsible for actin binding. One such mutation, S15A, caused the first helix to become extended at its C-terminal end to include the LKKT motif, leading to a 50% reduction in its actin-binding activity. Another mutation, K11P, disrupted the formation of the first helix, resulting in complete loss of actin binding. These data demonstrated that the presence of the first helix and its correct orientation in relation to the LKKT motif are essential for the Tβ4:actin interaction.

Thus, these structural data indicate that βTs are largely unstructured in solution, but have a latent propensity to form two regions of helix, which can be induced by aqueous-organic mixed solvents or low temperatures.

3. BETA-THYMOSINS IN COMPLEX WITH ACTIN

3.1 NMR-Based Tβ4:Actin Model

Chemical cross-linking data had established that Tβ4 adopts an extended conformation when bound to actin, with its N-terminal region in contact to subdomain 1 of actin and its C-terminal region in close vicinity to His40 and Gln41 of actin subdomain 2 (De La Cruz et al., 2000; Reichert, Heintz, Echner, Voelter, & Faulstich, 1996; Safer, Sosnick, & Elzinga, 1997). The first high-resolution model for the structural basis of the Tβ4:actin interaction was constructed from NMR data using selective labeling of Tβ4 (Domanski et al., 2004). The relatively large size of the Tβ4:actin complex (47 kDa) had precluded NMR spectroscopy studies of the whole complex. However, a complete assignment of amide protons and nitrogen resonances for Tβ4 was successfully achieved, while bound to actin, using a combination of selective labeling of Leu, Thr, and Lys of Tβ4 in a three-dimensional ^{1}H-^{15}N NOESY-HSQC experiment. The actin-bound model (PDB ID: 1UY5; Fig. 1B) was constrained in its positioning on actin by the distance restraints arising from the cross-linking data. This model predicts that the entire length of Tβ4 is engaged in binding to actin, causing the Tβ4 backbone to adopt a conformation, with the two terminal helices flanking the extended region. Cleavage of the actin DNase I binding loop by subtilisin did not modify the signal arising from Tβ4 in the complex, suggesting that the Tβ4 backbone is not in the vicinity of actin residues 42–47. These data established that Tβ4 terminal segments form helices that are induced and/or stabilized through binding to actin.

3.2 X-ray Structures

3.2.1 C-Terminal Segment of Tβ4

Further breakthroughs in understanding the Tβ4:actin interaction came with novel strategies to stabilize the complex for X-ray crystallography. Through creating a hybrid protein between domain 1 of gelsolin (G1), a protein that severs and caps actin filaments (Nag, Larsson, Robinson, & Burtnick, 2013), and the C-terminal segment of Tβ4 (hereafter referred to as G1-Tβ4), the structure of the G1-Tβ4:actin complex was elucidated at 2 Å resolution (PDB ID: 1T44; Fig. 1C) (Irobi et al., 2004). This structure revealed that the C-terminal segment of Tβ4 forms a helix which bridges, and narrows, the gap between subdomains 2 and 4 of actin, in contrast to the NMR model where it only contacts subdomain 2. The new conformation provided an explanation for the biochemical evidence that Lys38 of Tβ4, located at the Tβ4 C-terminus, could be crosslinked to Gln41 in the DNase I loop of actin (Safer et al., 1997). In the G1-Tβ4: actin structure, the flexible actin DNase I loop (residues 39–50) is, like in most other actin structures, disordered. The distance between the main chain Cα carbon atoms of Tβ4's Lys38 and actin's Pro38, which is the closest residue to Gln41 that was resolved in the G1-Tβ4:actin structure, is 15.1 Å. Given the length of the side chains of Lys (6.4 Å) and Gln (4.4 Å), the flexibility of actin's DNase I loop, and the unstructured nature of Tβ4 before becoming fully bound to actin, it is reasonable to envisage that these two residues may be accessed by a zero-length cross-linker. Therefore, this crystal structure revealed the binding geometry of Tβ4's C-terminal segment to actin that induces and/or stabilizes the helical conformation of Tβ4.

3.2.2 N-Terminal Segment of Ciboulot's First βT Repeat

In the same year as the determination of the G1-Tβ4:actin X-ray structure (2004), Hertzog et al. determined the X-ray structure of a 25-residue peptide from the three-repeat β-thymosin protein, ciboulot, in complex with actin at 2.5 Å (PDB ID: 1SQK; Fig. 1D) (Hertzog et al., 2004). The peptide in the structure corresponds to the N-terminal segment of ciboulot's first βT repeat (Asp10 to Ser34, hereafter referred to as Cib1N), which has moderate sequence similarity to Tβ4. This X-ray structure is of considerable significance because it was the first to reveal how the N-terminal helix and the central LKKT motif of the βT/WH2 superfamily of proteins interact with actin, and the mode of interaction was later confirmed repeatedly by other complex structures of actin-bound WH2 motifs (Chereau et al., 2005; Lee et al., 2007). The N-terminal α-helix of Cib1N in the structure was extended relative to later βT/WH2:actin structures. Later it became apparent that the helix was extended by stabilization through crystal packing, and that the actual actin-bound helix is one turn shorter (Didry et al., 2012). Nevertheless, this work demonstrated that actin binding induces and/or stabilizes the βT/WH2 N-terminal helix and revealed the binding geometry of this helix and the conserved LKKT motif.

3.2.3 C-Terminal Segment of Ciboulot's Second βT Repeat

Utilizing a similar strategy employed for the G1-Tβ4 hybrid, a hybrid protein comprised of G1 and ciboulot's second and third βT repeats (hereafter referred to as G1-Cib23), was used in an attempt to determine the structure of this hybrid bound to actin (PDB ID: 2FF6; Fig. 1E) (Aguda et al., 2006). Although the hybrid contained nearly half of ciboulot (Glu72 to Ala129), the structure revealed only a short stretch of 11 residues from the C-terminal segment of ciboulot's second βT repeat (Glu72 to Lys82). This visible portion closely resembled Tβ4, in that it bound to the same place on the surface of actin. The last visible ciboulot residue, Lys82, corresponded to Lys31 in Tβ4, the first residue of the C-terminal helix. The absence of an ordered C-terminal helix in the G1-Cib23:actin structure may be attributed to the sequence differences between Tβ4 and ciboulot: Ser31 and Thr34, in Tβ4, are Asp81 and Ala84 in ciboulot, respectively. These substitutions were reasoned to cause destabilization of the helix, leading to partial loss of the actin binding by the C-terminal segment of ciboulot's second βT repeat. In agreement with such a hypothesis, the second repeat in isolation is the weakest in binding to actin when compared with the first and the third ciboulot repeats (Chereau et al., 2005). Thus, this work established that actin stabilization of the C-terminal helix is βT/WH2 module specific.

3.2.4 Structure-Based Models of Tβ4

A structure-based model for the full-length Tβ4:actin complex was constructed by combining the G1-Tβ4:actin and Cib1N:actin structures (Irobi et al., 2004). The model relied on the conserved central LKKT motif common to the βT/WH2 superfamily. By simply superimposing the actins from these two structures, the N-terminal segment of ciboulot's first βT repeat, from the Cib1N:actin structure, naturally connected, via its LKKT motif, to the C-terminal segment of Tβ4 in the G1-Tβ4:actin structure. Ciboulot residues were then substituted with the corresponding residues from Tβ4 to yield a model of full-length Tβ4. Overall, this model is of good credibility, as its construction was based on X-ray structures and involved minimum structural manipulations. Certain details are, however, expected to be less reliable, due to the sequence differences in these two proteins. The model was later further refined to take into account the presence of a helix-breaking proline (Pro4) at the N-terminus of Tβ4 (Fig. 1F) (Xue, Aguda, & Robinson, 2007). This proline residue effectively restricted the length of Tβ4's N-terminal helix, making the helix sterically more compatible with the formation of a ternary complex composed of actin, Tβ4 and profilin (Xue et al., 2014; Yarmola, Parikh, & Bubb, 2001). Sequence alignment indicates that this proline is generally absent in multi-repeat βTs and WH2 motifs, but is common among the single-repeat βTs (Edwards, 2012; Paunola et al., 2002), implying similar functional roles of this residue in other single-repeat βTs.

3.2.5 Full-Length and N-Terminal Segment of Tβ4

Recently, we reported two novel X-ray structures of Tβ4:actin complexes, which provided structural information on the actin interactions of full-length Tβ4 and, independently, its N-terminal segment (Xue et al., 2014). The full-length Tβ4:actin structure was made possible by creating a hybrid protein consisting of full-length human Tβ4 fused to *Pichia* actin (PDB ID: 4PL7; Fig. 1G, hereafter referred to as Tβ4FL). In this structure there was interpretable electron density for the majority of Tβ4 (Asp5 to Ala40), but a lack of electron density for the mobile termini, residues 1–4 and 41–43, respectively. The second structure was a complex between rabbit skeletal actin and a hybrid peptide comprising Tβ4's N-terminal segment prior to the LKKT motif (residues 1–16), the lysine-rich region of cordon-bleu (residues 1095–1106), and full-length Tβ4 (residues 1–43). Clear electron density was apparent for residues 4–25 for the full-length Tβ4 (residues 34–55 in the hybrid, hereafter referred to as Tβ4N), which were unambiguously assigned (PDB ID: 4PL8; Fig. 1H).

In good agreement with the NMR studies and the structure-based models, full-length Tβ4 in the Tβ4FL:actin structure interacts extensively with actin through its two terminal helices and throughout the connecting linker region. The validity of the structure could be crosschecked with the segments of Tβ4 in the G1-Tβ4:actin and Tβ4N:actin structures. Despite the differences in hybrid design and crystallization conditions, the full-length Tβ4 in Tβ4FL:actin corresponded well with its independently determined N- and C-terminal segments taken from Tβ4N:actin and G1-Tβ4:actin, respectively (Fig. 2). Not only did the main chains of Tβ4 in these structures superimpose precisely, but many side chains also adopted nearly identical conformations, suggesting their critical roles in anchoring of Tβ4 to the surface of actin. The first-ordered residue in Tβ4FL is Asp5, which lies at the beginning of the N-terminal helix, in agreement with the NMR measurements in aqueous solution at 14°C (Czisch et al., 1993). The N-terminus in Tβ4N:actin was extended by one residue to Pro4; a residue that has a minimal contact with actin. Thus, these structures defined the full-length Tβ4

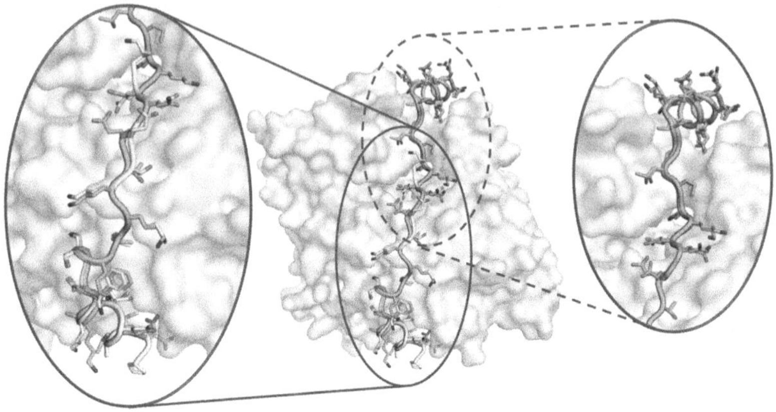

Fig. 2 Superimposition of the three known Tβ4:actin structures. Tβ4s in Tβ4FL:actin, Tβ4N:actin, and G1-Tβ4:actin are colored in *green*, *yellow*, and *violet*, respectively. Actin from Tβ4FL:actin is shown as a gray surface. (See the color plate.)

interaction with actin and demonstrated that many of the residues that could be induced to adopt helical conformations in aqueous-organic mixed solvents or at low temperatures are also functional as helices in interacting with actin.

4. Tβ4:ACTIN INTERFACE ANALYSIS

Here, we present details of the Tβ4:actin interface based on the analysis of the three available Tβ4:actin structures using PISA (Krissinel & Henrick, 2007). Of the 36 ordered Tβ4 residues in the Tβ4FL:actin structure, Met6, Ile9, Phe12, Leu17, Pro27, Leu28, Pro29, and Ile34 are the major contributors of hydrophobic interactions, whereas Asp5, Glu8, Glu10, Lys14, Lys18, Thr20, Thr22, Gln23, Lys25, Asn26, Lys31, Glu37, and Lys38 engage actin through hydrogen bonds and salt bridges (Fig. 3A); only eight residues (Ala7, Lys11, Asp13, Ser15, Glu32, Glu35, Gln39, and Ala40) have no contribution in actin binding, judged by the absence of buried surface area for these residues. However, as pointed out earlier (Xue et al., 2007), some of the noncontacting residues, for example, Asp13 and Ser15, contribute indirectly by stabilizing the actin-bound conformation of Tβ4.

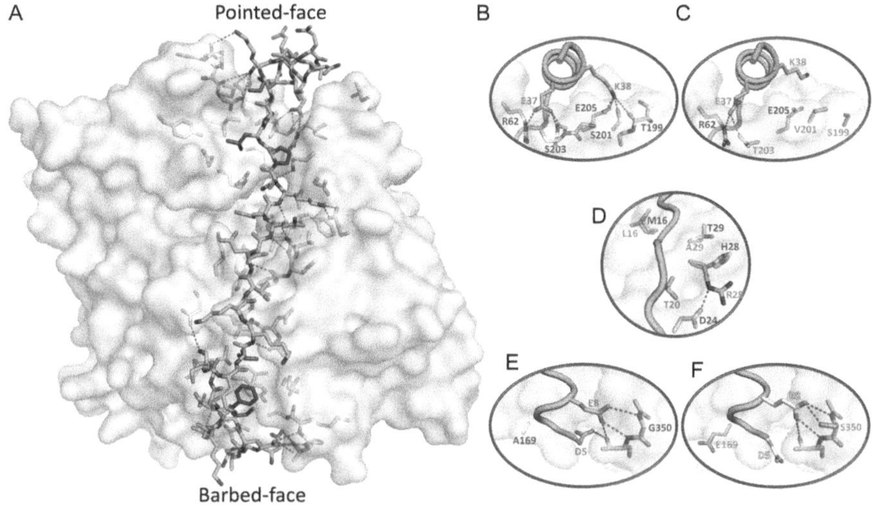

Fig. 3 Detailed interactions between Tβ4 and actin. (A) The Tβ4:actin interface based on the Tβ4FL:actin structure. Tβ4 is shown as *green* sticks except for its buried hydrophobic residues, which are colored in *chocolate*. Actin is shown as a gray surface, with its interface residues shown as *light blue* sticks. *Dashed lines* in *magenta* represent hydrogen bonds and/or salt bridges. (B–E) Differences between the *Pichia* and rabbit skeletal muscle isoforms on the pointed-face (B vs C), side (D), and the barbed-face (E vs F) of actin. Tβ4 residues from Tβ4FL, Tβ4N, and G1-Tβ4 are colored in *green*, *yellow*, and *violet*, respectively. Nonidentical actin residues from the *Pichia* and the muscle isoforms are colored in *light blue* and *orange*, respectively. (See the color plate.)

An interesting question arose when we considered the Tβ4-bound actins in these structures, which are isoforms from either rabbit skeletal muscle or from the yeast, *Pichia pastoris*. Measured tissue distributions of Tβ4 in mouse indicate that its concentration in muscle is much lower when compared with some other tissues such as spleen, brain, liver, and lung (Hannappel, 2007). Likewise, EST data, which have been shown to correlate with protein concentrations in human and mouse (Xue & Robinson, 2013), show similar patterns of Tβ4 enrichment in mammals and birds (interested readers may check the EST profiles for the individual entries under http://www.ncbi.nlm.nih.gov/unigene/?term=TMSB4X for more details). Therefore, the most biologically relevant binding partners for Tβ4 in mammals are the cytoplasmic β- and γ1-actins. *Pichia* actin is cytoplasmic. It is, however, only ~85% identical in sequence to mammalian cytoplasmic actins. Moreover, the *Pichia* actin–Tβ4 interaction is nonendogenous, as no βT family members are found in yeasts.

To address the above question, we performed a multiple sequence alignment of *Pichia*, rabbit skeletal muscle, and human cytoplasmic β- and γ1-actins. We then compiled all the interface residues according to the PISA analysis of the three Tβ4:actin structures, and indicated them alongside the aligned sequences (Fig. 4). Particular attention was paid to the residues involved in forming hydrogen bonds and salt bridges with Tβ4, because these interactions are critical in engaging the relatively hydrophilic Tβ4 peptide and in defining the specificity between Tβ4 and actin. There are altogether 59 interface residues on actin. Using the two human cytoplasmic actins as references, which have identical interfaces, the numbers of nonidentical interface residues are only 2 and 7 for the rabbit skeletal muscle and the *Pichia* actins, respectively, indicating the atomic details on Tβ4:actin interaction obtained from the rabbit skeletal muscle and *Pichia* isoforms are indeed highly relevant to the mammalian cytoplasmic actins. Moreover, the variations in these two isoforms do not completely overlap, but somewhat complement each other. For the two variations in the rabbit skeletal muscle isoform, M16L and T201V, the former is nonexistent in the *Pichia* isoform, whereas the latter becomes a more similar T201S substitution. Conversely, 6 out of the 7 variations found in the *Pichia* isoform are nonexistent in the rabbit skeletal muscle isoform. This complementarity allowed us to examine the interactions between Tβ4 and human cytoplasmic actin by combining the structural information from the three known structures.

We took the Tβ4FL:actin structure as the starting model and focused on the nonidentical interface residues on actin that were involved in hydrogen bonding with Tβ4. One region of interest is at the barbed-face of actin, in which Gly350 in *Pichia* actin corresponds to Ser350 in both rabbit skeletal muscle and human cytoplasmic actins. As shown by the comparison between the Tβ4N:actin and the Tβ4FL:actin structures (Fig. 3F vs E), the presence of the hydroxyl group on the side chain of Ser350 gave rise to one extra hydrogen bond to Tβ4's Glu8. Another region of interest lies at the opposite end, the pointed-face of actin, in which *Pichia* actin's Thr199, Ser201, and Ser203 are replaced by a serine and two threonines in human cytoplasmic actins. Given the similarity between serines and threonines, these variations are not expected to change the hydrogen-bonding patterns observed in the Tβ4FL:actin structure for this region (Fig. 3B). In fact, a threonine at residue 203 had been observed in the G1-Tβ4:actin structure to form a hydrogen bond to Tβ4's Glu37 (Fig. 3C). It should be pointed out that the overall hydrogen-bonding pattern for the muscle isoform is rather different in this region. The likely cause for the difference is the muscle

```
                       10        20        30        40        50        60
               ....|....|....|....|....|....|....|....|....|....|....|....|
P60709|hsB     -DDDIAALVVDNGSGMCKAGFAGDDAPRAVFPSIVGRPRHQGVMVGMGQKDSYVGDEAQS
P63261|hsG1    -EEE.....I..................................................
P68135|ocA1    DE.ETT...C.....LV...........................................
Q9P4D1|pp      DGE.V....I..........Y......HT....V..................F.......
Interface                     x     wMxBxMMxMxx                       x  Sx

                       70        80        90       100       110       120
               ....|....|....|....|....|....|....|....|....|....|....|....|
P60709|hsB     KRGILTLKYPIEHGIVTNWDDMEKIWHHTFYNELRVAPEEHPVLLTEAPLNPKANREKMT
P63261|hsG1    ............................................................
P68135|ocA1    ...............I..........................T.................
Q9P4D1|pp      .......R...........................L.............M...S......
Interface       Sxxx                          wS

                      130       140       150       160       170       180
               ....|....|....|....|....|....|....|....|....|....|....|....|
P60709|hsB     QIMFETFNTPAMYVAIQAVLSLYASGRTTGIVMDSGDGVTHTVPIYEGYALPHAILRLDL
P63261|hsG1    ............................................................
P68135|ocA1    ........V.......................L........N.............M....
Q9P4D1|pp      ........V..F..S.................L........V....A.FS.......I..
Interface                            xMxxxB                  xx

                      190       200       210       220       230       240
               ....|....|....|....|....|....|....|....|....|....|....|....|
P60709|hsB     AGRDLTDYLMKILTERGYSFTTTAEREIVRDIKEKLCYVALDFEQEMATAASSSSLEKSY
P63261|hsG1    ............................................................
P68135|ocA1    ....................V.......................N...............
Q9P4D1|pp      .............S....T.S.S....................D..LQ.SSSQ...I....
Interface                         M SxBMS xx xx  ww                        x

                      250       260       270       280       290       300
               ....|....|....|....|....|....|....|....|....|....|....|....|
P60709|hsB     ELPDGQVITIGNERFRCPEALFQPSFLGMESCGIHETTFNSIMKCDVDIRKDLYANTVLS
P63261|hsG1    ............................................................
P68135|ocA1    ...................T......I....A......Y.......I.........N.M.
Q9P4D1|pp      ................A.....H..V..L.AS..DQ..Y.........V..E..S.I.M.
Interface        xM

                      310       320       330       340       350       360
               ....|....|....|....|....|....|....|....|....|....|....|....|
P60709|hsB     GGTTMYPGIADRMQKEITALAPSTMKIKIIAPPERKYSVWIGGSILASLSTFQQMWISKQ
P63261|hsG1    ............................................................
P68135|ocA1    .........................................................T..
Q9P4D1|pp      .....F....E.....L......S..V..S...................G..........
Interface                                     w S xS   x  xxx xxBBw xx

                      370
               ....|....|....|
P60709|hsB     EYDESGPSIVHRKCF
P63261|hsG1    ...............
P68135|ocA1    ....A..........
Q9P4D1|pp      ...........L...
Interface
```

Fig. 4 Sequence alignment of actin isoforms. hsB and hsG1: *Homo sapiens* (human) β- and γ1-actins; ocA1: *Oryctolagus cuniculus* (rabbit) skeletal muscle actin; pp: *Pichia pastoris* (yeast) actin. Residues lying on the Tβ4:actin interface are generally denoted by an (x), except for those with less than 1.6 Å^2 in buried surface area (w) according to PISA (Krissinel & Henrick, 2007) and for those that are involved in hydrogen bonding with Tβ4 through their main chain (M), side chain (S), or both (B). *Red boxes* highlight nonidentical interface residues. (See the color plate.)

specific residue, Val201, which eliminates the possibility of a 2.5 Å hydrogen bond observed between *Pichia* actin's Ser201 and Tβ4's Lys38, leading to complete disengagement of Lys38 in G1-Tβ4:actin. A third region of interest is at residue 28 of actin, which harbors a histidine in *Pichia*, or an arginine in muscle and cytoplasmic actins (Fig. 3D). Despite having longer side chain and greater hydrogen-bonding potential, the arginine in Tβ4N:actin did not provide an extra hydrogen bond to Tβ4, because it was tied up by participating in an intramolecular hydrogen bond to Asp24 of actin. Taken together, the interactions between mammalian cytoplasmic actins and Tβ4 appear to be captured by the *Pichia* actin isoform structure despite the non-existence of Tβ4 in this species.

5. CONCLUSION

Tremendous progress has been made during the past two and half decades in elucidating the structures of βT family proteins, and in particular that of the family's archetypal member, Tβ4. A body of evidence has emerged, suggesting that two regions within the largely unstructured βT repeat have a higher propensity towards helix formation, which can be induced when the βT repeat executes its biological function in binding and sequestering monomeric actin. How well the C-terminal helix interacts with actin determines whether the βT repeat acts as a suppressor or facilitator of actin polymerization at the fast-growing barbed end of actin filaments. Advances in genomics have brought forth a large number of new members into the βT family, which await further structural and functional characterization.

In addition to sequestering actin, Tβ4 also participates in numerous biological functions such as stimulating wound healing and tissue repair (Malinda et al., 1999), suppressing inflammation (Young et al., 1999), and enhancing cardiomyocyte regeneration (Qian et al., 2012; Smart, Bollini, Dube, et al., 2011). However, the mechanisms behind many of Tβ4's hormonal activities remain elusive (Goldstein, Hannappel, & Kleinman, 2005). Once the pathways and receptors involved in these processes have been identified, new structural data will be required to compare how closely the disorder-to-order transitions mirror those determined for actin binding and to help drive drug development in these areas.

ACKNOWLEDGMENTS

B.X. and R.C.R. would like to thank the Agency for Science, Technology and Research (A*STAR), Singapore for support.

REFERENCES

Aguda, A. H., Xue, B., Irobi, E., Preat, T., & Robinson, R. C. (2006). The structural basis of actin interaction with multiple WH2/beta-thymosin motif-containing proteins. *Structure, 14*, 469–476.

Andreeva, A., Howorth, D., Chothia, C., Kulesha, E., & Murzin, A. G. (2014). SCOP2 prototype: A new approach to protein structure mining. *Nucleic Acids Research, 42*, D310–D314.

Chereau, D., Kerff, F., Graceffa, P., Grabarek, Z., Langsetmo, K., & Dominguez, R. (2005). Actin-bound structures of Wiskott-Aldrich syndrome protein (WASP)-homology domain 2 and the implications for filament assembly. *Proceedings of the National Academy of Sciences of the United States of America, 102*, 16644–16649.

Czisch, M., Schleicher, M., Horger, S., Voelter, W., & Holak, T. A. (1993). Conformation of thymosin beta 4 in water determined by NMR spectroscopy. *European Journal of Biochemistry, 218*, 335–344.

De La Cruz, E. M., Ostap, E. M., Brundage, R. A., Reddy, K. S., Sweeney, H. L., & Safer, D. (2000). Thymosin-beta(4) changes the conformation and dynamics of actin monomers. *Biophysical Journal, 78*, 2516–2527.

Dehal, P., Satou, Y., Campbell, R. K., et al. (2002). The draft genome of Ciona intestinalis: Insights into chordate and vertebrate origins. *Science, 298*, 2157–2167.

Dhaese, S., Vandepoele, K., Waterschoot, D., Vanloo, B., Vandekerckhove, J., Ampe, C., et al. (2009). The mouse thymosin beta15 gene family displays unique complexity and encodes a functional thymosin repeat. *Journal of Molecular Biology, 387*, 809–825.

Didry, D., Cantrelle, F. X., Husson, C., et al. (2012). How a single residue in individual beta-thymosin/WH2 domains controls their functions in actin assembly. *The EMBO Journal, 31*, 1000–1013.

Domanski, M., Hertzog, M., Coutant, J., Gutsche-Perelroizen, I., Bontems, F., Carlier, M. F., et al. (2004). Coupling of folding and binding of thymosin beta4 upon interaction with monomeric actin monitored by nuclear magnetic resonance. *The Journal of Biological Chemistry, 279*, 23637–23645.

Edwards, J. (2004). Are beta-thymosins WH2 domains? *FEBS Letters, 573*, 231–232. author reply 233.

Edwards, J. (2012). Single-domain β-thymosins: The family history. *Annals of the New York Academy of Sciences, 1269*, 7–15.

Goldstein, A. L. (2007). History of the discovery of the thymosins. *Annals of the New York Academy of Sciences, 1112*, 1–13.

Goldstein, A. L., Hannappel, E., & Kleinman, H. K. (2005). Thymosin beta4: Actin-sequestering protein moonlights to repair injured tissues. *Trends in Molecular Medicine, 11*, 421–429.

Hannappel, E. (2007). Beta-thymosins. *Annals of the New York Academy of Sciences, 1112*, 21–37.

Hertzog, M., van Heijenoort, C., Didry, D., et al. (2004). The beta-thymosin/WH2 domain: Structural basis for the switch from inhibition to promotion of actin assembly. *Cell, 117*, 611–623.

Irobi, E., Aguda, A. H., Larsson, M., Guerin, C., Yin, H. L., Burtnick, L. D., et al. (2004). Structural basis of actin sequestration by thymosin-beta4: Implications for WH2 proteins. *The EMBO Journal, 23*, 3599–3608.

Krissinel, E., & Henrick, K. (2007). Inference of macromolecular assemblies from crystalline state. *Journal of Molecular Biology, 372*, 774–797.

Lee, S. H., Kerff, F., Chereau, D., Ferron, F., Klug, A., & Dominguez, R. (2007). Structural basis for the actin-binding function of missing-in-metastasis. *Structure, 15*, 145–155.

Malinda, K. M., Sidhu, G. S., Mani, H., Banaudha, K., Maheshwari, R. K., Goldstein, A. L., et al. (1999). Thymosin beta 4 accelerates wound healing. *The Journal of Investigative Dermatology, 113*, 364–368.

Nag, S., Larsson, M., Robinson, R. C., & Burtnick, L. D. (2013). Gelsolin: The tail of a molecular gymnast. *Cytoskeleton*, *70*, 360–384.
Oldfield, C. J., & Dunker, A. K. (2014). Intrinsically disordered proteins and intrinsically disordered protein regions. *Annual Review of Biochemistry*, *83*, 553–584.
Paunola, E., Mattila, P. K., & Lappalainen, P. (2002). WH2 domain: A small, versatile adapter for actin monomers. *FEBS Letters*, *513*, 92–97.
Qian, L., Huang, Y., Spencer, C. I., Foley, A., Vedantham, V., Liu, L., et al. (2012). In vivo reprogramming of murine cardiac fibroblasts into induced cardiomyocytes. *Nature*, *485*, 593–598.
Reichert, A., Heintz, D., Echner, H., Voelter, W., & Faulstich, H. (1996). Identification of contact sites in the actin-thymosin beta 4 complex by distance-dependent thiol cross-linking. *The Journal of Biological Chemistry*, *271*, 1301–1308.
Safer, D., Elzinga, M., & Nachmias, V. T. (1991). Thymosin beta 4 and Fx, an actin-sequestering peptide, are indistinguishable. *The Journal of Biological Chemistry*, *266*, 4029–4032.
Safer, D., Golla, R., & Nachmias, V. T. (1990). Isolation of a 5-kilodalton actin-sequestering peptide from human blood platelets. *Proceedings of the National Academy of Sciences of the United States of America*, *87*, 2536–2540.
Safer, D., Sosnick, T. R., & Elzinga, M. (1997). Thymosin beta 4 binds actin in an extended conformation and contacts both the barbed and pointed ends. *Biochemistry*, *36*, 5806–5816.
Sickmeier, M., Hamilton, J. A., LeGall, T., et al. (2007). DisProt: The database of disordered proteins. *Nucleic Acids Research*, *35*, D786–D793.
Simenel, C., Van Troys, M., Vandekerckhove, J., Ampe, C., & Delepierre, M. (2000). Structural requirements for thymosin beta4 in its contact with actin. An NMR-analysis of thymosin beta4 mutants in solution and correlation with their biological activity. *European Journal of Biochemistry*, *267*, 3530–3538.
Smart, N., Bollini, S., Dube, K. N., et al. (2011). De novo cardiomyocytes from within the activated adult heart after injury. *Nature*, *474*, 640–644.
Stoll, R., Voelter, W., & Holak, T. A. (1997). Conformation of thymosin beta 9 in water/fluoroalcohol solution determined by NMR spectroscopy. *Biopolymers*, *41*, 623–634.
Uversky, V. N. (2002). Natively unfolded proteins: A point where biology waits for physics. *Protein Science*, *11*, 739–756.
van der Lee, R., Buljan, M., Lang, B., et al. (2014). Classification of intrinsically disordered regions and proteins. *Chemical Reviews*, *114*, 6589–6631.
Van Troys, M., Dhaese, S., Vandekerckhove, J., & Ampe, C. (2007). In P. Lappalainen (Ed.), *Multirepeat β-thymosins. Actin-monomer-binding proteins* (pp. 71–81). New York: Springer.
Van Troys, M., Ono, K., Dewitte, D., Jonckheere, V., De Ruyck, N., Vandekerckhove, J., et al. (2004). TetraThymosinbeta is required for actin dynamics in Caenorhabditis elegans and acts via functionally different actin-binding repeats. *Molecular Biology of the Cell*, *15*, 4735–4748.
Xue, B., Aguda, A. H., & Robinson, R. C. (2007). Models of the actin-bound forms of the beta-thymosins. *Annals of the New York Academy of Sciences*, *1112*, 56–66.
Xue, B., Leyrat, C., Grimes, J. M., & Robinson, R. C. (2014). Structural basis of thymosin-β4/profilin exchange leading to actin filament polymerization. *Proceedings of the National Academy of Sciences of the United States of America*, *111*, E4596–E4605. A201412271.
Xue, B., & Robinson, R. C. (2013). Guardians of the actin monomer. *European Journal of Biochemistry*, *92*, 316–332.

Yarmola, E. G., Parikh, S., & Bubb, M. R. (2001). Formation and implications of a ternary complex of profilin, thymosin beta 4, and actin. *The Journal of Biological Chemistry*, *276*, 45555–45563.

Young, J. D., Lawrence, A. J., MacLean, A. G., Leung, B. P., McInnes, I. B., Canas, B., et al. (1999). Thymosin beta 4 sulfoxide is an anti-inflammatory agent generated by monocytes in the presence of glucocorticoids. *Nature Medicine*, *5*, 1424–1427.

Zarbock, J., Oschkinat, H., Hannappel, E., Kalbacher, H., Voelter, W., & Holak, T. A. (1990). Solution conformation of thymosin beta 4: A nuclear magnetic resonance and simulated annealing study. *Biochemistry*, *29*, 7814–7821.

CHAPTER FOUR

Phosphorylation of Prothymosin α. An Approach to Its Biological Significance

M. Freire[1], C.S. Sarandeses, G. Covelo, C. Díaz-Jullien
Facultad de Biología, Universidade de Santiago de Compostela, Santiago de Compostela, Spain
[1]Corresponding author: e-mail address: manuel.freire@outlook.es

Contents

Abstract

Prothymosin α (ProTα), the precursor of the thymosin α_1 and thymosin α_{11}, is a 109–111 amino acids protein widely distributed in the mammalian tissues that is essential for the cell proliferation and survival through its implication on chromatin remodeling and in the proapoptotic activity. ProTα is phosphorylated at Thr residues by the M2 isoenzyme of the pyruvate kinase in a process that is dependent on the cell proliferation activity, which constitutes a novel dual functionality of this enzyme. The Thr residues phosphorylated are apparently dependent on the carcinogenic transformation of the cells. Thus, in normal lymphocytes residues Thr11 or Thr12 are phosphorylated in addition to a Thr7 residue, while in tumor cells Thr7 is the only residue phosphorylated. Phosphorylation of ProTα seems to be related to its antiapoptotic activity, although other possibilities cannot be discarded.

Vitamins and Hormones, Volume 102
ISSN 0083-6729
http://dx.doi.org/10.1016/bs.vh.2016.04.001

1. INTRODUCTION

The essential role of the thymosins in the development of the lymphoid system and in the maintenance of the immune balance has led to an intensive research aimed at the isolation and identification of the biologically active thymic factors that were responsible for the physiological function of the thymus. This research was especially intense in the 1960–70 decade, and in 1966 these efforts led Abraham White and Alan Goldstein's group at the Albert Einstein College of Medicine in New York, to isolate from calf thymus a fraction containing a mixture of peptides that showed immune-proliferating properties in several in vitro and in vivo systems, that they named thymosin (Goldstein, Slater, & White, 1966). The best-known thymosin preparation is thymosin fraction 5 (TF5), which contains about 30 different heat-stable peptides with molecular weight ranging from 1 to 15 kDa (Goldstein, Guha, Zatz, Hardy, & White, 1972; Goldstein et al., 1966). Isoelectric focusing separation of TF5 led to classify its components based on their isoelectric point (pI). So that, polypeptides with pI below 5 were termed α, those with pI between 5 and 7 were designed as β and γ for those with pI higher than 7.

TF5 was used as a parent fraction for the further purification of most of the thymosinic peptides, and thymosin α_1 ($T\alpha_1$), the main component of the α-thymosins, was the first to be purified and its structure stablished (Goldstein et al., 1977). $T\alpha_1$ is a 28-amino acid residues polypeptide (Fig. 1A) with pI 4.2, which has been found to be more active than TF5 in several in vitro and in vivo assays (Goldstein et al., 1977; Low et al., 1979). A less abundant (four times less than $T\alpha_1$) α-thymosinic component named thymosin α_{11} ($T\alpha_{11}$) was subsequently isolated and characterized (Caldarella et al., 1983). The structure of $T\alpha_{11}$ (also shown in Fig. 1A) comprises the $T\alpha_1$ sequence plus seven additional amino acid residues at its C-terminus and shows immunoregulatory properties similar to those of $T\alpha_1$ (Low & Goldstein, 1985).

1.1 The Precursor of the α-Thymosins

The apparent importance of $T\alpha_1$ and $T\alpha_{11}$ as effectors of the development and function of thymus-dependent lymphocytes encouraged a research directed to characterize their cellular origin. The structural characteristics of the α-thymosins prompted the group of Bernard L. Horecker at the Roche Institute of Molecular Biology in Nutley (New Jersey) to investigate,

A

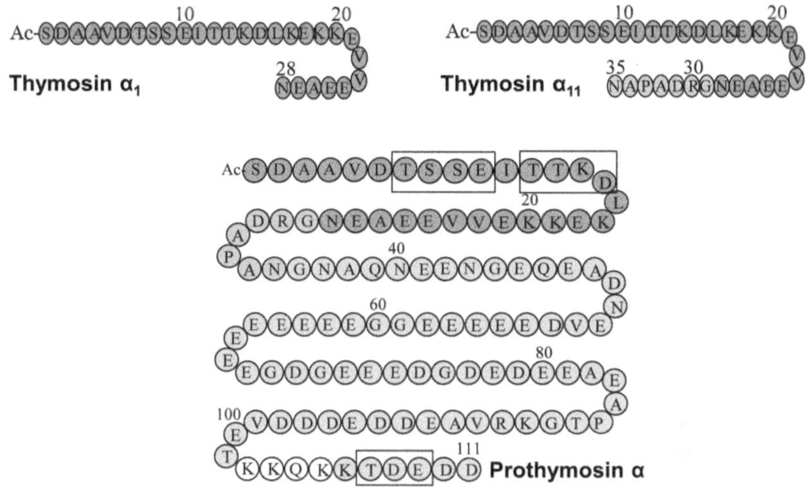

B

```
                    20                  40
Human: SDAAVDTSSEITTKDLKEKKEVVEEAENGRDAPANGNA-NEENGEQEADNEVDEEEEEG: 58
Rat  : SDAAVDTSSEITTKDLKEKKEVVEEAENGRDAPANGNAQNEENGEQEADNEVDEEEEEG: 59
Mouse: SDAAVDTSSEITTKDLKEKKEVVEEAENGRDAPANGNAQNEENGEQEADNEVDEEEEEG: 59
Calf : SDAAVDTSSEITTKDLKEKKEVVEEAENGREAPANGNA-NEENGEQEADNEVDEEEEEG: 58
Pig  : SDAAVDTSSEITTKDLKXXKEVVEEAENGREAPANGNA-NEENGEQEADNEVDEEEEEG: 58
Goat : SBAAVBTSSZITTKBLKEKKEVVEEAENGREAPANGNA-NEENGEQEABBZVBZZZZZG: 58

       60                80                  100
Human: GEEEEEEEEGDGEEEDGDEDEEAESATGKRAAEDDEDDDVDTKKQK-TDEDD: 109
Rat  : GEEEEEEEEGDGEEEDGDEDEEAEAPTGKRVAEDDEDDDVETKKQKKTDEDD: 111
Mouse: GEEEEEEEEGDGEEEDGDEDEEAEAPTGKRVAEDDEDDDVDTKKQK-TEEDD: 110
Calf : GEEEEEEEEGDGEEEDGDEDEEAEAATGKRAAEDDEDDDVDTKKQK-TDEDD: 109
Pig  : GEEEEEEEEGDGEEEDGDEDEEAEAATGKRAAEDDEDDDVDTKKQK-TDEDD: 109
Goat : GZZZZZZZZGBGZZZBGBZBZZAZAPTGKRAAEDDEDDDVDTKKQK-TDEDD: 109
```

Fig. 1 (A) Amino acid sequence of thymosin α_1 (Goldstein et al., 1977), thymosin α_{11} (Caldarella et al., 1983), and prothymosin α (Haritos, Blacher, Stein, Caldarella, & Horecker, 1985). *White circles* correspond to the NLS and *boxes* contain the CK-2 phosphorylation sites. (B) Alignment of the amino acid sequences of prothymosin α from human spleen (Goodall, Domínguez, & Horecker, 1986), rat spleen (Frangou-Lazaridis, Clinton, Goodall, & Horecker, 1988), mouse Ehrlich ascites cells (Schmidt & Werner, 1991), bovine thymus (Panneerselvam, Wellner, & Horecker, 1988), pig intestine (Wintero, Fredholm, & Davies, 1996), and goat spleen (Frillingos et al., 1991). (See the color plate.)

in the first months of 1977 for the origin of the α-thymosins, which was based on the possibility that both $T\alpha_1$ and $T\alpha_{11}$ might be synthesized as a larger precursor from which these peptides would be generated by proteolytic processing. With this purpose, a study was carried out to search for a component immunoprecipitable with antisera against TF5 or $T\alpha_1$ among the translation products of calf thymus mRNAs. This search led to the isolation and identification of a protein that included $T\alpha_1$ in its sequence

(Freire, Crivellaro, Isaacs, Moschera, & Horecker, 1978). The isolation of the mRNA coding for this protein (Freire et al., 1981) confirmed the existence of a precursor of the α-thymosins. The discovery of this precursor encouraged the study of thymic extracts prepared with procedures that minimize proteolytic activity by using strong denaturing conditions. Thus, in 1984 Horecker's group (Haritos, Goodall, & Horecker, 1984) isolated from rat thymus a protein showing the characteristics of the $T\alpha_1$ precursor that was named prothymosin α (ProTα). The sequence of this protein presented in Fig. 1A shown a number of important features, some of them discovered a few years later: (i) $T\alpha_1$ and $T\alpha_{11}$ correspond, respectively, with the first 28 and 35 amino acid residues of the ProTα sequence; (ii) ProTα possesses a very long acidic region of Glu and Asp residues, which are responsible for its highly acidic pI (3.55); (iii) no aromatic or sulfur amino acid is present in its sequence; (iv) ProTα has a nuclear localization signal (NLS) at its C-terminus (Gómez-Márquez & Segade, 1988; Manrow, Sburlati, Hanover, & Berger, 1991; Rubtsov et al., 1997); and (v) sequence of ProTα includes three putative casein kinase-2 (CK-2) phosphorylation sites (Barcia, Castro, Jullien, González, & Freire, 1992).

The primary structure of ProTα from other mammalians, deduced from both protein and cDNA sequencing, was later established (Fig. 1B) and from it can be deduced that ProTα is a 109–111 amino acid protein with a highly conserved sequence and widely distributed in mammalians. ProTα-like peptides have also been reported to be present in other vertebrates as chicken and trout (Yialouris et al., 1988) as well as in frog (Aniello et al., 2002), which contain 30 times lower concentration as that of ProTα in calf thymus.

Subsequent studies on ProTα revealed surprising results, firstly showing a generalized presence of this protein in other tissues, besides thymus. Thus, presence of ProTα was found in all the analyzed tissues from rat (Franco, Díaz, Barcia, & Freire, 1992; Haritos, Tsolas, & Horecker, 1984) or mouse (Clinton, Frangou-Lazaridis, Panneerselvam, & Horecker, 1989), at concentrations ranging from 40 to 116 μg/g of tissue, what means values in order of one-tenth of the concentration of a core histone. Moreover, the demonstration that acetylation of the N-terminal Ser in the ProTα sequence corresponds with a cotranslational modification of this protein (Nogueira & Freire, 1985) and the analysis of the ProTα gene structure (Eschenfeldt, Manrow, Krug, & Berger, 1989), showing the absence of a signal peptide in the ProTα sequence, seemed to rule out the possibility that ProTα can be secreted by a conventional procedure.

On the other hand, the failing to find Tα_1 and Tα_{11} in the thymic extracts prepared under highly denaturing conditions, led to suggest an artifactual origin of these peptides due to an uncontrolled proteolysis of ProTα generated during the isolation of TF5 (Haritos, Goodall, et al., 1984). However, subsequent results in our laboratory ruled out this possibility, since Tα_1 was isolated from thymus and other tissues, prepared in conditions in which proteolytic activity was prevented, in a concentration similar to that of ProTα (Franco et al., 1992; Freire, Rey-Méndez, Gómez-Márquez, & Arias, 1985). The natural origin of Tα_1 and Tα_{11} was later confirmed from our work which demonstrated that ProTα was processed in vivo by a lysosomal protease, identified as legumain, to generate Tα_1 and, transitorily, Tα_{11} (Sarandeses, Covelo, Díaz-Jullien, & Freire, 2003). Therefore these results proved the existence of native Tα_1 free in the cytoplasm and sharing its presence with that of ProTα in diverse mammalian tissues. The possibility that this cytoplasmic Tα_1 may have an intracellular role and/or that it can be secreted through an unconventional mechanism deserves future research.

Studies of the subcellular distribution of ProTα carried out by immunofluorescence microscopy (Conteas et al., 1990; Enkemann, Ward, & Berger, 2000; Moreira et al., 2013; Vareli, Frangou-Lazaridis, van der Kraan, Tsolas, & van Driel, 2000) or by fractionation of cell lysates (Covelo, Sarandeses, Díaz-Jullien, & Freire, 2006; Franco et al., 1992; Manrow et al., 1991; Tsitsiloni, Yialouris, Sekeri-Pataryas, & Haritos, 1989) gave contradictory results about the subcellular location of ProTα. Thus, while microscopic analysis indicated a preferential nuclear location, analysis of cell lysates showed that cytoplasm contained the major concentration of ProTα. However, after preventing protein leakage from the nucleus by treatment with cytochalasin B, a prominent ProTα presence was shown in the nucleoplasmic extracts (Manrow et al., 1991), proving that nuclear leaking of ProTα can occur during the preparation of the cell extracts. These studies indicated that ProTα had a preferred nuclear presence in the mammalian cells, which is in accordance with the presence of a NLS at its C-terminus and with its ability to interact with proteins involved in the nuclear import (Freire, Covelo, Sarandeses, Díaz-Jullien, & Freire, 2001) and export (Kubota, Adachi, Copeland, & Oroszlan, 1995).

Thus, a decade after the discovery of the α-thymosins and their precursor, experimental data ruled out the expectations of an exclusive implication of these polypeptides in the immunocompetence of lymphocytes and questioned their capacity to be exported as conventional hormones, indicating a possible intracellular role for Tα_1 and ProTα.

1.2 Progress in the Study of the Biological Function of ProTα

As indicated, preliminary research on the cellular distribution of ProTα demonstrated that this protein should have an intracellular function. The analysis of the ProTα gene expression was the first and more important step directed toward the elucidation of the ProTα function. The evaluation of the levels of ProTα and its mRNA in mitogen-activated lymphocytes (Eschenfeldt & Berger, 1986; Gómez-Márquez et al., 1989; Zalvide, Cancio, Alvarez, Regueiro, & Domínguez, 1992) and in regenerating tissues (Bustelo, Otero, Gómez-Márquez, & Freire, 1991; Wu, Boers, Reitsma, van Deventer, & Chamuleau, 1997) provided the first clues of a possible relationship between ProTα gene expression and cell proliferation. The decrease of cell growth in cells transfected with ProTα antisense oligomers contributes to confirm this relationship (Sburlati, Manrow, & Berger, 1991). On the other hand, studies to investigate possible links between ProTα function and cell differentiation processes (Franco del Amo & Freire, 1995; Smith et al., 1993) have not detected any direct relationship between the two. The preferential nuclear location of ProTα in proliferation cells (Conteas et al., 1990; Manrow et al., 1991; Moreira et al., 2013) seems to indicate that the nucleus should be its main site of action.

The research directed toward elucidating the putative mechanism by which ProTα can be involved in the cellular activity has been greatly assisted by the characterization of proteins that might interact with ProTα. The results from our group and others in the nineties, provided increasing evidence of a nuclear function of ProTα due to its ability to interact with histones (Covelo et al., 2006; Díaz-Jullien, Pérez-Estévez, Covelo, & Freire, 1996; Karetsou et al., 1998; Papamarcaki & Tsolas, 1994), with transcription factors (Cotter & Robertson, 2000; Karetsou, Kretsovali, Murphy, Tsolas, & Papamarcaki, 2002; Subramanian et al., 2002), and with the estrogen receptor (Martini, Delage-Mourroux, Kraichely, & Katzenellenbogen, 2000). Our results showing the influence of ProTα on nucleosome assembly (Díaz-Jullien et al., 1996), the affinity for nuclear-import and cell-cycle regulatory proteins (Freire et al., 2001) and the interaction with a nuclear multiprotein complex, in which are included core histones, H3 and H4 acetyltransferases, H3 methyltransferase, and proteins involved in chromatin remodeling (Covelo et al., 2006), contribute to demonstrate a nuclear role of ProTα related with the chromatin activity.

Notwithstanding the evidence of this nuclear role of ProTα, cytoplasmic functions of ProTα have also been reported. The cytoplasmic ProTα seems to have an antiapoptotic activity exerted through the prevention of the caspase 9 activation by blocking apoptosome formation, in a way that ProTα counteracts the pro-apoptotic activity of the protein ANP32A (Jiang et al., 2003; Qi, Wang, & Du, 2010).

With all described earlier, the ProTα behaves as multitask cytoplasmic–nuclear shuttling protein specially involved in cell proliferation and survival, of which important aspects of the molecular mechanism involved in its biological activity remain to be clarified. In this sense, studies of the posttranslational modifications of ProTα provide additional clues to its biological function. Thus, the aforementioned acetylation of the N-terminal Ser in the ProTα sequence has been demonstrated to be a cotranslational modification that, besides proving the inability of ProTα to be secreted by a conventional procedure, may constitute a signal for prevention of rapid degradation in the cell (Nogueira & Freire, 1985). On the other hand, proteolytic processing by legumain to generate free $T\alpha_1$ in the cytoplasm of mammalian cells, it is a process that contributes to diversify the study of the biological function of the α-thymosins, besides following the behavior of ProTα, a study on the possible intracellular role of $T\alpha_1$ should be addressed in the future.

The phosphorylation of ProTα, which occurs concomitantly with the cellular proliferation activity (Barcia, Castro, Jullien, & Freire, 1993; Pérez-Estévez, Díaz-Jullien, Covelo, Salgueiro, & Freire, 1997), constitutes another posttranslational modification that can shed light on the biological activity of this protein. A review of this process is presented in this chapter.

2. PHOSPHORYLATION IN VITRO vs IN VIVO

The reports concerning the involvement of CK-2 in the phosphorylation of acidic proteins (Hathaway & Traugh, 1982; Maelandsmo, Østvold, & Laland, 1989) led us to investigate whether ProTα could be included among the substrates of CK-2. In fact, the sequence of ProTα, as shown in Fig. 1A, has three putative CK-2 phosphorylation sites, two at the N-terminus, and one at the C-terminus. In agreement with this structural characteristic, ProTα isolated from calf thymus, previously treated with alkaline phosphatase, was phosphorylated in vitro by rat liver CK-2 using ATP or GTP as phosphate donors (Barcia et al., 1992). Analysis of the tryptic peptides derived from purified ^{32}P-ProTα indicated that phosphorylation sites were located in a unique peptide, which coincides with the first 14 amino acids of its sequence, with an equal proportion of Ser and Thr residues phosphorylated, as was demonstrated by phosphoamino acid analysis.

The efficiency of the ProTα phosphorylation by CK-2 in vitro was dependent on the main effectors of this enzyme, thus CK-2 activity was highly enhanced by polyamines (protamine and polylysine) and inhibited by heparin.

The next thing was to establish whether ProTα could also be phosphorylated in vivo; therefore, search for phosphorylated ProTα was accomplished in cells metabolically labeled with ^{32}P-orthophosphate. To this end, we used lymphocytes (thymocytes and splenic lymphocytes), because these cells have the highest concentration of ProTα in the mammalian tissues (about 100 ng/10^6 cells) (Franco et al., 1992; Sarandeses et al., 2003). Due to not finding phosphorylated ProTα in those extracts, we decided to investigate its presence in murine splenic lymphocytes mitogen-activated with ConA plus IL-2 (Barcia et al., 1993). The purification of ProTα from these cells labeled with ^{32}P-orthophosphate was achieved by a procedure which combines drastic denaturing conditions, ion-exchange chromatography and reverse phase HPLC, which have demonstrated to be very efficient in the isolation of this protein from diverse tissues (Franco et al., 1992; Haritos, Goodall, et al., 1984), as compared with others that use a procedure based in a phenol extraction of cell homogenates (Pavlov, Evstafieva, Rubtsov, & Vartapetian, 1995; Sburlati, Manrow, & Berger, 1990). Concerning the study of ProTα, it must be taken into account that the peculiar structural characteristics of this protein, the more acidic in the mammalian tissues (pI 3.55), although favored its purification as a thermostable protein, it constitutes a disadvantage for its immunological analysis, due to its low immunogenicity and its difficulty to interact with membranes in the western blotting analysis (Covelo et al., 2006).

A ^{32}P-labeled component copurifying with calf ProTα was isolated and immunologically identified as ^{32}P-ProTα. The HPLC analysis of the tryptic peptides derived from this component showed a pattern of phosphopeptides similar to that of ProTα phosphorylated by CK-2 (Fig. 2), with a unique phosphorylated peptide that corresponds to the 1–14 fragments of the ProTα sequence. However, the phosphoamino acid analysis, shown in the inset in Fig. 2, indicated that in vivo only Thr residues were phosphorylated, whereas CK-2 catalyzed the incorporation of ^{32}P into Thr and Ser residues in similar amounts. According to the sequence of the fragments 1–14 of ProTα, its phosphorylation in activated splenic lymphocytes should correspond with a total or partial phosphorylation of Thr residues located at positions 7, 12, or 13. A similar study of ProTα phosphorylation in vivo carried out in other cell types, as thymocytes and HeLa cells, showed identical results (Barcia et al., 1993). However, our results on the ProTα

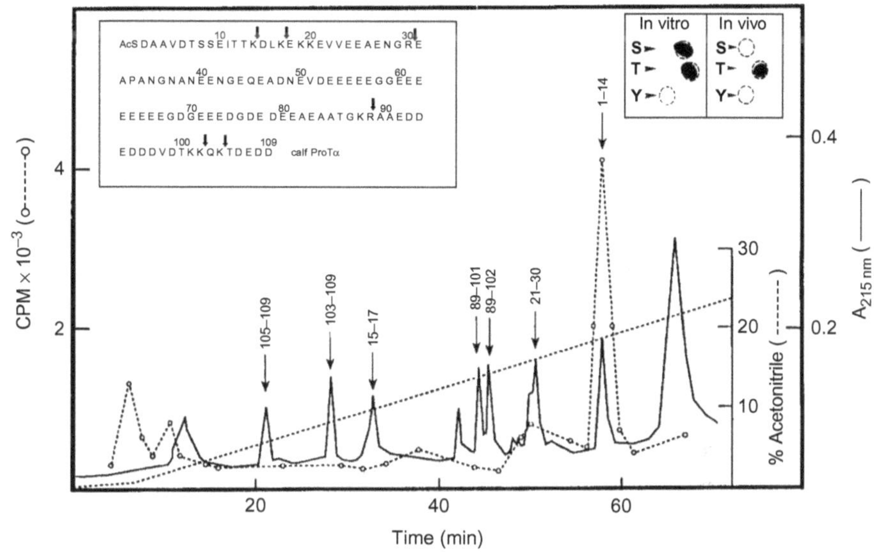

Fig. 2 Analysis by HPLC of the peptides derived from the tryptic digestion of ^{32}P-ProTα isolated from murine splenic lymphocytes labeled with ^{32}P-orthophosphate. *Insets* show the sequence of calf thymus ProTα used as carrier in the tryptic digestion, with arrows indicating sites of tryptic cleavage, and the phosphoamino acid analysis of the radioactive peak coeluting with the 1–14 fragments of calf ProTα derived from the ^{32}P-ProTα phosphorylated by CK-2 (in vitro) and from ^{32}P-ProTα isolated from murine splenic lymphocytes labeled with ^{32}P-orthophosphate (in vivo). *This research was originally published in Barcia, M. G., Castro, J. M., Jullien, C. D., & Freire, M. (1993). Prothymosin α is phosphorylated in proliferating stimulated cells.* Journal of Biological Chemistry, 268*(7), 4704–4708. ©The American Society for Biochemistry and Molecular Biology.*

phosphorylation sites were controversial with those reported by the Berger's group (Sburlati et al., 1993) that in a related paper claimed that in vivo ProTα was phosphorylated in the acetyl-Ser residue located at the N-terminus of its sequence. This possibility was also ruled out according to our results from further experiments designed to obtain a more detailed characterization of the ProTα phosphorylation sites (to be presented ahead).

The ProTα phosphorylation process, either in splenic lymphocytes, thymocytes, or HeLa cells, was highly dependent on proliferation activity of the cells, being the phosphorylation activity in activated splenic lymphocytes higher (about five times) than in the other cells. The analysis of the subcellular distribution of phosphorylated ProTα indicated that it was mainly recovered from the cytosolic extracts. Moreover, according to the ^{32}P-orthophosphate incorporated to the Thr residues of ProTα after 20 h of metabolically labeling mitogen-activated splenic lymphocytes, and due to the cellular concentration of ProTα in these cells, we estimated that in these lymphocytes only 18% of the cellular content of ProTα was phosphorylated (Pérez-Estévez et al., 1997).

With these results, it was demonstrated that phosphorylation of ProTα takes place in proliferating cells and, according to its phosphorylation sites, another protein kinase apparently different from CK-2 seemed to be responsible for this process. The next task should be the characterization of such enzyme.

3. THE PROTEIN KINASE WHICH PHOSPHORYLATES ProTα IN PROLIFERATING CELLS

The search for the enzyme responsible for the ProTα phosphorylation in vivo was accomplished in the same mitogen-activated splenic lymphocytes from which the phosphorylated ProTα was isolated (Pérez-Estévez et al., 1997). The investigation of the presence of an enzyme able to phosphorylate ProTα in different subcellular fractionates of the murine splenocytes, led us to isolate components in extracts from cytoplasm, nuclear envelopes, and nucleoplasm extracts showing affinity to ProTα as well as the ability to phosphorylate this protein, as indicated in Fig. 3. The ProTα phosphorylation activity was mainly present in the cytoplasmic components with a higher affinity for ProTα, while minor phosphorylation activities were found in the cytoplasmic fractions with moderate affinity to ProTα, and in the ProTα-affinity components from nuclear envelopes and nucleoplasm, which, respectively, represent 4.8%, 25%, and 5.9% of that found in the cytoplasmic fraction with higher affinity to ProTα. Further purification of these enzymes by ion-exchange HPLC revealed that ProTα phosphorylation activity from cytoplasm and nucleoplasm had similar elution patterns (although activities in the fractions from cytosolic components with low affinity to ProTα and from nucleoplasm were scarcely detected); while the kinase from nuclear envelopes showed a different elution pattern. This difference is in agreement with the different molecular size estimated for both ProTα kinases: 180 kDa for the cytoplasmic enzyme and 130 kDa for that from nuclear envelopes.

3.1 Characterization of the Purified ProTα Kinases

To establish whether the purified kinases from cytoplasm and nuclear envelopes were responsible for the phosphorylation of ProTα in vivo, the phosphorylation sites of ProTα by these enzymes were investigated

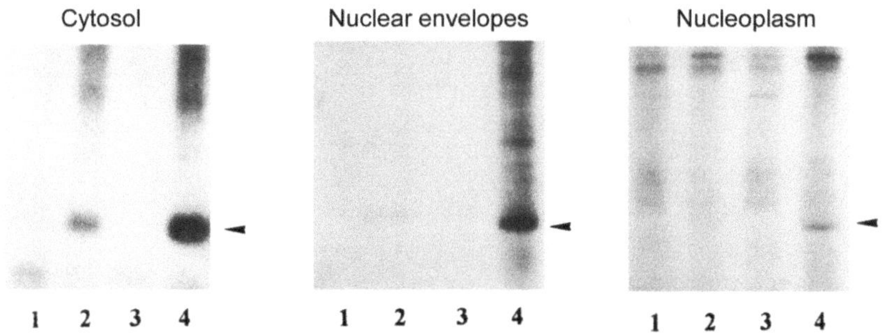

Fig. 3 ProTα phosphorylation activity in subcellular fractionates of mitogen-activated splenic lymphocytes. The ProTαK activity was assayed in the different eluates from the ProTα-Sepharose chromatography of the diverse subcellular fractionates. *Panels* show the autoradiographs of gels in which the components of the different reaction mixtures were separated by SDS-PAGE. *Lines* correspond to the analysis of the reactions mixtures from the different eluates: 1, flow-through; 2, eluted with 0.3 M NaCl (moderate affinity); 3, eluted with 1 M NaCl in which ProTα was omitted (control); 4, eluted with 1 M NaCl (high affinity). *Arrows* mark the position of dephosphorylated ProTα. *This research was originally published in Pérez-Estévez, A., Díaz-Jullien, C., Covelo, G., Salgueiro, M. T., & Freire, M. A. (1997). A 180 kDa protein kinase seems to be responsible for the phosphorylation of prothymosin α observed in proliferating cells.* Journal of Biological Chemistry, *272(16), 10506–10513. ©The American Society for Biochemistry and Molecular Biology.*

(Pérez-Estévez et al., 1997). The HPLC patterns of the tryptic peptides derived from ^{32}P-ProTα isolated, either from metabolically labeled murine splenic lymphocytes or from the reaction mixtures in which enzymatic activity of the purified enzymes was assayed, were identical and similar to that shown in Fig. 2: it was detected a unique radioactive peptide, which coelutes with the first 14 amino acids of the ProTα sequence. The phosphoamino acid analysis of this radioactive peptide from the diverse sources (insets in Fig. 4) indicates that the cytoplasmic kinase (named ProTαK), phosphorylates ProTα in Thr residues as that in vivo, while the enzyme from nuclear envelopes phosphorylates both Ser and Thr residues apparently in similar proportions. These results showed the cytoplasmic ProTαK as a preferred candidate to carry out the phosphorylation of ProTα in vivo. This hypothesis was supported by the results obtained in a structural analysis of the 1–14 fragments derived from the ProTα phosphorylated in vivo and in vitro by the isolated kinases. Thus, as shown in Fig. 4, the HPLC elution profiles of the ^{32}P-labeled 1–14 fragments treated with the protease V8, which cleaves it between residues Glu10 and Ile11 to generate a decapeptide and a tetrapeptide, provide evidence that the cytosolic ProTαK

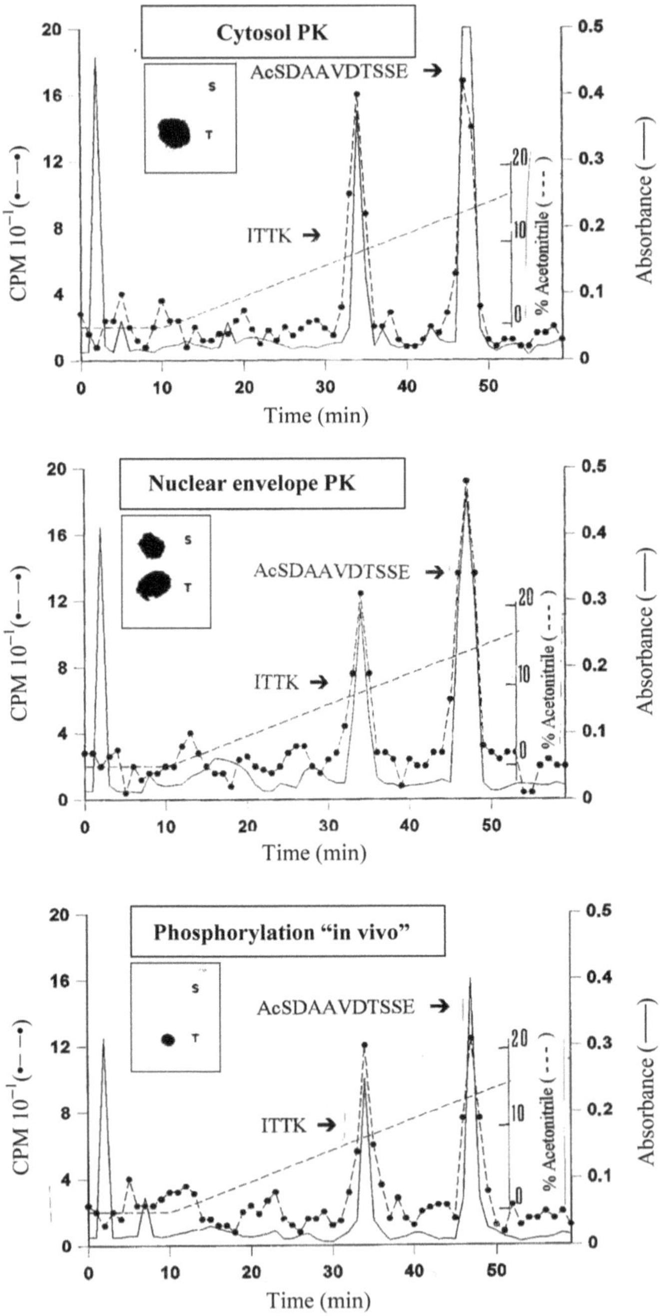

Fig. 4 See legend on opposite page.

phosphorylates ProTα in the same phosphorylation sites as in vivo. Specifically, Thr residues located at position 7 and 12 or 13 (or both partially). By contrast, the ProTαK activity from nuclear envelope phosphorylates ProTα at Thr and Ser residues that should correspond to Thr residues located at positions 7 and 12 or 13 (or both partially) and to one or two Ser residues located in the first 10 amino acid in the sequence of ProTα. A further study of the purified ProTα kinases using different substrates, as well as immunoassays, led us to identify the ProTαK from nuclear envelopes as the CK-2 (Pérez-Estévez et al., 1997, 2000). In fact, the phosphopeptide mapping and the phosphoamino acid analysis of the ProTα phosphorylated by this enzyme, shown in Fig. 4, match those observed for CK-2 (identical to that shown in Fig. 2). The analysis of the ProTα phosphorylation sites, besides confirming the identification of the cytosolic ProTαK as the kinase responsible for the phosphorylation of ProTα in vivo, also ruled out the former proposal claiming that acetyl-Ser would be the only amino acid residue of the ProTα sequence phosphorylated in proliferating cells (Sburlati et al., 1993).

Besides the ubiquitous distribution of the CK-2 in most of the cell types (Faust & Montenarh, 2000), we did not find any other ProTα-phosphorylation activity of this enzyme in the diverse fractionates of the murine splenic lymphocytes. It may be attributed to an exclusive location of this enzyme at the nuclear envelope of these cells, or that the CK-2, which may be present in other fractionates, had no affinity for ProTα. In any case, given the experimental data, CK-2 does not seem to be responsible for the phosphorylation in vivo of this protein in these cells.

Fig. 4 Identification of the sites phosphorylated in ProTα by the purified kinases in vitro and in vivo in proliferating splenic lymphocytes. *Insets* show the phosphoamino acid analysis of the radioactive peptide coeluting with the 1–14 fragments of ProTα derived from ^{32}P-ProTα isolated from the kinase activity assays of cytosolic protein kinase (Cytosol PK), nuclear envelope protein kinase (Nuclear envelope PK), and from splenic lymphocytes labeled with ^{32}P-orthophosphate (Phosphorylation "in vivo"). *Panels* show the separation by HPLC of the peptides resulting from the hydrolysis with V8 of the radioactive peptide coeluting with the 1–14 fragments of calf ProTα and derived of the ^{32}P-ProTα isolated from the three sources. Sequence of the peptides derived from the hydrolysis with V8 is indicated. *This research was originally published in Pérez-Estévez, A., Díaz-Jullien, C., Covelo, G., Salgueiro, M. T., & Freire, M. A. (1997). A 180 kDa protein kinase seems to be responsible for the phosphorylation of prothymosin α observed in proliferating cells.* Journal of Biological Chemistry, *272(16), 10506–10513. ©The American Society for Biochemistry and Molecular Biology.*

A similar study of the ProTαK activity performed in other cell types (Pérez-Estévez et al., 1997) demonstrated the presence of a protein kinase, similar to that found in murine splenocytes, in cytosolic extracts from mitogen-activated murine thymocytes, semiconfluent HeLa cells, murine hepatocytes, and in human lymphoma B cell (NC37 cells). A CK-2 activity was also detected in nuclear envelopes of NC37 cells, in a similar proportion to that found in activated murine splenic lymphocytes.

The results of the characterization of the kinase responsible for the phosphorylation of ProTα in vivo, revealed that it seems to be an 180 kDa cytosolic enzyme widely distributed in mammalian cells, whose levels of activity are dependent on the proliferation activity of the cells. In this sense, analysis of the gene expression in mitogen-activated murine splenic lymphocytes indicated a correspondence between the levels of the ProTαK activity and that of newly phosphorylated ProTα, which reach their maximum levels during the S-phase, and both begin a decrease as cells enter in G2-phase (Pérez-Estévez et al., 1997, 2000). On the other hand, this protein kinase might have a preferential specificity for ProTα, since in vitro we did not find other possible substrates, with the exception of the histones H2B and H3, which are also phosphorylated by this enzyme, although with phosphorylation activities, respectively, 2 and 4.5 times lower than ProTα phosphorylation activity (Pérez-Estévez et al., 1997). In reference to the catalytic properties of the cytoplasmic ProTαK, it is notable its inability to phosphorylate, either in vitro or in vivo, the peptides $T\alpha_1$ and $T\alpha_{11}$ (see Fig. 1), which include the ProTα phosphorylation sites in their sequence (Pérez-Estévez et al., 2000).

An important property of the ProTαK is that its activity seems to be regulated by a phosphorylation process. A component of the purified enzyme, in the range of 60–64 kDa was found to be phosphorylated as consequence of the mitogenic activation of the murine splenic lymphocytes to yield the active enzyme, whereas the dephosphorylation abolish the enzyme activity (Pérez-Estévez et al., 2000). Since the ProTαK has no autophosphorylation activity (Pérez-Estévez et al., 1997), it means that another protein kinase should be responsible for the phosphorylation that controls its activity. To this respect, a study on the effect of diverse protein kinases involved in regulatory circuits controlling the cell growth on both levels of the ProTαK activity and phosphorylated ProTα, demonstrated that the ProTα phosphorylation process was activated by a mitogen-initiated pathway that seem to be dependent of PKC, although this kinase by itself does not phosphorylate ProTα (Pérez-Estévez et al., 2000).

The next goals in the study of the ProTα phosphorylation process should be directed toward the structural characterization of the ProTαK and to the knowledge of the significance of this process in the biological function of ProTα.

3.2 Characterization of the Cytoplasmic Kinase that Phosphorylates ProTα

The cytosolic extracts from transformed human B-lymphocytes (NC37 cells), that contain high levels of ProTαK activity, were used to isolate this enzyme in quantitative amounts. The purification procedure combines ion-exchange and affinity chromatography with a final step in which fractions containing ProTα activity were separated by SDS-PAGE, and components holding the enzymatic activity searched among the diverse renatured slices in which the SDS-PAGE gel was fractionated (Díaz-Jullien et al., 2011). This procedure, whose results are summarized in Fig. 5, led us to isolate a protein in the range of 60 kDa identified by mass spectrometry as the M2-type isoenzyme of human pyruvate kinase (PK). The M2 is one of the four isoenzymes (L, R, M1, and M2) of the PK that is preferably expressed in proliferating mammalian cells through an alternative splicing of a common transcript of the PKM gene which also encoded the M1 isoenzyme (Clover et al., 2010; David, Chen, Assanah, Canoll, & Manley, 2010; Takenaka et al., 1991). This unexpected identification was confirmed by an immunological study of both ProTαK and PK activities in the NC37 cell extracts, which also proved the interaction in vitro and in vivo between both ProTα and M2 protein (Fig. 6A–C).

Further research on the characteristics of the ProTαK and PK activities in the NC37 cell extracts (Díaz-Jullien et al., 2011) demonstrated that only a small part of the cellular pool of the M2 protein (about 10% which is precipitated with 45% ammonium sulphate (AS) and has affinity for ProTα-Sepharose) corresponds to the M2 protein owning the ProTαK activity, which we named M2-ProTαK, while the main part of M2 (precipitable by 75% AS and without affinity for ProTα-Sepharose) maintains the PK activity in the NC37 cells. Moreover, we found that the M1 isoform has not ProTαK activity. The study of the phosphorylation sites of ProTα by the M2-ProTαK, carried out by mass spectrometry, demonstrated that it catalyzed the transference of orthophosphate exclusively to the residue Thr located at position 7 in the ProTα sequence, which is in agreement with the phosphoamino acid analysis of ProTα isolated from NC37 cells. Since in mitogen-activated murine splenic lymphocytes ProTα is phosphorylated at

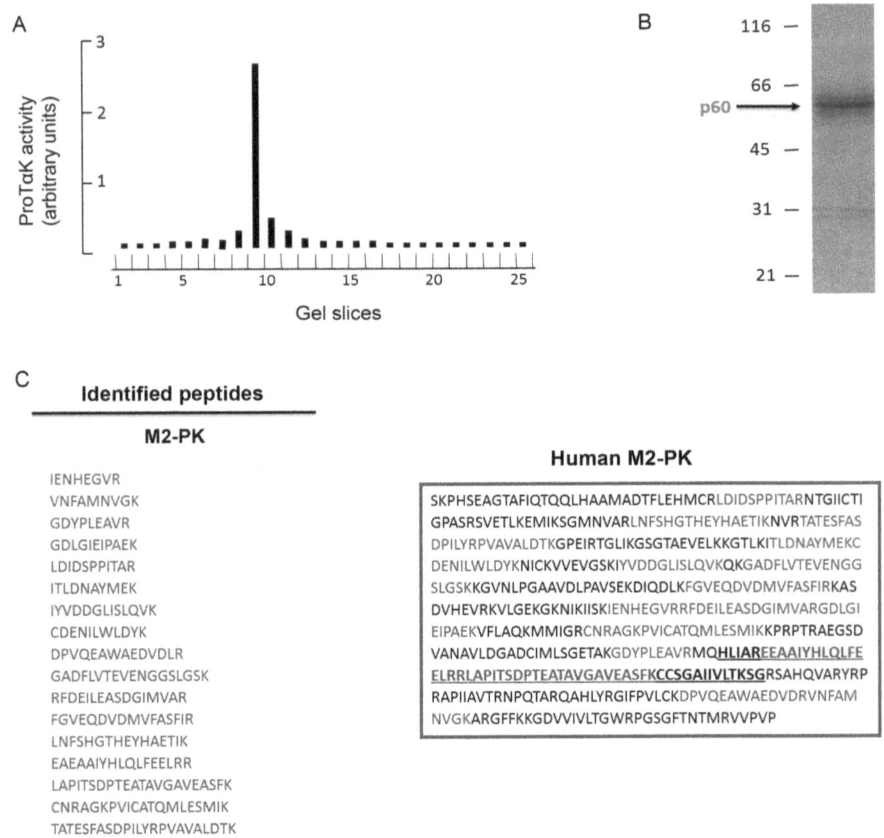

Fig. 5 Structural characterization of the protein with ProTαK activity obtained from NC37 cells. (A) ProTαK activity in the slices from the renatured SDS-PAGE gels in which chromatographically purified ProTαK was separated. (B) SDS-PAGE analysis of the components in the slice with maximal ProTαK activity. (C) Peptides identified in the mass spectrometry analysis of the p60 protein in (B). The identified peptides are displayed in the human M2 sequence showed in the *inset*, and the 56 amino acids from exon 10, which differentiate the M2 from the M1 isoenzyme, are *underlined. Díaz-Jullien, C., Moreira, D., Sarandeses, C. S., Covelo, G., Barbeito, P., & Freire, M. (2011). The M2-type isoenzyme of pyruvate kinase phosphorylates prothymosin α in proliferating lymphocytes.* Biochimica et Biophysica Acta, 1814, *355–365. Copyright (2011) Elsevier.* (See the color plate.)

Thr residues located at positions 7 and 12 or 13, it seems that ProTα is preferably phosphorylated at Thr7 and Thr12/13 might be additional phosphorylation sites accessible in nontumor cells, at least in lymphocytes. On the other hand, phosphorylation at residue Thr7 in tumor cells was confirmed by the analysis of ProTα expression in 293T cells transfected with a vector harboring recombinant ProTα in which Thr7 was substituted by Ala (Díaz-Jullien et al., 2011).

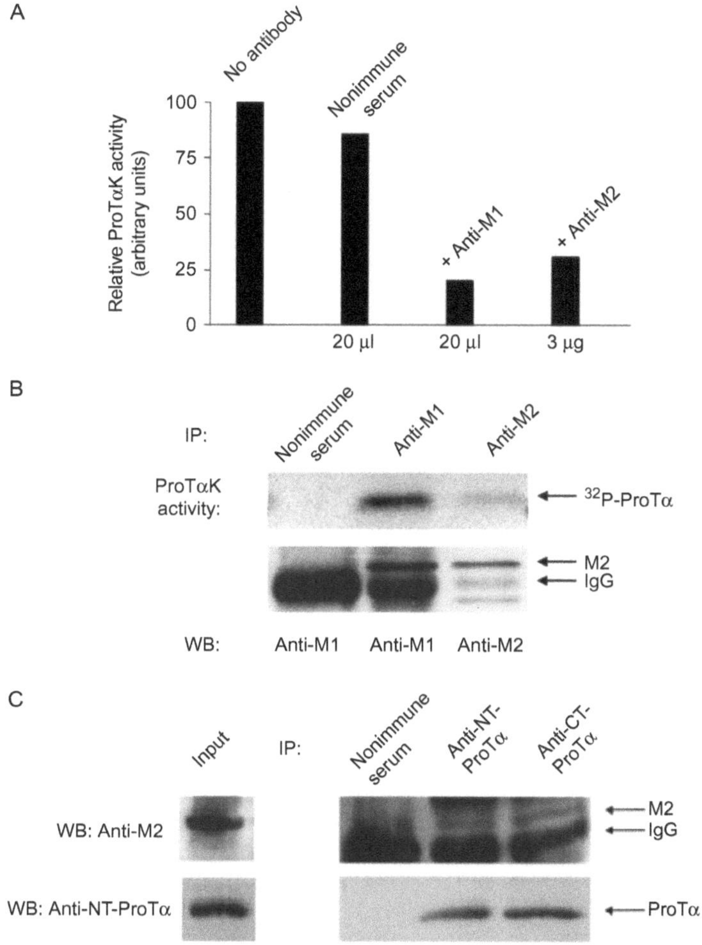

Fig. 6 Immunological analysis of ProTαK and PK activity in NC37 cells. (A) Immunodepletion of the ProTαK activity with anti-M1 and anti-M2 antibodies, both immunoreactive against the M2 protein, and with nonimmune serum. (B) Immunoprecipitation of purified ProTαK with anti-M1, anti-M2, or nonimmune serum. Aliquots of the diverse immunoprecipitates were used to determine the ProTαK activity (*upper row*) or to determine the presence of the M2 protein by western blotting (*lower row*). (C) Coimmunoprecipitation of ProTα and M2 protein by antibodies raised against segments of ProTα located at its N-terminus (anti-NT-ProTα), at its C-terminus (anti-CT-ProTα), or with nonimmune serum. Aliquots of the diverse immunoprecipitates were analyzed by western blotting to determine the presence of M2 (*upper row*) or ProTα (*lower row*). *Díaz-Jullien, C., Moreira, D., Sarandeses, C. S., Covelo, G., Barbeito, P., & Freire, M. (2011). The M2-type isoenzyme of pyruvate kinase phosphorylates prothymosin α in proliferating lymphocytes.* Biochimica et Biophysica Acta, 1814, *355–365. Copyright (2011) Elsevier.*

Next, an important question that should be solved was that related to the knowledge of the molecular mechanisms by which a small part of the cellular pool of the M2 protein acquires the novel ability to function as a protein kinase, while losing its ability to catalyze the synthesis of ATP. The demonstrated influence of a phosphorylation process regulating the activity of the kinase of ProTα in proliferating splenic lymphocytes, also identified as M2-ProTαK (Díaz-Jullien et al., 2011), led us to investigate if a similar process could be involved in the catalytic evolution of the M2 protein in the NC37 cells. This investigation (Díaz-Jullien et al., 2011) demonstrated that a phosphorylation process was implicated in the dual functioning of the M2 protein. So that, phosphorylation of a small part of this protein promotes its affinity for ProTα and its ability to catalyze the phosphorylation of ProTα, properties which are lost when it is dephosphorylated, whereas the main pool of M2 maintaining the PK activity (90%) is not phosphorylated and is unable to interact with ProTα in the NC37 cells (Fig. 7A). Moreover, partial phosphorylation of the M2 protein, which occurs at Ser and Thr residues (Fig. 7B), has also influence on the quaternary structure of this protein. Thus, phosphorylated M2-ProTαK corresponds with a main form of 180 kDa (as in murine splenic lymphocytes) and a 60 kDa form with lower enzymatic activity (five times lower) which, respectively, should correspond with a trimer and a monomer. Dephosphorylation of M2-ProTαK, besides abolishing its enzymatic activity, originates a monomeric (60 kDa) inactive form of this protein. The physiological concentration of fructose 1,6-bisphosphate (FBP) also affects the quaternary structure of the phosphorylated M2-ProTαK promoting the formation of a less active tetrameric structure (240 kDa), promotion which is abolished when the M2-ProTαK is dephosphorylated. These properties differ from that of the M2 protein with PK activity, since it is not affected by phosphatase treatment and FBP is indispensable for its activity and to maintain a quaternary structure, constituted by a tetrameric form (more active) and a monomeric form (less active), as occurs with the M2-PK in other proliferating cells (Muñoz & Ponce, 2003). With this, the promotion of the M2 protein dual functionality to have M2-ProTαK activity would be dependent of a phosphorylation process that, according to the results from the study on the phosphorylation of ProTα in splenic lymphocytes (Pérez-Estévez et al., 2000), would correspond with a protein kinase included in an activation pathway dependent on the PKC activity. The characterization of the protein kinase responsible for this selective and partial phosphorylation of the M2 isoform, therefore, constitutes an important task to be accomplished in the future.

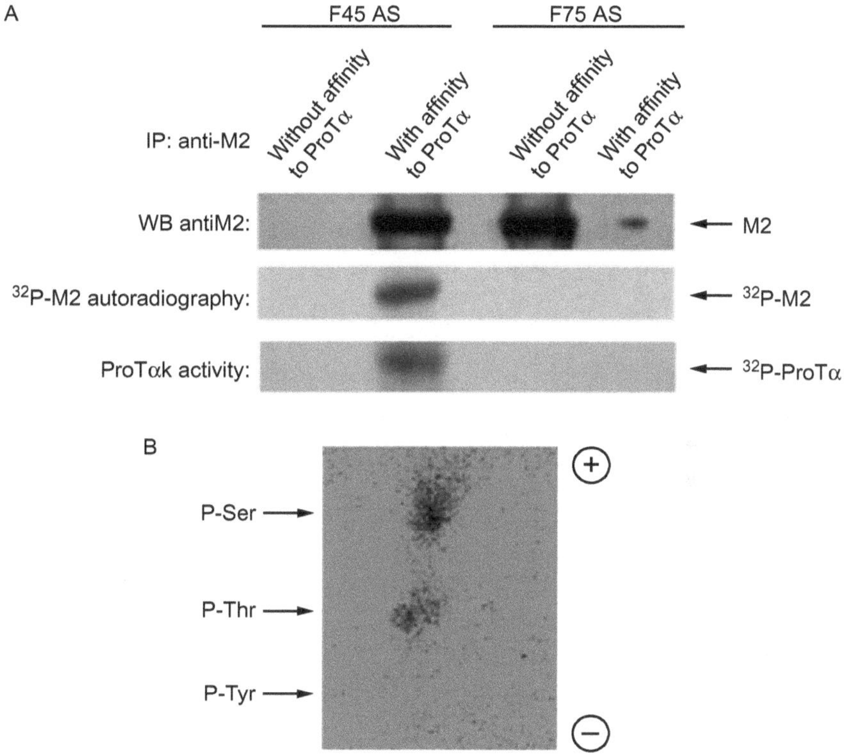

Fig. 7 Study of the dual functionality of the M2 protein. (A) Lysates of NC37 cells metabolically labeled with ^{32}P-orthophosphate were fractionated with 45% (F45AS) and with 75% (F75AS) ammonium sulfate and these fractionates chromatographed on a ProTα-Sepharose column. Components without affinity or with affinity to ProTα were immunoprecipitated with anti-M2 antibodies and aliquots of the diverse immunoprecipitates analyzed by western blotting to determine the presence of M2 (*upper row*) and autoradiographed to see ^{32}P-M2 (*middle row*) or used to assay the ProTαK activity (*bottom row*). (B) Phosphoamino acid analysis of the ^{32}P-M2 isolated from the F45AS fraction. *Díaz-Jullien, C., Moreira, D., Sarandeses, C. S., Covelo, G., Barbeito, P., & Freire, M. (2011) The M2-type isoenzyme of pyruvate kinase phosphorylates prothymosin α in proliferating lymphocytes.* Biochimica et Biophysica Acta, 1814, *355–365. Copyright (2011) Elsevier.*

The novel enzymatic activity found for M2 isoform of the PK was later corroborated by reports from different groups in which a protein kinase activity of this protein was also demonstrated. Thus, a protein kinase activity of the M2 protein that, using phosphoenolpyruvate (PEP) as phosphate donor, phosphorylates the transcription factor STAT3 at the nucleus of cancer cells, was reported a year later (Gao, Wang, Yang, Liu, & Liu, 2012).

This protein kinase activity corresponds to a dimeric form of the M2-PK favored by conformational changes in the M2 protein apparently due to interactions in which residues Arg399, Glu418, and Glu396 in its sequence are implicated. Another protein kinase activity of a nuclear form of M2 was later reported (Yang et al., 2012), which phosphorylates histone H3 at Thr11 residue, using PEP as phosphate donor, although the M2 protein responsible for such ability seems to have no structural modification. In a recent report from the same group (Jiang et al., 2014), they described another role of this nuclear M2-PK, which is involved in the regulation of the chromosome segregation and mitosis progression of tumor cells through its ability to phosphorylate the spindle checkpoint protein Bub3 at Tyr207 residue, in a process correlated with the histone H3 phosphorylation. In this sense, the ability of M2-ProTαK to phosphorylate histones H2B and H3 (Pérez-Estévez et al., 1997) would prove a capacity of the M2 protein to function as a histone kinase either in its cytoplasmic or nuclear location.

Thus, according to the indicated studies of the M2 protein it can be inferred that it has a dual functionality in proliferating and tumor cells, either working as a glycolytic enzyme or as a protein kinase. These abilities are propitiated by a cytoplasmic or nuclear location as well as by structural modifications of the M2 protein. So that, PK activity corresponds to tetrameric/ dimeric cytoplasmic forms, while the cytoplasmic protein kinase activity of M2, so far demonstrated, corresponds to the M2-ProTαK that uses ATP as phosphate donor. Whereas, the nuclear protein kinase activity of M2, with PEP as phosphate donor, seems to have several functions.

Once the phosphorylation of ProTα is characterized, the next goal in the ProTα research should be directed to stablish the biological significance of its phosphorylation process.

4. ABOUT THE BIOLOGICAL SIGNIFICANCE OF THE PHOSPHORYLATION OF ProTα

The predominant nuclear presence of ProTα affecting the chromatin activity through its interaction with histones and other nuclear factors, led us, firstly, to investigate the influence that the phosphorylation of ProTα might have on its nuclear interactions. Results of this research showed that interactions of ProTα with histones and other nuclear proteins (Covelo et al., 2006; Díaz-Jullien et al., 1996; Freire et al., 2001), as well as its implication in nucleosome assembly (Díaz-Jullien et al., 1996), are not affected by its phosphorylation. Moreover, the phosphorylation of ProTα has no influence on its import to the nucleus, as demonstrated in HeLa cells transfected with a vector harboring recombinant ProTα, either wild-type or nonphosphorylatable with Thr7 substituted by Ala (Moreira et al., 2013). Low cellular concentration of phosphorylated ProTα, which contrast with a high nuclear content of ProTα as deduced from the microscopic analysis of proliferating cells (Moreira et al., 2013), could justify a scanty influence of the phosphorylation process in the nuclear activity of ProTα.

The investigation on the biological significance of the ProTα phosphorylation was then addressed to test whether this process might have influence on the antiapoptotic activity of ProTα, since it has been reported that ProTα exerts a cytoplasmic antiapoptotic effect by blocking the apoptosome formation and, thence, the activation of caspase 9 and effector caspases 3 and 7 by an unrevealed mechanism that antagonizes the proapoptotic effect of the protein ANP32A (Jiang et al., 2003; Qi et al., 2010). Influence of the phosphorylated ProTα in this process was investigated in 293T cells transfected with recombinant ProTα to express wild-type or nonphosphorylatable ProTα, in which, evolution of the apoptosis induced by staurosporine treatment was determined following the caspases 9 and 3 activities. Results (Moreira et al., 2013) indicated that phosphorylation of ProTα at residue Thr7 seems to influence the antiapoptotic effect of ProTα, at least in proliferating tumor cells, since an overexpression of phosphorylatable ProTα provokes an important inhibition of the caspases 9 and 3 activation. However, the mechanism of this influence remains elusive. To this respect, recent results from our laboratory (manuscript in preparation) showing that in proliferating lymphocytes (Jurkat cells) a small part of the cellular content of ProTα (about 10%) interacts with proteins SET, ANP32A, and ANP32B forming cytoplasmic complexes, revealed possible mechanisms by which the cytoplasmic activity of ProTα could be achieved. The structure of these complexes, whose components have a function related to the regulation of the mitochondrial-dependent apoptotic pathway, make feasible that the characterized interactions of ProTα might be also related with its implication in this process. The high decrease in the concentration of these complexes observed when cell proliferation activity is diminished or apoptosis induced, could support this hypothesis. Future research should stablish the cellular function of these interactions of ProTα, and determine if phosphorylated ProTα is implicated in them.

Besides the apoptotic implication of phospho-ProTα, other possibilities of its cellular activity cannot be discarded. Thus, it should be taken into consideration the possibility that the ProTα phosphorylation process could be also related with the proteolytic processing of ProTα by the lysosomal protease legumain to generate $T\alpha_1$, or even with its total proteolytic degradation. On the other hand, although the phospho-ProTα seems not to be implicated in the chromatin-relative functions of ProTα, other phosphorylation-dependent nuclear function of ProTα could be possible.

5. CONCLUDING REMARKS AND PERSPECTIVES

The study of the phosphorylation process of ProTα has led to deepen in the mechanisms of the cellular function of this protein. Thus, dependence of this process on the cellular proliferation activity seem to indicate that phosphorylation of ProTα could be involved in the mechanisms that implicate ProTα in the growth of the mammalian cells.

A special circumstance in the ProTα phosphorylation process is the peculiar interaction of ProTα with a special isoform of the M2 protein that has acquired a novel protein kinase activity that propitiates the phosphorylation of ProTα in normal and tumor cells. This dual functionality of the M2 protein, caused by the phosphorylation of a small part of the cellular content of this protein, represents a novel ability of the M2-PK, and its phosphorylation process constitutes a key for the regulation of the ProTα phosphorylation in the cytoplasm of proliferating cells. The characterization of the protein kinase responsible for the partial phosphorylation of the cellular pool of M2 constitutes and important goal toward addressing the future research on the phosphorylation of ProTα.

An interesting property of the M2 protein kinase activity toward ProTα is the modification of the specificity on the phosphorylated residues of ProTα, which apparently depends on the carcinogenic transformation of the cells, at least in lymphocytes. So that, the incorporation of additional orthophosphate at Thr12 and/or Thr13 residues of ProTα by the M2-ProTαK in normal cells, seems to be lost when the cells become cancerous. The characterization of the mechanisms involved in the phosphorylation process by which the M2-ProTαK activity is promoted, may explain this change in its specificity, what at the same time, might denote the appearance of a marker of the carcinogenic transformation of lymphocytes and, perhaps, other mammalian cell types.

Concerning the biological significance of the phosphorylation of ProTα, our findings suggest that it seems related to an antiapoptotic activity of ProTα. In this sense, the low cellular content of phospho-ProTα would be in agreement with a participation in the cellular activity, which should concern to a small part of the cellular pool of ProTα. In this sense, cytoplasmic location of an antiapoptotic activity attributed to ProTα, which coincides with a scanty content of this protein, would be appropriated for a possible implication of phosphorylated ProTα in this activity. Recent results from our laboratory showing the interaction of ProTα with the proteins SET, ANP32A, ANP32B, and others, implicated in the regulation of the mitochondrial-induced apoptosis, might be indicative of an implication of ProTα in this process. Future research should assess this point, at the time that has to corroborate the involvement of phospho-ProTα in these interactions.

However, in the future research on biological significance of the phosphorylation of ProTα cannot be ruled out a study on the possibility that this event can influence other cytoplasmic activities, as the proteolytic processing of ProTα. Moreover, although phospho-ProTα seems not to be involved in a chromatin-related function of ProTα, it should be considered that some other nuclear phosphorylation-dependent functions of ProTα are possible.

REFERENCES

Aniello, F., Branno, M., De Rienzo, G., Ferrara, D., Palmiero, C., & Minucci, S. (2002). First evidence of prothymosin α in a non-mammalian vertebrate and its involvement in the spermatogenesis of the frog Rana esculenta. *Mechanisms of Development*, *110*(1–2), 213–217.

Barcia, M. G., Castro, J. M., Jullien, C. D., & Freire, M. (1993). Prothymosin α is phosphorylated in proliferating stimulated cells. *Journal of Biological Chemistry*, *268*(7), 4704–4708.

Barcia, M. G., Castro, J. M., Jullien, C. D., González, C. G., & Freire, M. (1992). Prothymosin α is phosphorylated by casein kinase-2. *FEBS Letters*, *312*(2,3), 152–156.

Bustelo, X. R., Otero, A., Gómez-Márquez, J., & Freire, M. (1991). Expression of the rat prothymosin α gene during T-lymphocyte proliferation and liver regeneration. *Journal of Biological Chemistry*, *266*(3), 1443–1447.

Caldarella, J., Goodall, G. J., Felix, A. M., Heimer, E. P., Salvin, S. B., & Horecker, B. L. (1983). Thymosin α_{11}: A peptide related to thymosin α_1 isolated from calf thymosin fraction 5. *Proceedings of the National Academy of Sciences of the United States of America*, *80*(24), 7424–7427.

Clinton, M., Frangou-Lazaridis, M., Panneerselvam, C., & Horecker, B. L. (1989). Prothymosin α and parathymosin: mRNA and polypeptide levels in rodent tissues. *Archives of Biochemistry and Biophysics*, *269*(1), 256–263.

Clover, C. V., Chatterjee, D., Wang, Z., Cantley, L. C., Vander Heiden, M. G., & Krainer, A. R. (2010). The alternative splicing repressors hnRNPA1/A2 and PTB influence pyruvate kinase isoform expression and cell metabolism. *Proceedings of the National Academy of Sciences of the United States of America*, *107*(5), 1894–1899.

Conteas, C. N., Mutchnick, M. G., Palmer, K. C., Weller, F. E., Luk, G. D., Naylor, P. H., et al. (1990). Cellular levels of thymosin immunoreactive peptides are linked to proliferative events: Evidence for a nuclear site of action. *Proceedings of the National Academy of Sciences of the United States of America, 87*(9), 3269–3273.

Cotter, M. A., & Robertson, E. S. (2000). Modulation of histone acetyltransferase activity through interaction of Epstein-Barr nuclear antigen 3C with prothymosin α. *Molecular and Cellular Biology, 20*(15), 5722–5735.

Covelo, G., Sarandeses, C. S., Díaz-Jullien, C., & Freire, M. (2006). Prothymosin α interacts with free core histones in the nucleus of dividing cells. *Journal of Biochemistry, 140*(5), 627–637.

David, C. J., Chen, M., Assanah, M., Canoll, P., & Manley, J. L. (2010). HnRNP proteins controlled by c-Myc deregulate pyruvate kinase mRNA splicing in cancer. *Nature, 463*, 364–368.

Díaz-Jullien, C., Moreira, D., Sarandeses, C. S., Covelo, G., Barbeito, P., & Freire, M. (2011). The M2-type isoenzyme of pyruvate kinase phosphorylates prothymosin α in proliferating lymphocytes. *Biochimica et Biophysica Acta, 1814*, 355–365.

Díaz-Jullien, C., Pérez-Estévez, A., Covelo, G., & Freire, M. (1996). Prothymosin α binds histones in vitro and shows activity in nucleosome assembly assay. *Biochimica et Biophysica Acta, 1296*, 219–227.

Enkemann, S. A., Ward, R. D., & Berger, S. L. (2000). Mobility within the nucleus and neighboring cytosol is a key feature of prothymosin α. *Journal of Histochemistry and Cytochemistry, 48*(10), 1341–1355.

Eschenfeldt, W. H., & Berger, S. L. (1986). The human prothymosin α gene is polymorphic and induced upon growth stimulation: Evidence using a cloned cDNA. *Proceedings of the National Academy of Sciences of the United States of America, 83*(24), 9403–9407.

Eschenfeldt, W. H., Manrow, R. E., Krug, M. S., & Berger, S. L. (1989). Isolation and partial sequencing of the human prothymosin α gene family. Evidence against export of the gene products. *Journal of Biological Chemistry, 264*(13), 7546–7555.

Faust, M., & Montenarh, M. (2000). Subcellular localization of protein kinase CK2: A key to its function. *Cell and Tissue Research, 301*, 329–340.

Franco del Amo, F., & Freire, M. (1995). The prothymosin α gene is specifically expressed in ectodermal and mesodermal regions during early postimplantation mouse embryogenesis. *FEBS Letters, 359*(1), 15–19.

Franco, F. J., Díaz, C., Barcia, M., & Freire, M. (1992). Thymosin α_1 is a native peptide in several tissues. *Biochimica et Biophysica Acta, 1120*(1), 43–48.

Frangou-Lazaridis, M., Clinton, M., Goodall, G. J., & Horecker, B. L. (1988). Prothymosin α and parathymosin: Amino acid sequences deduced from the cloned rat spleen cDNAs. *Archives of Biochemistry and Biophysics, 263*(2), 305–310.

Freire, J., Covelo, G., Sarandeses, C., Díaz-Jullien, C., & Freire, M. (2001). Identification of nuclear-import and cell-cycle regulatory proteins that bind to prothymosin α. *Biochemistry and Cell Biology, 79*, 123–131.

Freire, M., Crivellaro, O., Isaacs, C., Moschera, S., & Horecker, B. L. (1978). Translation of mRNA from calf thymus in the wheat germ system: Evidence for a precursor of thymosin α_1. *Proceedings of the National Academy of Sciences of the United States of America, 75*(12), 6007–6011.

Freire, M., Hannappel, E., Rey, M., Freire, J. M., Kibo, H. K., & Horecker, B. L. (1981). Purification of thymus mRNA coding for a 16,000 dalton polypeptide containing the thymosin α_1 sequence. *Proceedings of the National Academy of Sciences of the United States of America, 78*(1), 192–195.

Freire, M., Rey-Méndez, M., Gómez-Márquez, J., & Arias, P. (1985). Evidence for the synthesis of thymosin α_1 by calf thymocytes and for the production of this peptide by natural processing. *Archives of Biochemistry and Biophysics, 239*(2), 480–485.

Frillingos, S., Frangou-Lazaridis, M., Seferiadis, K., Hulmes, J. D., Pan, Y. C., & Tsolas, O. (1991). Isolation and partial sequence of goat spleen prothymosin α. *Molecular and Cellular Biochemistry, 108*(1), 85–94.

Gao, X., Wang, H., Yang, J. J., Liu, X., & Liu, Z. R. (2012). Pyruvate kinase M2 regulates gene transcription by acting as a protein kinase. *Molecular Cell, 45*, 598–609.

Goldstein, A. L., Guha, A., Zatz, M., Hardy, M. A., & White, A. (1972). Purification and biological activity of thymosin, a hormone of the thymus gland. *Proceedings of the National Academy of Sciences of the United States of America, 69*(7), 1800–1803.

Goldstein, A. L., Low, T. L. K., McAdoo, M., McClure, J., Thurman, G. B., Rossio, J., et al. (1977). Thymosin α_1: Isolation and sequence analysis of an immunologically active thymic polypeptide. *Proceedings of the National Academy of Sciences of the United States of America, 74*(2), 725–729.

Goldstein, A. L., Slater, F. D., & White, A. (1966). Preparation, assay and partial purification of a thymic lymphocytopoietic factor (thymosin). *Proceedings of the National Academy of Sciences of the United States of America, 56*(3), 1010–1017.

Gómez-Márquez, J., & Segade, F. (1988). Prothymosin α is a nuclear protein. *FEBS Letters, 226*(2), 217–219.

Gómez-Márquez, J., Segade, F., Dosil, M., Pichel, J. G., Bustelo, X. R., & Freire, M. (1989). The expression of prothymosin α gene in T lymphocytes and leukemic lymphoid cells is tied to lymphocyte proliferation. *Journal of Biological Chemistry, 264*(15), 8451–8454.

Goodall, G. J., Domínguez, F., & Horecker, B. L. (1986). Molecular cloning of cDNA for human prothymosin α. *Proceedings of the National Academy of Sciences of the United States of America, 83*(23), 8926–8928.

Haritos, A. A., Blacher, R., Stein, S., Caldarella, J., & Horecker, B. L. (1985). Primary structure of rat thymus prothymosin α. *Proceedings of the National Academy of Sciences of the United States of America, 82*(2), 343–346.

Haritos, A. A., Goodall, G. J., & Horecker, B. L. (1984). Prothymosin α isolation and properties of the major immunoreactive form of thymosin α_1 in rat thymus. *Proceedings of the National Academy of Sciences of the United States of America, 81*(4), 1008–1011.

Haritos, A. A., Tsolas, O., & Horecker, B. L. (1984). Distribution of prothymosin α in rat tissues. *Proceedings of the National Academy of Sciences of the United States of America, 81*(5), 1391–1393.

Hathaway, G. M., & Traugh, J. A. (1982). Casein kinases—Multipotential protein kinases. *Current Topics in Cellular Regulation, 21*, 101–127.

Jiang, X., Kim, H. E., Shu, H., Zhao, Y., Zhang, H., Kofron, J., et al. (2003). Distinctive roles of PHAP proteins and prothymosin α in a death regulatory pathway. *Science, 299*, 223–226.

Jiang, Y., Li, X., Yang, W., Hawke, D. H., Zheng, Y., Xia, Y., et al. (2014). PKM2 regulates chromosome segregation and mitosis progression of tumor cells. *Molecular Cell, 53*, 1–13.

Karetsou, Z., Kretsovali, A., Murphy, C., Tsolas, O., & Papamarcaki, T. (2002). Prothymosin α interacts with the CREB-binding protein and potentiates transcription. *EMBO Reports, 3*(4), 361–366.

Karetsou, Z., Sandaltzopoulos, R., Frangou-Lazaridis, M., Lai, C. Y., Tsolas, O., Becker, P. B., et al. (1998). Prothymosin α modulates the interaction of histone H1 with chromatin. *Nucleic Acids Research, 26*(13), 3111–3118.

Kubota, S., Adachi, Y., Copeland, T. D., & Oroszlan, S. (1995). Binding of human prothymosin α to the leucine-motif/activation domains of HTLV-I Rex and HIV-1 Rev. *European Journal of Biochemistry, 233*(1), 48–54.

Low, T. L. K., & Goldstein, A. L. (1985). Thymic hormones: An overview. In G. D. Sabato, J. J. Langone, & H. V. Vunakis (Eds.), *Methods in enzymology: Vol. 116*. (pp. 213–219). San Diego, CA: Academic Press, Inc.

Low, T. L. K., Thurman, G. B., McAdoo, M., McClure, J., Rossio, J. L., Naylor, P. H., et al. (1979). The chemistry and biology of thymosin. I. Isolation, characterization, and biological activities of thymosin α_1 and polypeptide β_1 from calf thymus. *Journal of Biological Chemistry*, *254*(3), 981–986.

Maelandsmo, G. M., Østvold, A. C., & Laland, S. G. (1989). Phosphorylation of the high-mobility-group-like protein P1 by casein kinase-2. *European Journal of Biochemistry*, *184*(3), 529–534.

Manrow, R. E., Sburlati, A. R., Hanover, J. A., & Berger, S. L. (1991). Nuclear targeting of prothymosin α. *Journal of Biological Chemistry*, *266*(6), 3916–3924.

Martini, P. G., Delage-Mourroux, R., Kraichely, D. M., & Katzenellenbogen, B. S. (2000). Prothymosin α selectively enhances estrogen receptor transcriptional activity by interacting with a repressor of estrogen receptor activity. *Molecular and Cellular Biology*, *20*(17), 6224–6232.

Moreira, D., Díaz-Jullien, C., Sarandeses, C. S., Covelo, G., Barbeito, P., & Freire, M. (2013). The influence of phosphorylation of prothymosin α on its nuclear import and antiapoptotic activity. *Biochemistry and Cell Biology*, *91*, 265–269.

Muñoz, E. M., & Ponce, E. (2003). Pyruvate kinase: Current status of regulatory and functional properties. *Comparative Biochemistry and Physiology. Part B: Biochemistry & Molecular Biology*, *135*, 197–218.

Nogueira, M., & Freire, M. (1985). Characteristics of the translation of thymosin α_1 precursor mRNA by cell-free wheat germ system. Evidence for the acetylation of thymosin α_1 precursor. *International Journal of Biochemistry*, *17*(4), 533–536.

Panneerselvam, C., Wellner, D., & Horecker, B. L. (1988). The amino acid sequence of bovine thymus prothymosin α. *Archives of Biochemistry and Biophysics*, *265*(2), 454–457.

Papamarcaki, T., & Tsolas, O. (1994). Prothymosin α binds to histone H1 in vitro. *FEBS Letters*, *345*(1), 71–75.

Pavlov, N., Evstafieva, A., Rubtsov, Y., & Vartapetian, A. (1995). Human prothymosin α inhibits division of yeast Saccharomyces cerevisiae cells, while its mutant lacking nuclear localization signal does not. *FEBS Letters*, *366*, 43–45.

Pérez-Estévez, A., Díaz-Jullien, C., Covelo, G., Salgueiro, M. T., & Freire, M. (1997). A 180-kDa protein kinase seems to be responsible for the phosphorylation of prothymosin α observed in proliferating cells. *Journal of Biological Chemistry*, *272*(16), 10506–10513.

Pérez-Estévez, A., Freire, J., Sarandeses, C., Covelo, G., Díaz-Jullien, C., & Freire, M. (2000). Properties of the protein kinase that phosphorylates prothymosin α. *Molecular and Cellular Biochemistry*, *208*, 111–118.

Qi, X., Wang, L., & Du, F. (2010). Novel small molecules relieve prothymosin α-mediated inhibition of apoptosome. *Biochemistry*, *49*, 1923–1930.

Rubtsov, Y. P., Zolotukhin, A. S., Vorobjev, I. A., Chichkova, N. V., Pavlov, N. A., Karger, E. M., et al. (1997). Mutational analysis of human prothymosin α reveals a bipartite nuclear localization signal. *FEBS Letters*, *413*(1), 135–141.

Sarandeses, C. S., Covelo, G., Díaz-Jullien, C., & Freire, M. (2003). Prothymosin α is processed to thymosin α_1 and thymosin α_{11} by a lysosomal asparaginyl endopeptidase. *Journal of Biological Chemistry*, *278*(15), 13286–13293.

Sburlati, A. R., De La Rosa, A., Batey, D. W., Kurys, G. L., Manrow, R. E., Pannell, L. K., et al. (1993). Phosphorylation of human and bovine prothymosin α in vivo. *Biochemistry*, *32*(17), 4587–4596.

Sburlati, A. R., Manrow, R. E., & Berger, S. L. (1990). Human prothymosin α: Purification of a highly acidic nuclear protein by means of a phenol extraction. *Protein Expression and Purification*, *1*(2), 184–190.

Sburlati, A. R., Manrow, R. E., & Berger, S. L. (1991). Prothymosin α antisense oligomers inhibit myeloma cell division. *Proceedings of the National Academy of Sciences of the United States of America*, *88*(1), 253–257.

Schmidt, G., & Werner, D. (1991). Nucleotide sequence of the murine prothymosin α cDNA and its deduced primary and secondary protein structure. *Biochimica et Biophysica Acta*, *1088*(3), 442–444.
Smith, M. R., al-Katib, A., Mohammad, R., Silverman, A., Szabo, P., Khilnani, S., et al. (1993). Prothymosin α gene expression correlates with proliferation, not differentiation, of HL-60 cells. *Blood*, *82*(4), 1127–1132.
Subramanian, C., Hasan, S., Rowe, M., Hottiger, M., Orre, R., & Robertson, E. S. (2002). Epstein-Barr virus nuclear antigen 3C and prothymosin α interact with the p300 transcriptional coactivator at the CH1 and CH3/HAT domains and cooperate in regulation of transcription and histone acetylation. *Journal of Virology*, *76*(10), 4699–4708.
Takenaka, M., Noguchi, T., Sadahiro, S., Hirai, H., Yamada, K., Matsuda, T., et al. (1991). Isolation and characterization of the human pyruvate kinase M gene. *European Journal of Biochemistry*, *198*(1), 101–106.
Tsitsiloni, O. E., Yialouris, P. P., Sekeri-Pataryas, K., & Haritos, A. A. (1989). Prothymosin α is not a nuclear polypeptide. *Experientia*, *45*(4), 332–334.
Vareli, K., Frangou-Lazaridis, M., van der Kraan, I., Tsolas, O., & van Driel, R. (2000). Nuclear distribution of prothymosin α and parathymosin: Evidence that prothymosin α is associated with RNA synthesis processing and parathymosin with early DNA replication. *Experimental Cell Research*, *257*(1), 152–161.
Wintero, A. K., Fredholm, M., & Davies, W. (1996). Evaluation and characterization of a porcine small intestine cDNA library: Analysis of 839 clones. *Mammalian Genome*, *7*(7), 509–517.
Wu, C. G., Boers, W., Reitsma, P. R., van Deventer, S. J., & Chamuleau, R. A. (1997). Overexpression of prothymosin α, concomitant with c-myc, during rat hepatic carcinogenesis. *Biochemical and Biophysical Research Communications*, *232*(3), 817–821.
Yang, W., Xia, Y., Hawke, D., Li, X., Liang, J., Xing, D., et al. (2012). PKM2 phosphorylates histone H3 and promotes gene transcription and tumorigenesis. *Cell*, *150*, 685–696.
Yialouris, P. P., Evangelatos, G. P., Soteriadis-Vlahos, C., Heimer, E. P., Félix, A. M., Tsitsiloni, O. E., et al. (1988). The identification of prothymosin α-like material in vertebrate lymphoid organs by a radioimmunoassay for the N-terminal decapeptide. *Journal of Immunological Methods*, *106*(2), 267–275.
Zalvide, J. B., Cancio, E., Alvarez, C. V., Regueiro, B. J., & Domínguez, F. (1992). Prothymosin α mRNA levels are invariant throughout the cell cycle. *Journal of Biological Chemistry*, *267*(12), 8692–8695.

CHAPTER FIVE

Mechanism of Action of Thymosinα1: Does It Interact with Membrane by Recognition of Exposed Phosphatidylserine on Cell Surface? A Structural Approach

R. Nepravishta*,†, W. Mandaliti*, P.S. Vallebona*, F. Pica*, E. Garaci*,‡, M. Paci*,[1]

*University of Rome "Tor Vergata", Rome, Italy
†Faculty of Pharmacy Catholic University "Our Lady of Good Counsel", Tirane, Albania
‡San Raffaele Pisana Scientific Institute for Research, Hospitalization and Health Care, Rome, Italy
[1]Corresponding author: e-mail address: paci@uniroma2.it

Contents

Abstract

Thymosinα1 is a peptidic hormone with pleiotropic activity, which is used in the therapy of several diseases. It is unstructured in water solution and interacts with negative regions of micelles and vesicles assuming two tracts of helical conformation with a

Vitamins and Hormones, Volume 102
ISSN 0083-6729
http://dx.doi.org/10.1016/bs.vh.2016.04.002

structural flexible break in between. The studies of the interaction of Thymosinα1 with micelles of mixed dipalmitoylphosphatidylcholine and sodium dodecylsulfate and vesicles with mixed dipalmitoylphosphatidylcholine/dipalmitoylphosphatidylserine, the latter the negative component of the membranes, by ^{1}H and natural abundance ^{15}N NMR are herewith reported, reviewed, and discussed. The results indicate that the preferred interactions are those where the surface is negatively charged due to sodium dodecylsulfate or due to the presence of dipalmitoylphosphatidylserine exposed on the surface. In fact the unbalance of dipalmitoylphosphatidylserine on the cellular surface is an important phenomenon present in pathological conditions of cells. Moreover, the direct interaction of Thymosinα1 with K562 cells presenting an overexposure of phosphatidylserine as a consequence of resveratrol-induced apoptosis was carried out.

1. INTRODUCTION

The α-thymosins belong to a group of peptides isolated for the first time from calf thymus extracts, denominated Thymosin fraction 5 (TF5) (Goldstein, Guha, Zatz, Hardy, & White, 1972) which showed both in vitro and in vivo systems immune regulatory properties (Low, Thurman, McAdoo, et al., 1979).

Thymosinα1, the most abundant α-Thymosin found in TF5, was the first isolated and sequenced (Low & Goldstein, 1979). It is a 28-amino acid peptide with sequence: Ac-Ser-Asp-Ala-Ala-Val-Asp-Thr-Ser-Ser-Glu-Ile-Thr-Thr-Lys-Asp-Leu-Lys-Glu-Lys-Lys-Glu-Val-Val-Glu-Glu-Ala-Glu-Asn-OH, where Ac indicates that is acetylated on its N-terminus group. The peptide derives from the N-terminus tract of ProThymosinα (ProTα) from where is cleaved by legumain, a lysosomal asparagine endopeptidase (AEP) present also in mammals (Sarandeses, Covelo, Díaz-Jullien, & Freire, 2003). Both Thymosinα1 (Franco, Diaz, Barcia, & Freire, 1992) and legumain (Chen, Dando, Rawlings, et al., 1997) share a wide distribution in different tissues, suggesting that the ProTα processing to yield Thymosinα1 represents a generalized process in mammalian tissues. Moreover, high legumain activity has been found in lymphoid tissues that also show high levels of Thymosinα1 (Franco et al., 1992), which argue for important biological function of the peptide in these cells.

Thymosinα1 cellular location, in the absence of secretion signals and the incapacity for nuclear migration, is in accord for a cytosolic localization differently from its precursor ProTα (Eschenfeldt, Manrow, Krug, & Berger, 1989; Manrow, Sburlati, Hanover, & Berger, 1991).

Actually, Thymosinα1 is approved in different countries for treatment of several viral infections (Andreone, Cursaro, Gramenzi, et al., 2001; Iino, Toyota, Kumada, et al., 2005; Kullavanuava, Treeprasertsuk, Thong-Ngam, et al., 2001; You, Zhuang, Cheng, et al., 2006) and as an adjuvant for immune enhancement (Carraro, Naso, Montomoli, et al., 2012; Goldstein, 2009). Moreover, it is also developed for treatment of non-small cell lung cancer, hepatocellular carcinoma, AIDS, and malignant melanoma (Billich, 2002; Maio, Mackiewicz, Testori, et al., 2010). An univocal mechanism of action of Thymosinα1is still unknown since no specific receptors have been identified at the level of the lymphocyte membrane and a growing body of evidence obtained in the last decades attests the clear-cut pleiotropy of this peptide, which targets both normal and tumor cells interacting with multiple cellular components (Garaci, Pica, Serafino, et al., 2012; Romani, Moretti, Fallarino, et al., 2012; Serafino, Pica, Andreola, et al., 2014).

Previous research showed that Thymosinα1 upregulates the expression of the major histocompatibility complex class I in murine and human tumor cell lines and in primary cultures of human monocyte-enriched peripheral blood mononuclear cells (Giuliani, Napolitano, Mastino, et al., 2000). Thymosinα1 induces T-cell and dendritic cell (DC) maturation and interleukin (IL-2) expression (Knutsen, Freeman, Mueller, et al., 1999; Romani, Bistoni, Gaziano, et al., 2004); it is also capable of up-regulating the expression of Toll-like receptors (TLRs) 2 and 9 greater than for 5 and 8 in murine DCs and protects mice from challenge by invasive aspergillosis in the MyD88 (myeloid differentiate factor 88)-dependent way (Romani et al., 2004). Thymosinα1 activates a TRAF6- atypical protein kinase C (PKC)-IkB kinase signaling pathway that leads to the activation of nuclear factor-kB, which in turn initiates cytokine gene expression in murine bone marrow-derived macrophages (Zhang, Chan, Dragoi, et al., 2005). The complex of these findings has generated a renewed interest of the research on this hormone peptide, which is aimed to a better knowledge of its therapeutic potential and to an improvement of its delivery to the target cells, thus implementing its own effectiveness. In this regard, the conjugation of Thymosinα1 with a RGD motif has been attempted very recently to implement Thymosinα1 interaction with tumor cells (Lao, Liu, Chen, & Zheng, 2013). In fact, the amino acidic RGD sequence (Arg-Gly-Asp) is able to recognize integrins avb3, which is specifically associated with upregulated expression of tumor vessels and tumor cells of the endothelium (Zetter, 1997). All these researches would benefit of a greater understanding of the mechanism(s) involved in the Thymosinα1-cell interaction.

2. THE NMR STRUCTURAL STUDY IN TRIFLUOROETHANOL SOLUTION

In this regard, the knowledge of the Thymosinα1 conformation when interacting with membranes may represent a useful step toward a better understanding of all the processes involved in its action. In the past, investigations by NMR has been carried out (Elizondo-Riojas, Chamow, Tuthill, et al., 2011; Grottesi, Sette, Palamara, et al., 1998) the latter with a limited instrumental NMR resolution and, the former, at much higher magnetic field. Both studies were performed using trifluoroethanol (TFE) as solvent. This solvent was for long time considered as a "membrane-like" environment and, then, as an optimal tool for determining the structure of biologically important peptides (Elizondo-Riojas et al., 2011; Grottesi et al., 1998). The addition of TFE to peptides which often are completely unstructured in water solution acts as a structuring medium in some case forcing and propagating the secondary structure.

The structures found for Thymosinα1 identifies two secondary structure elements: two helix tracts with a central region flexible in between with few negligible differences (Elizondo-Riojas et al., 2011; Grottesi et al., 1998).

3. THE STRUCTURAL STUDY BY NMR IN MICELLAR ENVIRONMENT

The limitation of the above reported studies is that the chaotropic solvent (TFE) wraps around the peptide structuring it into a helical conformation. Considering that Thymosinα1 may interact with cellular membrane by partial insertion, the determination of these structures in TFE did not give sufficient information to support a feasible hypothesis on the mechanism of action. Thus, the hypothesis that an interaction with membrane might be the possible first step of the mechanism of action was exploited. As discussed below this possibility has been invoked for several peptide, hormones and proteins to elucidate the biological mechanism of interaction and the signaling cascade due to this interaction (Nepravishta et al., 2015).

Micelles of negatively charged sodium dodecylsulfate (SDS) were used to test the possibility and therefore the modality of interaction of this hormone peptide with a model membrane. In the structural studies of peptides the use of SDS micelles is considered a good tool for the investigation of the transmembrane domain of some membrane proteins or of isolated helixes belonging to their transmembrane domains. This medium ensures a physico-chemical similarity with the biological membranes and, in addition, shares a high stability of samples upon time. This represents an absolute requirement for NMR spectroscopy, due to its long experimental time, which is not fulfilled by the phospholipid assembly like phospholipid vesicles, which are, generally, stable only for short times. Results of the investigation on the conformation of Thymosinα1 by the use of NMR and circular dichroism (CD) spectroscopy techniques also indicate that Thymosinα1 is substantially unstructured in water solution and that, interacting with micelles, it assumes a definite conformation with two helical domains and, in between, a short flexible tract (residues 9–14) that divides the two helical domains. Moreover by paramagnetic NMR relaxation it was found that Thymosinα1 is inserted in the hydrophobic region of the micelles by the residues 1–5 of the N-terminus end. This result further clarifies the modality of insertion that was not possible to obtain in the previous NMR studies in TFE. These findings suggest the hypothesis that, however, has to be experimentally verified, the Thymosinα1 folds upon the membrane insertion and, when inserted, may be capable of floating on the membrane due to its fluidity and, eventually, may interact with nearby proteins and/or receptors acting as effector and causing a biological signaling cascade. The structure obtained is reported in Fig. 1A–C where the Thymosinα1 structure alone (A and B) and the scheme of its insertion in SDS micelle (C) is depicted.

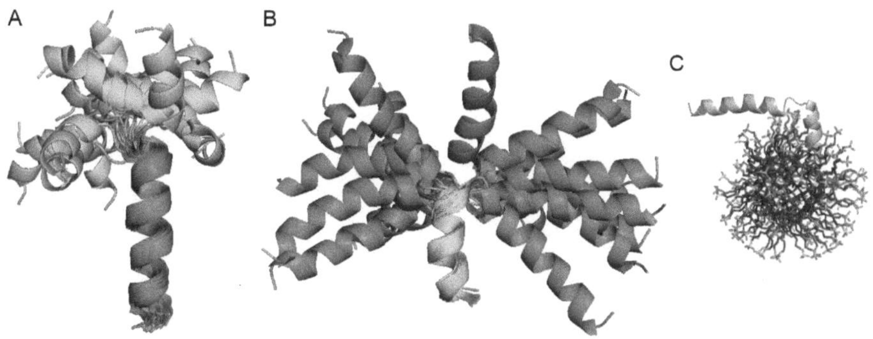

Fig. 1 (A, B, and C) Thymosinα1 structured upon binding to SDS micelles: (A) the Thymosinα1 in SDS micelles aligned with the N-terminus tract, (B) the same as in (A) aligning the C-terminus tract, and (C) representing its insertion on the SDS micelle. (See the color plate.)

4. THE INTERACTIONS WITH PHOSPHOLIPIDIC MEMBRANES WITH NEGATIVE REGIONS

These findings led us to investigate the behavior of Thymosinα1 in the presence of membrane models containing phospholipids that are the natural constituents of the membrane. The interaction either with vesicles of dodecylphosphocholine alone or with vesicles of dodecylphosphocholine-sodium dodecylsulfate was performed. The results of the NMR spectroscopy experiments indicated that Thymosinα1 interacts by an aspecific modality with phospholipidic membrane model exposing choline polar heads on the surface interacts by an aspecific modality (not shown). The NMR spectrum appears broad and all the resonance shows that a large aspecific binding occurring on the surface (not shown). By contrast, in the mixed dodecylphosphocholine-sodium dodecylsulfate micelles an insertion in the micellar hydrophobic region was observed and a conformation was assumed which was similar to that found in sodium dodecylsulfate micelles (SDS). In the presence of these mixed micelles, both the insertion and structuration of the peptide occur in preferred regions, particularly where the membrane models are negatively charged.

It is important to note that, when increasing the concentration of the SDS-d_{25} in the micelles, the resonance attributed to the OH of Ser1 of Thymosinα1 (Nepravishta et al., 2015) appears clearly visible in the proximity of the resonance of water protons. This effect had been already observed and assigned in the TOCSY spectra of Thymosinα1 performed in micelles of SDS-d_{25} (Nepravishta et al., 2015). We can consider its observation as diagnostic for the insertion of the N-terminus region of the Thymosinα1 in the micelle as already found for SDS-d_{25} micelles (Nepravishta et al., 2015). Increasing the amount of SDS-d_{25} in the mixed micelles, these resonances increase their intensity and the TOCSY spectrum of this region clearly confirms this observation showing the scalar correlation of this resonance with CHα and CH_2 protons of Ser1. It is important to note that this resonance is absent in the micelle of DPC-d_{38} alone (reported above).

5. THE CONFORMATION OF THYMOSINα1 IN MIXED DPC-D_{38}/SDS-D_{25} MICELLES

The collection of NOEs reported above allowed us to perform a restrained molecular dynamics simulation of Thymosinα1 inserted in the mixed micelles. The results allowed us to report the final structure in Fig. 2 where

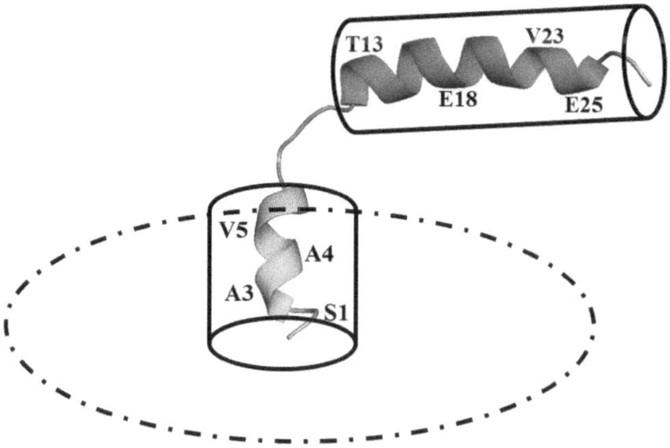

Fig. 2 Thymosinα1 inserted in mixed micelles of DPC and perdeuterated SDS. (See the color plate.)

Thymosinα1 appears structured in helical conformation in two tracts, the first one from residue 1 to 6 which corresponds to the region inserted in mixed micelles and the second one from residue 15 to about 26. A flexible tract in between gives the mobility of the two secondary structure elements.

6. THE INTERACTION OF THYMOSINα1 WITH PERDEUTERATED DPC AND PERDEUTERATED DPC–SDS MICELLES

The determination of the nuclear magnetic spin-lattice relaxation times of the NMR resonances of Thymosinα1 in the presence of mixed micelles of DPC-d_{38}/SDS-d_{25} gave for the aliphatic resonances a longitudinal relaxation time (T_1) of about 570 ms. The same spectral region in the sample of Thymosinα1 in water the value of the spin-lattice relaxation time obtained was about 813 ms. According to the Bloembergen–Pound–Purcell theory the difference of correlation times of Thymosinα1 in water or in DPC–SDS micelles corresponds to a change of the tumbling rate of different molecular weights in the limits of the Stoke's law of spherical molecule approximation, indicating clearly a tight binding of Thymosinα1 to DPC–SDS micelles. The measure of the diffusion of Thymosinα1 has been performed in DPC-d_{38} and DPC-d_{38}/SDS-d_{25} using the NMR DOSY technique which essentially gave results similar to those reported for Thymosinα1 in the previous paper in SDS-d_{25} micelles (Nepravishta et al., 2015). The results indicated that the diffusion coefficient of

Thymosinα1 is influenced by the binding to both DPC-d_{38} and DPC-d_{38}/SDS-d_{25} micelles. In fact, the diffusion coefficients of Thymosinα1 in these micellar systems evaluated by the diffusion fronts in DOSY experiments (not shown) differ from those observed in water. Particularly, the diffusion slows down in presence of DPC-d_{38} and, even markedly more, in the presence of DPC-d_{38}/SDS-d_{25} micelles. These results indicate the presence of a diffuse interaction with the former micellar system in dynamical exchange which is typical of a widespread nonspecific interaction and of another more important interaction with the last one as revealed by the reduction of the mean average diffusion value. It is important to note that the most dramatic change of diffusion coefficient was found for the interaction with SDS micelles where beyond the structuring mechanism and insertion of the N-terminus (1–5 N-terminus residues) in the hydrophobic region of the micelle was revealed (Nepravishta et al., 2015).

7. THE CIRCULAR DICHROISM STUDY OF THYMOSINα1 IN PERDEUTERATED DPC AND PERDEUTERATED DPC–SDS MICELLES

The circular dichroism spectroscopy of Thymosinα1 in perdeuterated micelles of DPC and DPC–SDS showed spectra with very different shapes. Particularly the spectrum of Thymosinα1 in DPC micelles is similar to that obtained in water with a spectral shape diagnostic of a poorly stable structuration if any. On the other hand the spectrum of Thymosinα1 in DPC–SDS mixed micelles can be attributed to a helical conformation. On the basis of the results of previous studies on circular dichroism spectroscopy of peptides and the results related to Thymosinα1 in SDS micelles (Nepravishta et al., 2015) a possible interpretation of results obtained can be drawn attributing the conformation to a 3_{10} helix. In fact, in our case, the intensities of the bands at 208 nm and at 222 nm are markedly different with the 208 nm band much more negative than that at 222 nm. The ratio between these two absorption values, that at 222 nm respect to that at 208 nm, was found to be 0.30. This value is in complete agreement with studies previously published on 3_{10} helix conformation in peptides (Toniolo et al., 2004; Toniolo, Polese, Formaggio, Crisma, & Kamphius, 1996).

The different behavior of Thymosinα1 in the circular dichroism spectra in the presence of DPC and of mixed DPC–SDS micelle appears as the effect of the difference in the electrostatic nature of the two lipidic surfaces. In fact,

DPC has an amphipathic nature due to the phosphocholine group (which has a positive and the negative charge linked together in the phosphoric and choline groups) different from the negative character of the only sulfate group in SDS. This indicates that this mixed electrostatic surface charge(s) in DPC micelles binds the Thymosinα1 with a rather nonspecific modality. By contrast the SDS negative groups in mixed micelles attract the protein to the negative surface causing its insertion similarly to that found and reported previously (Nepravishta et al., 2015). These results are in line with the results obtained by the DOSY NMR diffusion measurements (Mandaliti et al., 2016). The change of hydrodynamic radius from the value of diffusion coefficient D for Thymosinα1 alone appears lower for the presence of DPC than in DPC–SDS mixed micelles. The lowest values were observed for Thymosinα1 in the presence of SDS micelles (Nepravishta et al., 2015). This indicates that a weak nonspecific interaction occurs with a fast dynamic exchange with DPC micelles. The change observed in mixed micelles indicated that increasing the negative composition of the model membrane increases the interaction of Thymosinα1 with structuration process and the inclusion of Thymosinα1 into the micelles.

These findings suggest the hypothesis that the Thymosinα1 folds on the membrane with an interaction scheme hypothesized previously as possible for hormone peptides (Sargent & Schwyzer, 1986). This similarity suggests the possible influence of the characteristics of the surface of the lipid bilayer of target cell membranes on rate and equilibria of peptide–receptor interactions and qualify the interaction peptide/lipid bilayer as the initial steps of the hormone-cell contact. The Thymosinα1 linked to the membrane surface assumes a helical conformation in two helical tracts in presence of mixed micelles, one from 1 to 6 inserted in hydrophobic region and the second one from residue 15 to 26 as reported above in Fig. 2. This conformation likely behaves as a determinant factor for the interaction with effectors nearby present on the membrane that can be considered as a starting event of a signaling cascade. In conclusion, the novelty of these results is indicative of a membrane insertion of Thymosinα1 as shown in the scheme of Fig. 2, which may represent a sort of preferentiality of Thymosinα1 toward negative regions of phospholipid membrane in the mechanism of lymphocyte activation. This appears rather important in comparison with the low specificity of the interaction of Thymosinα1 with DPC. The effects of the interaction of Thymosinα1 with the nonphysiological nature of SDS, lead to hypothesize that, in the cell membrane environment, the negative phospholipids like phosphatidylserine may be the target of the interaction of

Thymosinα1. In fact, from the point of view of the mechanism of action, insertion of its N-terminus in negative regions similar to those membranes with exposure of phosphatidylserine is likely to occur. The recent growing interest in studying the physio-pathological significance of the phosphatidylserine exposure in cells may enforce the interest for these findings.

8. THE STRUCTURAL STUDY BY NMR OF THE INTERACTION OF THYMOSINα1 WITH PHOSPHATIDYLSERINE IN MEMBRANES

Thymosinα1, unstructured in water solution interacts with negative regions of vesicles assuming two tracts of helical conformation with a structural break in between. The study of the interaction of Thymosinα1 with vesicles with mixed dipalmitoylphosphatidylcholine and dipalmitoylphosphatidylserine, the negative component of the membranes by ^{1}H and natural abundance ^{15}N NMR is herewith reported. The results indicate that the preferred interactions are those where the membrane is negatively charged.

The summary of the results reported in Fig. 3 indicates the presence of the helical tracts (1–10 and 15–24) detected by the sole NH–NH NOEs, in agreement with results previously published (Nepravishta et al., 2015) of Thymosinα1 in presence of SDS and in mixed DPC/SDS vesicles (Mandaliti et al., 2016).

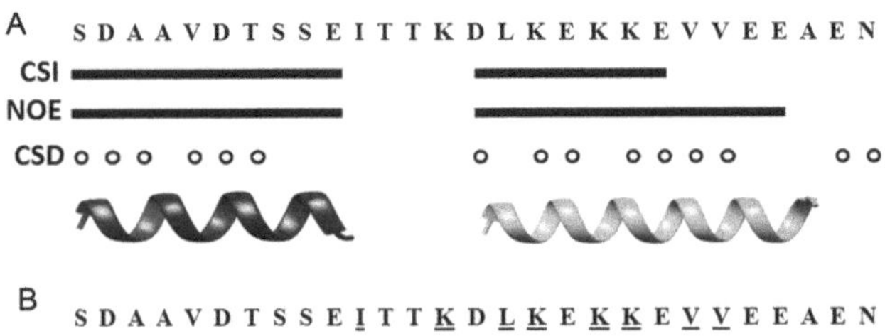

Fig. 3 Structural experimental details of the study of the interaction of Thymosinα1 with vesicles with mixed dipalmitoylphosphatidylcholine and dipalmitoylphosphatidylserine. (A) The sequence of residues in Thymosinα1 and the tracts assigned in the NMR spectrum together with the secondary structure indication. CSI are those attributed by chemical shift index (CSI), NOE are those attributed by NOEs crosspeaks diagnostic for the helix. (B) The results of the perturbation by ^{15}N NMR spectra upon binding to PC–PS vesicles; *open dots* refer to the results of Fig. 4. (C) Reports the results of the experiments of binding with human serum albumin (HSA) by STD experiments and NOE transfer, not reported in this chapter (paper submitted). The *underlined* residues are those that likely represent the region of binding to the plasmatic macromolecules.

9. ^{15}N NMR SPECTROSCOPY STUDY OF THE INTERACTION

To obtain a clear description of the mechanism of interaction of Thymosinα1 with negative regions of the vesicle, a natural abundance ^{15}N NMR study of the interaction was performed. The possibility of obtaining the ^{15}N HSQC spectra without interference of the numerous phospholipid protons allowed us to study in detail the perturbation of the residues involved in PS interaction. In fact ^{15}N resonances also at natural abundance are sensitive of the change of physico-chemical environment and this approach gave several important results in the study of protein structure and interaction. The ^{15}N HSQC spectrum of Tα1 at the concentration of 4 mM in the presence of PC/PS at a concentration of 20 mM (PC + PS); is reported in Fig. 4.

The assignments were performed considering the ^{1}H NMR resonances; all the resonances of Thymosinα1 have been identified.

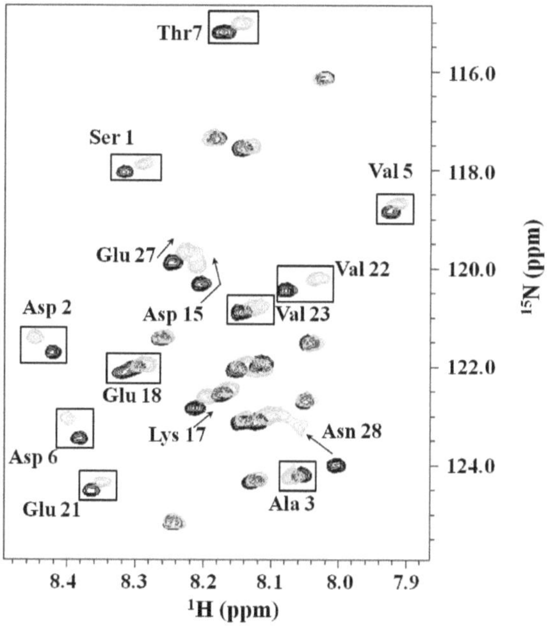

Fig. 4 Assignment of the resonances of natural abundance ^{15}N HSQC NMR of Thymosinα1 in the presence of 20 mM of PC–PS (4:1) (*black*) and in the presence of increasing PC/PS 4:1 *gray*. *Black* PC–PS 5 mM; *gray* PC–PS 20 mM. (See the color plate.)

The assignments allowed us to use the chemical shift index (Wishart, Sykes, & Richards, 2002) corrected for short peptide (Schwarzinger, Kroon, Foss, Wright, & Dyson, 2000) which indicated the existence of two tracts of helix in the conformation of Thymosinα1 in the interaction with vesicles. These results are reported in comparison with the NH–NH NOEs detected and reported above along the Thymosinα1 sequence in Fig. 3. Upon increasing PC–PS micellar concentration from 5 to 20 mM, it was possible to follow in the natural abundance ^{15}N HSQC spectrum the shifts of the resonances which are more perturbed by the interaction with an increase of the concentration of the PC–PS vesicles as reported in Fig. 4 (gray).

Particularly, in the natural abundance ^{15}N HSQC spectra of Thymosinα1 the black crosspeaks are due to Thymosinα1 in presence of PC/PS 4:1 at 5 mM concentration; the gray crosspeaks are those with Thymosinα1 in presence of PC/PS 4:1 at 20 mM concentration. One can see that some of these resonances (circled and arrows) are markedly shifted upon increasing PC + PS concentration, this indicating a major interaction of the peptide with the vesicle. In Fig. 4, the arrows are help to the eye to observe the shift induced by the increase of vesicles concentration. The dots along the sequence reported in Fig. 3 indicate the regions that are markedly interested by the interaction with PC/PS vesicles. It is important to remember that the previously reported results (Nepravishta et al., 2015) indicated that it is the N-terminal part which inserts in the negative region of the membrane. This result indicated that in the membranes the regions of exposure of PS can be the target of the interaction with Thymosinα1.

The results indicated that Thymosinα1 polypeptidic chain is induced to fold into helical conformations during the interaction with membranes exposing negative charges on the surface. Thymosinα1 inserts its N-terminus into the hydrophobic region of the membrane. On the other hand, the NMR experiments performed on K562 cells activated with resveratrol and, then, induced in apoptosis (with exposure of PS) did not give a clear-cut answer on this point. In fact, the difference in concentration between the Thymosinα1 sample (10 mM) necessary to detect the ^{15}N natural abundance NMR spectra in these conditions and the surfaces of a few million cells available to the interaction did not give enough effect to be detected by NMR. This is also due to the not so large difference of exposure of PS in the K562 cells upon activation by resveratrol induction of apoptosis (the unbalance of PS in the outer is about 20%).

Moreover in other experiments performed in our laboratory that are not presented here, it was found that human serum albumin (HSA) binds to Thymosinα1 in the C-terminal region acting as a transporter of the peptide to the target districts for the insertion in negative membrane.

It is important to remember that the lack of receptors for the thymic hormone Thymosinα1 was one of the most important points where the investigation about the therapeutic role of Thymosinα1 failed until now. It was also found that Thymosinα1 binds HSA (at least in the hypothesis of only one binding site which is a very unlikely limitation). The value of the binding constant found (paper submitted) gives the idea that the peptide can bind to HSA strongly enough to be effectively transported until it reaches the target membrane where is to be inserted. The target membrane is one presenting the regions negatively charged by the PS exposure. This condition is present in a large variety of biological processes.

10. IMPLICATIONS OF THYMOSINα1 BINDING TO PS EXPOSURE

Phosphatidylserine (PS) is the negative phospholipid that is generally mainly localized in the inner leaflet of membranes and its exposure is due to the action of enzymes like scramblases (An et al., 2001). Recent studies shed light on the potential function of PS interaction with cytoskeletal proteins to mediate anchorage of actin filaments to the phospholipid bilayer. Moreover, the erythrocyte protein 4.1R that binds to phosphatidylserine interacts with the negative head-group and, subsequently forms a tight hydrophobic interaction with the fatty acid moieties of the membrane.

Also annexin binds to PS (Lee, Pollard, & Arispe, 2002; van Engeland, Nieland, Ramaekers, Schutte, & Reutelingsperger, 1998) and this discovery was used as an affinity assay to test PS exposure in cells targeting the loss of phospholipid asymmetry of the plasma membrane, an early event in apoptosis, independent of the cell type, resulting in the exposure of PS at the outer plasma membrane.

Moreover, the modification of distribution of phospholipids in membrane (Denisov, Wanaski, Glaser, & McLaughlin, 1998) upon addition of positive peptide was reported. The binding of proteins to exposed PS (Denisov et al., 1998; Lee et al., 2002; van Engeland et al., 1998) has been studied theoretically and experimentally as found also in coagulation process.

In this case, the translocation of PS from the inner to outer membrane leaflet of activated blood platelets and platelet-derived microvesicles provides a catalytic surface for interacting coagulation factors (Lentz, 2003; Ortel, Devore-Carter, Quinn-Allen, & Kanes, 1992). This process is impaired in the Scott syndrome, a rare congenital bleeding disorder, underscoring the indispensable role of PS in hemostasis. This also reveals a defect of a protein-catalyzed scrambling of membrane phospholipids in such pathological states as chronic ethanol administration (Sun & Sun, 1983) and after oxidative stress (Freikman, Amer, Cohen, Ringel, & Fibach, 2008). These evidences has been reviewed (Lemmon, 2008) where the conclusions are that the engulfment due to the apoptotic process exhibits a unique pharmacological profile, where the sensitivity to blockers correlates with the occurrence of phosphatidylserine exposure on the phagocytic prey. In particular, phosphatidylserine exposure occurs on both the prey and the phagocyte. While the phospholipid scrambling exposes phosphatidylserine on the dying cells, lipid randomization is required on the phagocyte (Verhoven, Schlegel, & Williamson, 1995).

The specificity for PS exposure is a biologically important response and its appearance on the cell surface during apoptosis in thymocytes and cytotoxic T lymphocyte cell lines provokes PS-dependent recognition by activated macrophages (Mariño & Kroemer, 2013).

Activating a translocase and a nonspecific lipid scramblase are responsible for PS reaching the surface from its intracellular location before DNA degradation, meiosis, and cell lysis in the apoptotic pathway (Mariño & Kroemer, 2013). The finding that Thymosinα1 interacts with regions of membrane surface where an exposure of PS assuming partial helical conformation(s) prior the interaction with receptors can be an indication of the possible action pathway for its pharmacological action.

11. IMPLICATION OF THYMOSINα1 BINDING TO MEMBRANE AND CELLS

The binding to membrane of Thymosinα1 together with the assumption of structural elements absent in water solution led us to hypothesize that this can be included among the many pathways proposed for the action of these peptidic hormones. After the proposal of Sargent and Schwyzer (1986), several cases have been reported in literature where an insertion in the membrane, also partial, can be considered an initial step of a biological cascade. Many examples have been proposed of binding to model membranes to peptides; ie, antimicrobic, peptidic hormones, transmembrane tracts (mainly helicoidal), and neurotransmitters (Castanho & Fernandes, 2006; Langelaan & Rainey, 2010; Maler, 2013; Vauquelin & Packeu, 2009; Ying, Ahn, Jacobsen, Brown, & Hruby, 2003). Examples have been reported as oxytocin (Pliska, Folkers, & Spiwok, 2014), neurohypophyseal hormone-like peptides (Iłowska et al., 2012). In these cases it was proven that the direct interaction with membrane is the way of peptide to find receptors as, also, in case of ghrelin (Staes et al., 2010), of the neuropeptide PPY-3 (Pedersen et al., 2012) for an analog of kyotorphin (Machuqueiro, Campos, Soares, & Baptista, 2010) or about the role of the membrane itself in the interaction to reach hERG potassium channel (Chartrand, Arnold, Gravel, Jenna, & Marcotte, 2010) and for apelin (Langelaan & Rainey, 2009). In some cases the interaction was stepwise and in other cases the interaction with membrane is considered as a step of the action at the molecular level on the HIV inhibitor enfuvirtide and of inhibitor peptide T-1249 revealing a correlation with improved efficiency (Veiga, Henriques, Santos, & Castanho, 2004; Veiga, Santos, Loura, Fedorov, & Castanho, 2004). In other cases, an improvement of the efficacy can be attributed to membrane interaction (Vauquelin, 2010) by a reduction of their diffusion able to contribute to their kinetics of action just increasing the permanence of drug upon time and, thus, the duration of effects (Axelrod & Wang, 1994; Wang, Gou, & Axelrod, 1992).

This particular mode of action of peptides has been recently reviewed (Lesovoy et al., 2009) in the role of receptor(s) activation(s) upon binding to biological membranes (daCosta, Wagg, Marlene, McKay, & Baenziger, 2004; Lee et al., 2014; Sykes et al., 2014).

In particular, the role (and the differences) of lipids involved in model membranes and in cell membranes has been studied (Castanho & Fernandes, 2006; Maler, 2012, 2013; Vauquelin & Packeu, 2009). Thus, the idea that Thymosinα1 can undergo a direct interaction with membrane in negative regions, namely exposed PS due to cell phospholipid unbalance, and, after assuming partial structural elements, can be considered a model for the scheme of activity mechanism of Thymosinα1 as activator of biological cascade(s) upon interaction with different receptors.

ACKNOWLEDGMENTS

The technical assistance of Fabio Bertocchi in the skillful maintaining and checking the performances of the NMR instrumentation is gratefully acknowledged.

REFERENCES

An, X., Takakuwa, Y., Manno, S., Han, B., Gascard, B., & Mohandas, N. (2001). Structural and functional characterization of protein 4.1R-phosphatidylserine interaction potential role in 4.1R sorting within cells. *The Journal of Biological Chemistry*, *276*, 35778–35785.

Andreone, P., Cursaro, C., Gramenzi, A., et al. (2001). In vitro effect of thymosin-alpha1 and interferon-alpha on Th1 and Th2 cytokine synthesis in patients with chronic hepatitis C. *Journal of Viral Hepatitis*, *8*, 194–201.

Axelrod, D., & Wang, D. (1994). Reduction-of-dimensionality kinetics at reaction-limited cell surface receptors. *Biophysical Journal*, *66*, 588–600.

Billich, A. (2002). Thymosin alpha1 SciClone Pharmaceuticals. *Current Opinion in Investigational Drugs*, *3*, 698–707.

Carraro, G., Naso, A., Montomoli, E., et al. (2012). Thymosin-alpha 1 (Zadaxin) enhances the immunogenicity of an adjuvated pandemic H1N1v influenza vaccine (Focetria) in hemodialyzed patients: A pilot study. *Vaccine*, *30*, 1170–1180.

Castanho, M. A. R. B., & Fernandes, M. X. (2006). Lipid membrane-induced optimization for ligand–receptor docking: Recent tools and insights for the "membrane catalysis" model. *European Biophysical Journal*, *35*, 92–103.

Chartrand, E., Arnold, A. A., Gravel, A., Jenna, S., & Marcotte, I. (2010). Potential role of the membrane in hERG channel functioning and drug-induced long QT syndrome. *Biochimica et Biophysica Acta*, *1798*, 1651–1662.

Chen, J. M., Dando, P. M., Rawlings, N. D., et al. (1997). Cloning, isolation, and characterization of mammalian legumain, an asparaginyl endopeptidase. *The Journal of Biological Chemistry*, *272*, 8090–8098.

daCosta, C. J. B., Wagg, I. D., Marlene, E., McKay, M. E., & Baenziger, J. E. (2004). Phosphatidic acid and phosphatidylserine have distinct structural and functional interactions with the nicotinic acetylcholine receptor. *The Journal of Biological Chemistry*, *279*, 14967–14974.

Denisov, G., Wanaski, S., Glaser, M. L., & McLaughlin, S. (1998). Binding of basic peptides to membranes produces lateral domains enriched in the acidic lipids phosphatidylserine and phosphatidylinositol 4,5-bisphosphate: An electrostatic model and experimental results. *Biophysical Journal*, *74*, 731–744.

Elizondo-Riojas, M. A., Chamow, S. M., Tuthill, C. W., et al. (2011). NMR structure of human Thymosinalpha-1. *Biochemical and Biophysical Research Communications*, *416*, 356–361.

Eschenfeldt, W. H., Manrow, R. E., Krug, M. S., & Berger, S. L. (1989). Isolation and partial sequencing of the human proThymosinalpha gene family. Evidence against export of the gene products. *The Journal of Biological Chemistry*, *264*, 7546–7555.

Franco, F. J., Diaz, C., Barcia, M., & Freire, M. (1992). Thymosinalpha 1 is a native peptide in several tissues. *Biochimica et Biophysica Acta*, *1120*, 43–48.

Freikman, I., Amer, J., Cohen, J. S., Ringel, I., & Fibach, E. (2008). Oxidative stress causes membrane phospholipid rearrangement and shedding from RBC membranes—An NMR study. *Biochimica et Biophysica Acta*, *1778*, 2388–2394.

Garaci, E., Pica, F., Serafino, A., et al. (2012). Thymosinα1 and cancer: Action on immune effector and tumor target cells. *The Annals of the New York Academy of Sciences*, *1269*, 26–33.

Giuliani, C., Napolitano, G., Mastino, A., et al. (2000). Thymosin-alpha1 regulates MHC class I expression in FRTL-5 cells at transcriptional level. *European Journal of Immunology*, *30*, 778–786.

Goldstein, A. L. (2009). From lab to bedside: Emerging clinical applications of Thymosinalpha 1. *Expert Opinion on Biological Therapy*, *9*, 593–608.

Goldstein, A. L., Guha, A., Zatz, M. M., Hardy, M. A., & White, A. (1972). Purification and biological activity of thymosin, a hormone of the thymus gland. *Proceedings of the National Academy of Sciences of the United States of America*, *69*, 1800–1803.

Grottesi, A., Sette, M., Palamara, T., et al. (1998). The conformation of peptide Thymosinalpha 1 in solution and in a membrane like environment by circular dichroism and NMR spectroscopy. A possible model for its interaction with the lymphocyte membrane. *Peptides*, *19*, 1731–1738.

Iino, S., Toyota, J., Kumada, H., et al. (2005). The efficacy and safety of Thymosinalpha-1 in Japanese patients with chronic hepatitis B; results from a randomized clinical trial. *Journal of Viral Hepatitis*, *12*, 300–306.

Iłowska, E., Wyrzykowski, D., & Kwiatkowska, A. (2012). Membrane structure and interactions of peptide hormones with model lipid bilayers. *Biochimica et Biophysica Acta*, *1818*, 2982–2991.

Knutsen, A. P., Freeman, J. J., Mueller, K. R., et al. (1999). Thymosin-alpha1 stimulates maturation of CD34 + stem cells into CD3+4 + cells in an in vitro thymic epithelia organ coculture model. *International Journal of Immunopharmacology*, *21*, 15–26.

Kullavanuava, P., Treeprasertsuk, S., Thong-Ngam, D., et al. (2001). The combined treatment of interferon alpha-2a and Thymosinalpha 1 for chronic hepatitis C: The 48 weeks end of treatment results. *Journal of the Medical Association of Thailand*, *84*, 462–468.

Langelaan, D. N., & Rainey, J. K. (2009). Headgroup-dependent membrane catalysis of apelin-receptor interactions is likely. *The Journal of Physical Chemistry B*, *113*, 10465–10476.

Langelaan, D. N., & Rainey, K. N. (2010). Membrane catalysis of peptide–receptor binding. *Biochemistry and Cell Biology*, *88*, 203–210.

Lao, X., Liu, M., Chen, J., & Zheng, H. (2013). A tumor-penetrating peptide modification enhances the antitumor activity of Thymosinalpha 1. *PLoS One*, *8*, 722–742.

Lee, S. Y., Muthu, D., Hall, S. M., Ramos-Colon, C., Rankin, D., Hu, J., et al. (2014). Discovery of amphipathic Dynorphin A analogues to inhibit the neuroexcitatory effects of Dynorphin A through Bradykinin receptors in the spinal cord. *Journal of the American Chemical Society*, *136*, 6608–6616.

Lee, G., Pollard, H. B., & Arispe, N. (2002). Annexin 5 and apolipoprotein E2 protect against Alzheimer's amyloid-beta-peptide cytotoxicity by competitive inhibition at a common phosphatidylserine interaction site. *Peptides*, *23*, 1249–1263.

Lemmon, M. A. (2008). Engulfment of apoptotic cells involves membrane recognition by phospholipid-binding domains. *Nature Cell Biology*, *1*, 456–461.

Lentz, B. R. (2003). Exposure of platelet membrane phosphatidylserine regulates blood coagulation. *Progress in Lipid Research*, *42*, 423–438.

Lesovoy, D. M., Bocharov, E. V., Lyukmanova, E. N., Kosinsky, Y. A., Shulepko, M. A., Dolgikh, D. A., et al. (2009). Specific membrane binding of neurotoxin II can facilitate its delivery to acetylcholine receptor. *Biophysical Journal*, *97*, 2089–2097.

Low, T. L., & Goldstein, A. L. (1979). The chemistry and biology of thymosin. II. Amino acid sequence analysis of Thymosinalpha1 and polypeptide beta1. *The Journal of Biological Chemistry*, *254*, 987–995.

Low, T. L., Thurman, G. B., McAdoo, M., et al. (1979). The chemistry and biology of thymosin. I. Isolation, characterization, and biological activities of Thymosinalpha1 and polypeptide beta1 from calf thymus. *The Journal of Biological Chemistry*, *254*, 981–986.

Machuqueiro, M., Campos, S. R. R., Soares, C. M., & Baptista, A. M. (2010). Membrane-induced conformational changes of kyotorphin revealed by molecular dynamics simulations. *The Journal of Physical Chemistry B*, *114*, 11659–11667.

Maio, M., Mackiewicz, A., Testori, A., et al. (2010). Large randomized study of Thymosinalpha 1, interferon alfa, or both in combination with dacarbazine in patients with metastatic melanoma. *Journal of Clinical Oncology*, *28*, 1780–1787.

Maler, L. (2012). Solution NMR studies of peptide-lipid interactions in model membranes. *Molecular Membrane Biology*, *5*, 155–176.

Maler, L. (2013). Solution NMR studies of cell-penetrating peptides in model membrane systems. *Advanced Drug Delivery Reviews*, *65*, 1002–1011.

Mandaliti, W., Nepravishta, R., Sinibaldi Vallebona, P., Pica, F., Garaci, E., & Paci, M. (2016). New studies about the insertion mechanism of Thymosinα1 in negative regions of model membranes as starting point of the bioactivity. *Amino acids*, *48*, 1231–1239.

Manrow, R. E., Sburlati, A. R., Hanover, J. A., & Berger, S. L. (1991). Nuclear targeting of proThymosinalpha. *The Journal of Biological Chemistry*, *266*, 3916–3924.

Mariño, G., & Kroemer, G. (2013). Mechanisms of apoptotic phosphatidylserine exposure. *Cell Research*, *23*, 1247–1248.

Nepravishta, R., Mandaliti, W., Eliseo, T., Sinibaldi-Vallebona, P., Pica, F., Garaci, E., et al. (2015). Thymosinα1 inserts N-terminus into model membranes assuming a helical conformation. *Expert Opinion on Biological Therapy*, *2*, 11–15.

Ortel, T. L., Devore-Carter, D., Quinn-Allen, M. A., & Kanes, W. H. (1992). Evidence for a phosphatidylserine binding site in factor v; the second c-type domain. *The Journal of Biological Chemistry*, *267*, 4189–4198.

Pedersen, S. L., Bhatia, V. K., Jurt, S., Paulsson, J. F., Pedersen, M. H., Jorgensen, R., et al. (2012). Improving membrane binding as a design strategy for amphipathic peptide hormones: 2-helix variants of PYY3-36. *Journal of Peptide Science*, *18*, 579–587.

Pliska, V., Folkers, G., & Spiwok, V. (2014). Thermodynamics of the interaction between oxytocin and its myometrial receptor in sheep: A stepwise binding mechanism. *Biochemical Pharmacology*, *91*, 119–127.

Romani, L., Bistoni, F., Gaziano, R., et al. (2004). Thymosinalpha 1 activates dendritic cells for antifungal Th1 resistance through toll-like receptor signaling. *Blood*, *103*, 4232–4239.

Romani, L., Moretti, S., Fallarino, F., et al. (2012). Jack of all trades: Thymosinα1 and its pleiotropy. *Annals of the New York Academy of Sciences*, *1269*, 1–6.

Sarandeses, C. S., Covelo, G., Díaz-Jullien, C., & Freire, M. (2003). ProThymosinalpha is processed to Thymosinalpha 1 and Thymosinalpha 11 by a lysosomal asparaginyl endopeptidase. *The Journal of Biological Chemistry*, *278*, 13286–13293.

Sargent, D. F., & Schwyzer, R. (1986). Membrane lipid phase as catalyst for peptide receptor interactions. *Proceedings of the National Academy of Sciences of the United States of America*, *83*, 5774–5778.

Schwarzinger, S., Kroon, G., Foss, T. R., Wright, P. E., & Dyson, H. J. (2000). Random coil chemical shifts in acidic 8 M urea: Implementation of random coil shift data in NMRView. *Journal of Biomolecular NMR*, *18*, 43–48.

Serafino, A., Pica, F., Andreola, F., et al. (2014). Thymosinα1 activates complement receptor-mediated phagocytosis in human monocyte-derived macrophages. *Journal of Innate Immunity*, *6*, 72–88.

Staes, E., Absil, P. A., Lins, L., Brasseur, R., Deleu, M., Lecouturier, N., et al. (2010). Acylated and unacylated ghrelin binding to membranes and to ghrelin receptor: Towards a better understanding of the underlying mechanisms. *Biochimica et Biophysica Acta, Biomembranes*, *1798*, 2102–2113.

Sun, G. Y., & Sun, A. Y. (1983). Chronic ethanol administration induced an increase in phosphatidylserine in guinea pig synaptic plasma membranes. *Biochemical and Biophysical Research Communications*, *113*, 262–268.

Sykes, D. A., Parry, C., Reilly, J., Wright, P., Fairhurst, R. A., & Charlton, S. J. (2014). Observed drug-receptor association rates are governed by membrane affinity: The importance of establishing "Micro-Pharmacokinetic/Pharmacodynamic Relationships" at the b2-Adrenoceptor. *Molecular Pharmacology*, *85*, 608–617.

Toniolo, C., Formaggio, F., Tognon, S., Broxterman, Q. B., Kaptein, B., Huang, R., et al. (2004). The complete chirospectroscopic signature of the peptide 3_{10}-helix in aqueous solution. *Biopolymers*, *75*, 32–45.

Toniolo, C., Polese, A., Formaggio, F., Crisma, M., & Kamphius, J. (1996). Circular dichroism spectrum of a peptide 3_{10}-helix. *Journal of the American Chemical Society*, *118*, 2744–2745.

van Engeland, M., Nieland, L. J. W., Ramaekers, F. C. S., Schutte, B., & Reutelingsperger, C. P. M. (1998). Annexin V-affinity assay: A review on an apoptosis detection system based on phosphatidylserine exposure. *Cytometry*, *31*, 1–9.

Vauquelin, G. (2010). Rebinding: Or why drugs may act longer in vivo than expected from their in vitro target residence time. *Expert Opinion on Drug Discovery*, *5*, 927–941.

Vauquelin, G., & Packeu, A. (2009). Ligands, their receptors and plasma membranes: A review. *Molecular and Cellular Endocrinology*, *311*, 1–10.

Veiga, S., Henriques, S., Santos, N., & Castanho, M. (2004). Putative role of membranes in the HIV fusion inhibitor enfuvirtide mode of action at the molecular level. *The Biochemical Journal*, *377*, 107–110.

Veiga, S., Santos, N., Loura, L., Fedorov, A., & Castanho, M. (2004). HIV Fusion Inhibitor Peptide T-1249 is able to insert or adsorb to lipidic bilayers. Putative correlation with improved efficiency. *Journal of the American Chemical Society*, *126*, 14758–14763.

Verhoven, B., Schlegel, R. A., & Williamson, P. (1995). Mechanisms of phosphatidylserine exposure: A phagocyte recognition signal, on apoptotic T lymphocytes. *The Journal of Experimental Medicine*, *182*, 1597–1601.

Wang, D., Gou, S. Y., & Axelrod, D. (1992). Reaction rate enhancement by surface diffusion of adsorbates. *Biophysical Chemistry*, *43*, 117–137.

Wishart, D. S., Sykes, B. D., & Richards, F. M. (2002). The chemical shift index: A fast and simple method for the assignment of protein secondary structure through NMR spectroscopy. *Biochemistry*, *31*, 1647–1651.

Ying, J., Ahn, J. M., Jacobsen, N. E., Brown, M. F., & Hruby, V. J. (2003). NMR solution structure of the glucagon antagonist(desHis1, desPhe6, Glu9)glucagon amide in the presence of perdeuterateddodecylphosphocholine micelles. *Biochemistry*, *4*, 2825–2835.

You, J., Zhuang, L., Cheng, H. Y., et al. (2006). Efficacy of Thymosinalpha-1 and interferon alpha in treatment of chronic viral hepatitis B: A randomized controlled study. *World Journal of Gastroenterology*, *12*, 6715–6721.

Zetter, B. R. (1997). On target with tumor blood vessel markers. *Nature Biotechnology*, *15*(12), 1243–1244.

Zhang, P., Chan, J., Dragoi, A. M., et al. (2005). Activation of IKK by thymosinα1 requires the TRAF6 signaling pathway. *EMBO Reports*, *6*, 531–537.

CHAPTER SIX

Thymosin Beta 4 Is a Potential Regulator of Hepatic Stellate Cells

J. Kim, Y. Jung[1]
Pusan National University, Pusan, Republic of Korea
[1]Corresponding author: e-mail address: y.jung@pusan.ac.kr

Contents

Abstract

Liver fibrosis, a major characteristic of chronic liver disease, is inappropriate tissue remodeling caused by prolonged parenchymal cell injury and inflammation. During liver injury, hepatic stellate cells (HSCs) undergo transdifferentiation from quiescent HSCs into activated HSCs, which promote the deposition of extracellular matrix proteins, leading to liver fibrosis. Thymosin beta 4 (Tβ4), a major actin-sequestering protein, is the most abundant member of the highly conserved β-thymosin family and controls cell morphogenesis and motility by regulating the dynamics of the actin cytoskeleton. Tβ4 is known to be involved in various cellular responses, including antiinflammation, wound healing, angiogenesis, and cancer progression. Emerging evidence suggests that Tβ4 is expressed in the liver; however, its biological roles are poorly understood. Herein, we introduce liver fibrogenesis and recent findings regarding the function of Tβ4 in various tissues and discuss the potential role of Tβ4 in liver fibrosis with a special focus on the effects of exogenous and endogenous Tβ4. Recent studies have revealed that activated HSCs express Tβ4 *in vivo* and *in vitro*. Treatment with the exogenous Tβ4 peptide inhibits the proliferation and migration of activated HSCs and reduces liver

Vitamins and Hormones, Volume 102
ISSN 0083-6729
http://dx.doi.org/10.1016/bs.vh.2016.04.011

fibrosis, indicating it has an antifibrotic action. Meanwhile, the endogenously expressed Tβ4 in activated HSCs is shown to promote HSCs activation. Although the role of Tβ4 has not been elucidated, it is apparent that Tβ4 is associated with HSC activation. Therefore, understanding the potential roles and regulatory mechanisms of Tβ4 in liver fibrosis may provide a novel treatment for patients.

1. INTRODUCTION

The liver has the remarkable ability to self-repair and restore its structure and function after liver injury; as little as 25% of a liver can regenerate the whole organ (Higgins, 1931; Michalopoulos & DeFrances, 1997; Taub, 2004). This remarkable regenerative ability of liver is evidenced by a Greek myth. Prometheus was chained to a rock where his liver was devoured by an eagle every day, only to grow back each night. This myth suggests that the ancient Greeks already knew about the liver's remarkable capacity for self-regeneration. Even so, many types of hepatic injury result in some degree of scarring, despite the exceptional regenerative potential. Progressive replacement of functional hepatic parenchyma with fibrotic scar harms the tissue and results in liver dysfunction (Bataller & Brenner, 2005; Schuppan & Afdhal, 2008). In wound healing processes, hepatic stellate cells (HSCs) are activated and produce extracellular matrix (ECM) proteins. However, excessive collagen deposition by overproliferation of the activated HSCs replaces the parenchyma with fibrosis which distorts normal tissue architecture with hepatocellular dysfunction, eventually contributing to liver fibrosis (Friedman, 2008a). Because HSCs play an essential role in the pathogenesis of liver fibrosis, research has focused on elucidating the molecular mechanisms; how to initiate, maintain, and inhibit the activation of HSCs.

Thymosin beta 4 (Tβ4), the most abundantly expressed member of the beta-thymosins, mainly acts as a G-actin-sequestering peptide in eukaryotes (Safer, Elzinga, & Nachmias, 1991; Sanders, Goldstein, & Wang, 1992). Recently, Tβ4 was reported to be involved in many biological processes including angiogenesis, tissue remodeling, and wound repair (Goldstein, Hannappel, & Kleinman, 2005). Growing evidence shows that Tβ4 is involved in the pathogenesis of fibrotic disease in various organs, such as the heart, kidney, lung, skin, brain, and liver (De Santis et al., 2011; Gupta et al., 2012; Kim et al., 2015; Vartiainen, Pyykönen, Hökfelt, & Koistinaho, 1996; Zuo et al., 2013). Interestingly, Tβ4 was also reported to influence the proliferation, migration, and activation of HSCs in liver fibrosis (Barnaeva, Nadezhda, Hannappel, Sjogren, & Rojkind, 2007; Kim et al., 2015; Reyes-Gordillo et al., 2011; Xiao et al., 2014). However, it remains unclear how Tβ4 is related to hepatic fibrogenesis and what types of cells express Tβ4 in the liver. In addition, conflicting results in the regulation of Tβ4 for HSC activation, inhibition, or promotion have been reported. Hence, in this chapter, we review liver fibrosis, HSCs, and the functions and related pathways of Tβ4 in several types of tissue and/or cells, and present in detail the endogenous role and exogenous actions of Tβ4 in liver, to suggest the possible involvement of Tβ4 in the pathogenesis of liver fibrosis.

2. LIVER CELLS AND DISEASES

The liver, one of the essential organs, is the largest internal organ and gland in the human body. The liver regulates a wide range of reactions necessary for normal vital functions, including regulation of metabolism, hormone production, and detoxification (Boyer, Wright, & Manns, 2011). Two major types of hepatic cell populations reside in the liver: parenchymal cells and nonparenchymal cells (NPCs) (Kmiec, 2001). Hepatocytes are a major type of liver cells and are classified as parenchymal cells. These parenchymal hepatocytes occupy nearly 80% of the liver volume and carry out the major functions of liver, such as glycolysis, bile production, detoxification, and regeneration. Although hepatocytes stay in a quiescent state, hepatocytes do enter the cell cycle and proliferate to repopulate damaged regions when healthy liver is injured (Grisham, 1962; Higgins, 1931; Michalopoulos & DeFrances, 1997; Stocker & Heine, 1971). However, when liver damage is severe, and the capability of hepatocytes to divide and replace damaged tissues is compromised, NPCs participate in liver reconstitution. In healthy liver, NPCs constitute 40% of the total number of hepatic cells but only 6.5% of its volume (Kmiec, 2001; Wisse et al., 1996). Liver sinusoids are lined with two types of cells, sinusoidal endothelial cells and phagocytic Kupffer cells (Bouwens, De Bleser, Vanderkerken, Geerts, & Wisse, 1992). Liver sinusoidal endothelial cells line the sinuses (the space between hepatocytes) and modulate the transport of molecules between blood and liver. The Kupffer cells, hepatic macrophages, protect the liver from bacterial infections and other harmful antigens (Bilzer, Roggel, & Gerbes, 2006; Sung, Costerton, & Shaffer, 1992). The Kupffer cells also secrete various proinflammatory mediators and cytokines (Tsukamoto, 2002). HSCs, one of the NPCs, are located in the space of

Disse and are in close contact with hepatocytes and sinusoidal endothelial cells (Wisse et al., 1996). In a normal liver, HSCs are quiescent and function as the major storage facility for retinoids (vitamin A). However, when a liver is injured, HSCs undergo transdifferentiation from "quiescent" HSCs into "activated/myofibroblastic" HSC, which is the major fibrogenic cell type producing collagen in the injured liver (Bataller & Brenner, 2005; Reeves & Friedman, 2002). Because the liver is composed of these specialized hepatic cell populations, it is important to understand the cellular pathophysiological process of these cells in liver diseases.

Liver disease is a major cause of morbidity and mortality in the world (Bica et al., 2001; Fattovich et al., 1997; Wong, McQuillan, McHutchison, & Poynard, 2000). Liver disease can be inherited or caused by a variety of factors which trigger persistent and/or severe damages, such as viral infections, autoimmune disorders, ischemia, and several xenobiotics, including drugs, alcohols, or toxins (Pinzani & Rombouts, 2004). These repetitive and/or severe injuries lead to the excessive death of hepatocytes surpassing the proliferation of hepatocytes, resulting in abnormal tissue remodeling and hepatic dysfunction (Friedman, 2008b). This is especially seen in chronic liver diseases, where disease progression in the liver gradually destructs and regenerates the parenchyma to fibrosis (Pinzani & Rombouts, 2004). It includes a wide range of liver pathologies, such as hepatitis, cirrhosis, and hepatocellular carcinoma. The most common characteristic of chronic liver disease is fibrosis (Bataller & Brenner, 2005). Hepatic fibrosis is the immoderate wound healing, where excessive connective tissue builds up in the liver (Reeves & Friedman, 2002). Apoptosis of hepatocytes accompanying inflammatory infiltrates is a pathological hallmark in the early stages of fibrosis (Faouzi et al., 2001; Jaeschke, 2002), and excessive deposition of the ECM in parenchyma is prevalent in the later stages of fibrosis (Bataller & Brenner, 2005; Pinzani & Rombouts, 2004). Fibrosis itself causes no symptoms but can lead to portal hypertension or cirrhosis due to disruption of the physiological tissue architecture, which leads to hepatic dysfunction and death. Thus, loss of liver function caused by a decrease in parenchyma and replacement by fibrosis is the major cause for the high morbidity and mortality of these diseases.

3. HEPATIC STELLATE CELLS

The main fibrogenic cell is the activated or myofibroblastic (MF) HSC. In the quiescent stage, inactivated HSCs store perinuclear retinoid droplets and synthesize glial fibrillary acidic protein (GFAP) (Gard, White, & Dutton, 1985). As HSCs are activated, the retinoid droplets and GFAP are gradually lost, followed by development into myofibroblast-like cells with increased synthesis of ECM proteins and expression of alpha-smooth muscle actin (α-SMA) and profibrogenic factors, such as vimentin and transforming growth factor-beta (TGF-β) (Friedman, 2008b). These activated or MF-HSCs increasingly proliferate and generate collagen fibrils leading to liver fibrosis. HSCs contribute to >80% of the myofibroblasts in a fibrotic liver (Friedman, Roll, Boyles, & Bissell, 1985). The ECM components, mainly consisting of collagen types I, III, and VI and fibronectins, are released by persistently activated HSCs (Knittel et al., 1999). As the disease progresses, collagen bundles form bridges that start the development of regenerative nodules and the architectural distortion with impairment of liver function, becoming typical cirrhosis (Friedman, 2008b). Therefore, development of the effective strategies to prevent these events could reduce fibrosis or suppress the progression into chronic liver disease. Many studies have focused on the mechanisms underlying transdifferentiation of quiescent HSCs into activated HSCs because this step has been considered for a possible therapeutic target to inhibit liver fibrosis (Bataller & Brenner, 2001; Li & Friedman, 1999). HSCs are activated by autocrine/paracrine signaling (Friedman, 2000; Gressner, 1995). Activated HSCs or neighboring cells, including hepatocytes, immune cells, and Kupffer cells, release several factors, such as cytokines, chemokines, and growth factors, in response to liver damage (Gressner, 1995). In addition, oxidative stress, as well as extensive changes in composition and organization of ECM components, contributes to the activation of HSCs (Nieto, Friedman, & Cederbaum, 2002a, 2002b; Nieto, Friedman, Greenwel, & Cederbaum, 1999; Nieto et al., 2000). Injured hepatocytes secret reactive oxygen species (ROS) and ROS cytokines, such as tumor necrosis factor-alpha and vascular endothelial growth factor (VEGF) (Bataller & Brenner, 2005; Nieto & Cederbaum, 2003; Svegliati Baroni et al., 1998). Both the endogenous and exogenous ROS are capable of enhancing activation of and collagen synthesis in HSCs (Siegmund & Brenner, 2005). Kupffer cells release TGF-β and ROS intermediates, which stimulate secretion of retinoids from inactivated HSCs and matrix synthesis in and proliferation of activated HSCs (Bilzer et al., 2006; Imamura, Ogawa, Sasaguri, Chayama, & Ueno, 2005; Matsuoka & Tsukamoto, 1990). They also recruit inflammatory cells by secreting ROS, TGF-β1, platelet-derived growth factor (PDGF), and other profibrogenic cytokines (Nieto, 2006; Tsukamoto, 1999). Similarly, activated

HSCs increase the secretion of inflammatory cytokines (Friedman, 1998, 2008a). TGF-β, a well-known fibrosis-stimulating factor, notably promotes HSC activation and ECM production in HSCs (Gressner, Weiskirchen, Breitkopf, & Dooley, 2002; Inagaki & Okazaki, 2007). Other signaling molecules, such as PDGF, phosphatidylinositol 3-kinase (PI3K)/protein kinase B (AKT), and hedgehog (Hh), are known to regulate hepatic fibrogenesis. These molecules are highly expressed in damaged liver and stimulate the transition of quiescent HSCs and/or promote the proliferation of newly generated HSCs. The PI3K/AKT pathway is activated by PDGF, a well-known mitogen, and required for the survival and proliferation of HSCs (Pinzani, 2002; Reif et al., 2003; Wang, Jiang, Liu, & Jiang, 2008). Hh signaling regulates the proliferation and activation of HSCs, leading to liver fibrosis (Choi et al., 2009; Sicklick et al., 2005). In addition, TGF-β and Hh promote the epithelial-to-mesenchymal transition (EMT), which is an important pathogenic pathway that contributes to the excessive accumulation of activated HSCs during liver fibrosis (Bi, Yang, & Shi, 2012; Choi et al., 2009).

Clearance of activated HSCs brings to the regression of fibrosis (Elsharkawy, Oakley, & Mann, 2005; Kong, Zhang, Zhang, Lu, & Zheng, 2013). Therefore, many studies have been conducted to identify which molecular mechanisms underlie the persistent activation of HSCs and the progression of liver fibrosis (Li & Friedman, 1999). However, there is no effective therapeutics against hepatic fibrogenesis, yet. Nevertheless, further study of the complex network involved in the activation of HSCs is necessary to terminate the fibrogenic reaction, eventually achieving the successful control of chronic liver disease.

4. THYMOSIN BETA 4

Beta-thymosins were first isolated from bovine thymus and thought to be thymic hormones (Goldstein, Slater, & White, 1966). Among the known family members, Tβ4, Tβ10, and Tβ15 are found in humans (Goldstein et al., 2005; Huff, Muller, Otto, Netzker, & Hannappel, 2001). Tβ4 is the most abundantly expressed member of the beta-thymosin family (Low, Hu, & Goldstein, 1981). Because Tβ4 is mainly involved in maintaining the large pool of unpolymerized G-actin (actin monomer) as intracellular actin-sequestering molecules, it regulates actin polymerization, which is necessary for cell motility and organogenesis (Sanders et al., 1992). Growing evidence suggests that Tβ4 is associated with several physiological processes including angiogenesis, apoptosis, development, differentiation, hair growth, inflammation, and wound healing (Cha, Jeong, & Kleinman, 2003; Grant et al., 1999; Huff et al., 2001; Philp, Goldstein, & Kleinman, 2004). Expression of Tβ4 is detected in the nervous system and in cardiac-differentiating cells during embryogenesis (Gomez-Marquez, Franco del Amo, Carpintero, & Anadon, 1996; Lugo et al., 1991; Yamamoto, Yamagishi, Yaginuma, Murakami, & Ueno, 1994). Transgenic mice with Tβ4 overexpression show abnormal tooth development (Cha et al., 2010). Tβ4 also influences T cell and endothelial cell differentiation (Grant et al., 1999; Hu, Low, & Goldstein, 1981; Low et al., 1981). In dermal wounds, Tβ4 treatment promotes wound healing through multiple mechanisms including increased angiogenesis, keratinocyte migration, collagen deposition, and wound contraction (Malinda et al., 1999; Philp et al., 2003). Tβ4 elevates TGF-β which is important for collagen synthesis (Sosne et al., 2004). The upregulation of TGF-β by Tβ4 seems to play a key role in accelerating collagen deposition in dermal wounds. In heart injuries after coronary ligation, Tβ4 endogenously increases and promotes repair processes resulting in improved cardiac function (Bock-Marquette, Saxena, White, DiMaio, & Srivastava, 2004). These findings indicate that Tβ4 is involved in the development and differentiation of several types of cells, as well as an injury response in multiple organs.

Tβ4 expression has been observed in many pathological conditions in the human body. In damaged tissue, Tβ4 is produced and released by various cell types, such as fibroblasts, granulocytes, platelets, macrophages, vascular endothelial cells, and corneal epithelial cells (Goldstein et al., 2005; Grant et al., 1995; Huff et al., 2001; Huff, Otto, Muller, Meier, & Hannappel, 2002; Xu, Hannappel, Morgan, Hempstead, & Horecker, 1982). The upregulation of Tβ4 is observed in scleroderma lung, hypoxic brain ischemia, and renal fibrosis (De Santis et al., 2011; Vartiainen et al., 1996; Zuo et al., 2013). In the fibrotic lesions, Tβ4 directly controls the production of ECM components by stimulating gene expression and synthesis of collagen and fibronectin (Kumar & Gupta, 2011; Malinda et al., 1999; Sosne et al., 2004; Zuo et al., 2013). Increased expression of Tβ4 is also shown in several types of tumors, such as metastatic melanoma, fibrosarcoma, malignant gliomas, breast cancers, ovarian cancers, and colorectal cancers (Clark, Golub, Lander, & Hynes, 2000; Ji et al., 2013; Kobayashi et al., 2002; Larsson & Holck, 2007; Wang et al., 2004; Wirsching et al., 2014). In tumors, Tβ4 promotes EMT by stimulating synthesis of many factors, including VEGF and fibroblast growth factor, which contribute to angiogenesis and tumor metastasis (Cha et al., 2003; Kim et al., 2011). Taken together, these results suggest that Tβ4 participates in both normal and pathological physiology, and regulates a variety of biological processes. Next, we discuss the expression and potential role of Tβ4 in the liver.

5. Tβ4 IN LIVER

Liver fibrosis is the main characteristic of most chronic liver diseases, and HSCs are known to be the major source of fibrous matrix production (Li & Friedman, 1999). Scientific progress has been made in understanding the pathophysiology of hepatic fibrosis and several forms of therapy evolved in the attempts to inhibit or reverse the disease. However, there is no effective therapeutics against liver fibrosis. Hence, studying the regulators that influence the activation of HSCs will provide important clues for developing therapeutics to inhibit liver fibrosis (Bataller & Brenner, 2001). Tβ4 accelerates wound healing and modulates inflammatory mediators in several damaged tissues, such as cornea, skin, kidney, and heart (Malinda et al., 1999; Sosne, Qiu, & Kurpakus-Wheater, 2007; Sosne et al., 2002). In corneal wound disorders, Tβ4 promotes rapid wound healing and reduces both leukocyte infiltration and production of inflammatory cytokine and/or chemokines (Bock-Marquette et al., 2004; Sosne et al., 2007, 2002). In the full-thickness wound model of rats, increased collagen deposition and angiogenesis were observed in Tβ4-treated wounds (Malinda et al., 1999). Therefore, liver researchers are now paying attention to the beneficial and healing ability of Tβ4 and have studied the potential role of Tβ4 in liver pathophysiology. In this section, we discuss in depth the accumulated reports about the hepatic expression and function of Tβ4 in the liver.

5.1 Hepatic Expression of Tβ4

Tβ4, a highly conserved peptide across species, is known to be distributed ubiquitously and is found at concentrations of 1×10^{-5} to 5.6×10^{-1} M in various types of tissues (Hannappel & Leibold, 1985; Hannappel & van Kampen, 1987; Hannappel, Xu, Morgan, Hempstead, & Horecker, 1982). In the early stage, researchers examined the endogenous level of Tβ4 in every organ using an enzymatic immunoassay. In 1997, Mora et al. estimated the basal concentrations of Tβ4 in major organs (brain, thymus heart, lungs, liver, spleen, kidneys, ovaries, fat, and muscle) of 20-week-old female Swiss Webster mice (Mora, Baumann, Paino, Goldstein, & Badamchian, 1997). Among these organs, the basal level of

Tβ4 in the liver was the lowest (9 μg/g), except muscle had no Tβ4. Even in mice treated intraperitoneally with 400 μg of synthetic Tβ4, the concentration of Tβ4 only slightly increased in liver (15 μg/g), whereas its concentration in muscle and kidney tissue showed a twofold increase. Goodall et al. also found similar results in Sprague–Dawley rats (Goodall, Hempstead, & Morgan, 1983). Compared with the thymus and spleen, the liver and muscle contain very low levels of Tβ4, indicating low levels of endogenous Tβ4 in healthy murine liver.

Although the concentration of Tβ4 in the liver has been estimated, the types of hepatic cells that specifically express Tβ4 have not been well established. In 2011, Nemolato et al. reported that hepatocytes expressed Tβ4 in healthy human liver tissue, as examined by immunohistochemical (IHC) staining (Fig. 1A) (Nemolato et al., 2011). They showed the localization of Tβ4 in the cytoplasm of hepatocytes, but not in Kupffer cells or HSCs. However, in 2013, they found the complete absence of Tβ4 in parenchyma areas including the hepatocytes of rat liver (Nemolato et al., 2013). These data differed from their previous results in human liver. They explained that species-specific expression of Tβ4 seemed to result in this

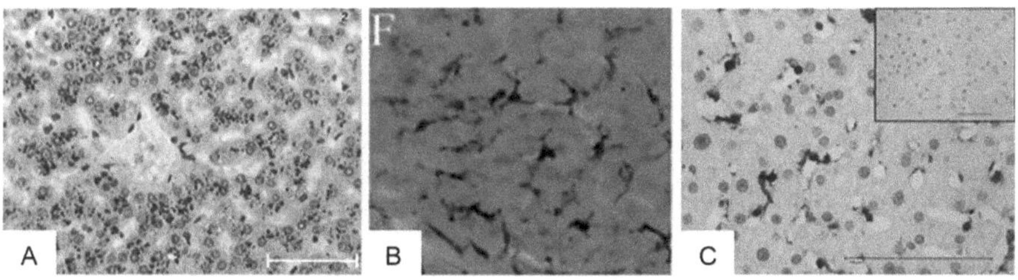

Fig. 1 Immunohistochemistry analysis of Tβ4 in the liver. (A) A Tβ4-immunoreactivity was observed in the cytoplasm of hepatocyte in healthy human liver. Scale bars represent 400 μm. (B) Kupffer-like cells are positive for Tβ4 in healthy liver of mice. These cells have small cell bodies with long branched processes. (C) Hepatic expression of Tβ4 was observed in Kupffer cells in liver of HCV-infected patients. The inserted image shows a negative control omitting primary antibody. Scale bars represent 100 μm. *Panel A: Figure from Nemolato, S., Van Eyken, P., Cabras, T., Cau, F., Fanari, M., Locci, A., …, Castagnola, M. (2011). Expression pattern of thymosin beta 4 in the adult human liver.* European Journal of Histochemistry, 55*(3), e25, PagePress, Italy, ©2011 Nemolato et al. Panel B: Figure from Paulussen, M., Landuyt, B., Schoofs, L., Luyten, W., & Arckens, L. (2009). Thymosin beta 4 mRNA and peptide expression in phagocytic cells of different mouse tissues.* Peptides, 30*(10), 1822–1832 ©2009 Elsevier Inc. Panel C: Figure from Henkel, C., Schwamborn, K., Zimmermann, H. W., Tacke, F., Kuhnen, E., Odenthal, M., …, Weiskirchen, R. (2011). From proteomic multimarker profiling to interesting proteins: Thymosin-beta(4) and kininogen-1 as new potential biomarkers for inflammatory hepatic lesions.* Journal of Cellular and Molecular Medicine, 15*(10), 2176–2188, ©2011 John Wiley and Sons Inc.* (See the color plate.)

conflicting result. In contrast with Nemolato's findings, Paulussen et al. reported that Kupffer cells express Tβ4 in healthy livers of mice, which was assessed by IHC (Fig. 1B) (Paulussen, Landuyt, Schoofs, Luyten, & Arckens, 2009). They proposed that Tβ4 indirectly controls the migration of Kupffer cells based on its ability to regulate actin polymerization, but they did not show any direct evidence to support their suggestion. Henkel et al. also showed that Kupffer cells expressed Tβ4 in liver sections of patients with the hepatitis C virus (HCV) through IHC staining (Henkel et al., 2011), suggesting that Tβ4 is a potential biomarker for inflammatory hepatic lesions (Fig. 1C). However, Tβ4 expression in healthy liver was unknown because they presented Tβ4 expression in HCV-infected liver only without examining it in healthy human liver. In addition, the effect of Tβ4 expression in Kupffer cells was not investigated. Recent studies have revealed that activated HSCs express Tβ4 in chronically damaged livers. We discuss the recent findings in more detail in Section 5.3.

5.2 The Potential Role of Exogenous Tβ4 in Liver

Based on previous findings for the antiinflammatory and antifibrotic actions of the Tβ4 peptide in extra-hepatic tissues, such as the lung and heart (Conte et al., 2012, 2014; Gupta et al., 2012; Sopko et al., 2011), researchers have investigated whether it exerts an antifibrotic effect in liver. In 2007, Barnaeva et al. suggested that Tβ4 has an antifibrotic effect on HSCs (Barnaeva et al., 2007). They demonstrated that using a Tβ4 treatment changed the expression of the genes responsible for ECM accumulation and transdifferentiation of HSCs, as assessed by reverse transcription polymerase chain reaction (RT-PCR) analysis. In this study, Tβ4-treated HSCs showed an upregulation of hepatocyte growth factor (HGF) and a downregulation of PDGF-β receptor at the RNA levels. Given that the level of HGF expression was downregulated in fully activated HSCs (Schirmacher, Geerts, Pietrangelo, Dienes, & Rogler, 1992) whereas the PDGF-β receptor was required for the activation of HSCs (Pinzani et al., 1995), this suggests that exogenous Tβ4 has an antifibrogenic function in human HSCs. However, α-SMA, a well-known marker of activated HSCs (Rockey, Boyles, Gabbiani, & Friedman, 1992), β-catenin, and glycogen synthase kinase (GSK)-3β are members of the Wnt pathway, which is involved in the transdifferentiation of HSCs (Cheng et al., 2008; Jiang, Parsons, & Stefanovic, 2006), were upregulated in a Tβ4-dose-dependent manner. In addition, the expression of other ECM components, such as collagen type1α2, tissue inhibitor of metalloproteinase (TIMP)-1, and matrix metalloproteinase (MMP)-2, was not changed after administration of Tβ4. Although these results suggest that downregulation of the PDGF-β receptor by Tβ4 prevents the proliferation of HSC-derived myofibroblasts, there has been no explanation for the increased expression of other ECM members, especially α-SMA, by Tβ4. To investigate this, further studies are required to demonstrate the disparity in the actions of Tβ4 with the PDGF-β receptor and α-SMA, as both have been considered hallmarks of HSC differentiation into myofibroblasts. In 2011, Reyes-Gordillo et al. presented that Tβ4 treatment prevented PDGF-ββ-dependent proliferation and the migration of cultured human HSCs by inhibiting PDGF-ββ-dependent phosphorylation of AKT (Reyes-Gordillo et al., 2011). In these studies, the researchers demonstrated that the binding of Tβ4 to actin interrupted the movement of AKT into PI3K, inhibiting the phosphorylation of AKT by PI3K in HSCs treated with PDGF-ββ. While treatment with exogenous Tβ4 successfully reduced the proliferation and migration of HSCs stimulated by PDGF-ββ, when the activated HSCs were treated with Tβ4, these cells also showed a decreased or unchanged expression of the genes under examination. Hence, it is possible that endogenous Tβ4 is combined with the exogenous Tβ4 and that relatively high concentrations of this combined Tβ4 desensitize the Tβ4 signaling pathway, eventually blocking the proliferation of HSCs. In these two studies, it is noted that Barnaeva and Reyes-Gordillo did not examine the endogenous expression or role of Tβ4 in HSCs. Nevertheless, Reyes-Gordillo et al. demonstrated in 2012 that the exogenous Tβ4 treatment attenuated acute liver damage induced by a single injection of carbon tetrachloride (CCl_4) in rats (Reyes-Gordillo, Shah, Arellanes-Robledo, Rojkind, & Lakshman, 2012). The liver morphology is improved in the CCl_4 + Tβ4-injected group compared with the CCl_4-injected group, as was assessed by hematoxylin and eosin staining. The hepatic protective effect of exogenous Tβ4 was mediated by inhibiting the upregulation of fibrotic markers (PDGF-β receptor, α-SMA, collagen type1α2, and fibronectin) and methyl-CpG-binding protein 2 (MeCP2), and the downregulation of adipogenic transcription factor peroxisome proliferator-activated receptor gamma (PPAR-γ) at the RNA level. The PPAR-γ expressed by quiescent HSCs is known to be reduced by MeCP2 in activated HSCs (Mann et al., 2007). The researchers suggested that Tβ4 maintained the expression levels of PPAR-γ by reducing MeCP2 expression, thus contributing to maintaining the quiescent phenotypic state of HSCs in the rat livers. This is a clinically significant, given that Tβ4 has therapeutic potential for liver fibrosis. However, the repair response, including activation of HSCs and inflammation, is a necessary process in damaged livers (Bataller & Brenner, 2005). The liver's health is dictated by a balance between the factors that regulate injury and repair. Liver injury stimulates a repair response, and when the repair response is inadequate, tissue damage persists and can progress. Therefore, defective repair responses permit mild liver injury to progress to cirrhosis (Friedman, 2008b). Liver damage caused by a single CCl_4 injection stimulated a repair response, and treatment with Tβ4 clearly showed an improved morphology in the livers injected with CCl_4, indicating that Tβ4 promotes liver regeneration. Therefore, the mechanism by which the repair process occurs without the activation of HSCs should be identified.

Taken together, these findings suggest that exogenous Tβ4 has a potential role in reducing liver fibrosis, a process that appears to be medicated by suppressing the proliferation and migration of HSCs (Barnaeva et al., 2007; Reyes-Gordillo et al., 2012, 2011). However, it is poorly understood how exogenous Tβ4 is taken up by the intracellular part and interacts with signaling molecules to inactivate HSCs. In addition, it is necessary to investigate what types of cells express endogenous Tβ4 in the liver, as well as whether endogenous Tβ4 is actually expressed in activated or inactivated HSCs, in order to understand the role and mechanism underlying the effects of Tβ4 in the liver.

5.3 The Potential Role of Endogenous Tβ4

As the expression and function of Tβ4 have been investigated recently in the liver, research on the types of cells that express Tβ4 has attracted attention. However, hepatic expression of Tβ4 is still controversial, as was mentioned earlier. Recently, Xiao et al. and Kim et al. provide direct evidence that activated HSCs express Tβ4 in vivo and in vitro (Kim et al., 2015; Xiao et al., 2014). Xiao et al. demonstrated that LX-2 cells, the activated human HSC line, strongly expressed Tβ4, as examined by immunofluorescent (IF) staining and Western blot (Fig. 2A). In line with their finding, Kim et al. showed that Tβ4 is expressed in both LX-2 cells and activated primary HSCs using quantitative RT-PCR (QRT-PCR), Western blot, IHC and IF staining assays. However, they suggested directly opposite interpretations of the role of Tβ4 in the liver. Xiao et al. reported that level of Tβ4 was downregulated in the serum of patients with cirrhosis compared with healthy humans (Xiao et al., 2014). They also showed that the expression

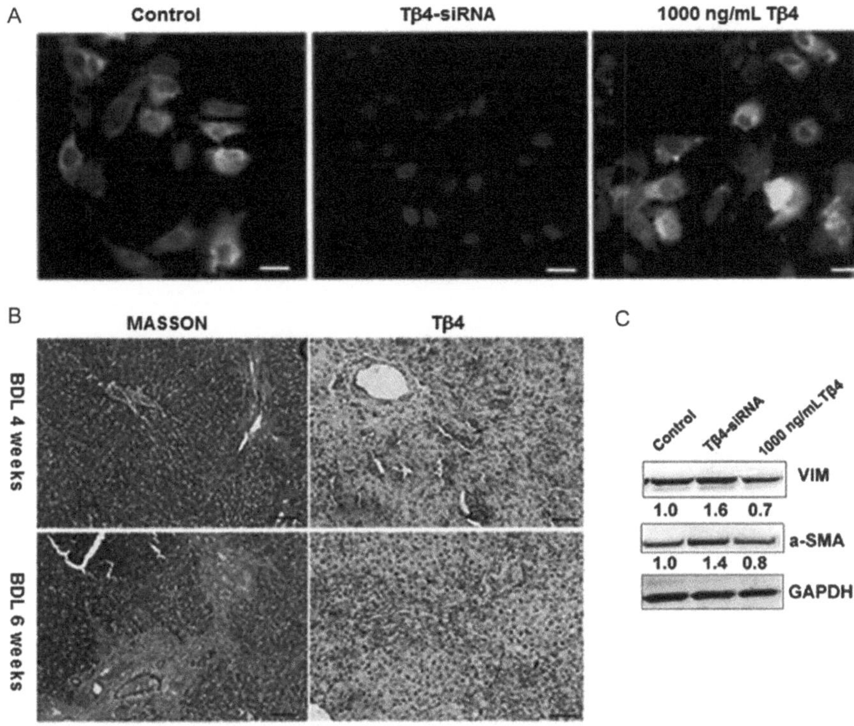

Fig. 2 Tβ4 is upregulated in LX-2 cells but downregulated in fibrotic liver. (A) Immunofluorescence analysis shows that Tβ4 is expressed in LX-2 cells treated with or without Tβ4 peptide. Tβ4-siRNA downregulates the endogenous expression of Tβ4 in these cells (×20). (B) Immunohistochemical staining for Tβ4 in liver of rats at 4 and 6 weeks after bile duct ligation. (C) Expression of α-SMA and vimentin in Tβ4-siRNA or Tβ4 peptide-treated LX-2 cells, as examined by Western blot assay. The expression of GAPDH was used as an internal control. *Figure modified from Xiao, Y., Qu, C., Ge, W., Wang, B., Wu, J., Xu, L., & Chen, Y. (2014). Depletion of thymosin β4 promotes the proliferation, migration, and activation of human hepatic stellate cells.* Cellular Physiology and Biochemistry: International Journal of Experimental Cellular Physiology, Biochemistry, and Pharmacology, 34*(2), 356–367, ©2014 S. Karger AG, Basel.* (See the color plate.)

of Tβ4 declined in the liver of bile duct-ligated (BDL) rats (Fig. 2B). Bile duct ligation is a well-established model, in which the accumulated myofibroblasts or the activated HSCs produce excessive ECMs, leading to fibrosis (Kountouras, Billing, & Scheuer, 1984). Therefore, the livers of those rats should have shown increased Tβ4 expression because the activated HSCs proliferate abundantly in those livers (Chen et al., 2008; Giebeler et al., 2009; Luk et al., 2007) and Xiao et al. presented an upregulation of Tβ4 in these activated HSCs. However, Tβ4 levels decreased in the BDL livers, although it was difficult to compare Tβ4 expression in the BDL liver with that of a healthy liver without evidence of Tβ4 expression in the livers of control rats. On the other hand, Kim et al. showed an upregulation of Tβ4 in human livers with stage 4 fibrosis compared with healthy livers (Fig. 3A) (Kim et al., 2015). In addition, they demonstrated that

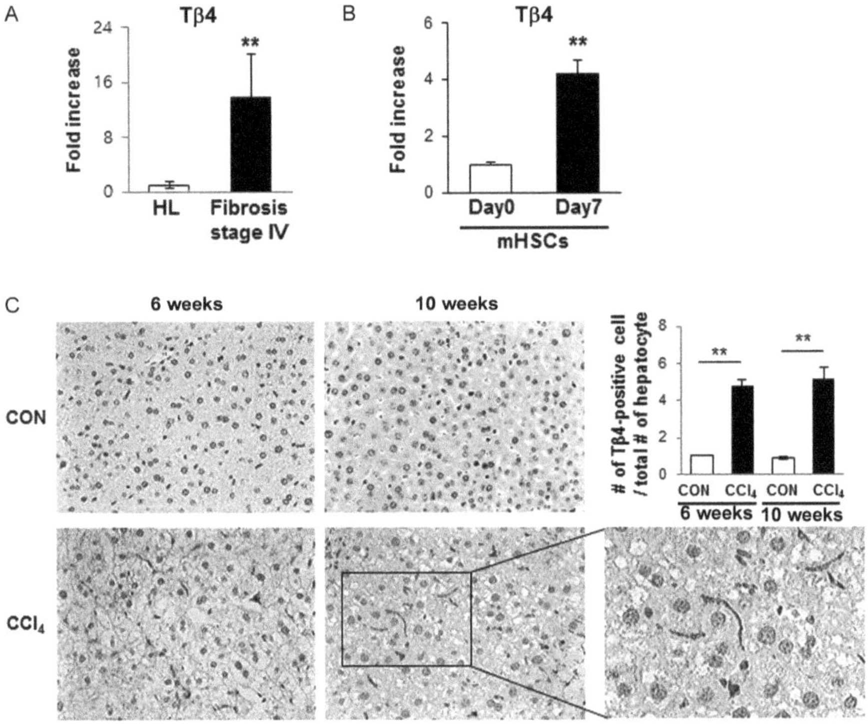

Fig. 3 Tβ4 increases during hepatic fibrogenesis and HSC activation. (A) QRT-PCR analysis for Tβ4 in human liver obtained from healthy (HL) and fibrotic livers with four stage of fibrosis (fibrosis stage IV). Mean ± SD results are graphed and data represent the mean ± SD of three independent experiments (**$p < 0.005$). (B) Tβ4 is upregulated during HSC culture, as assessed by QRT-PCR (Day 0: freshly isolated and not being cultured primary HSCs/Day 7: cultured primary HSCs for 7 days). Mean ± SD results are graphed. Data represent the mean ± SD of three independent experiments (**$p < 0.005$). (C) Immunohistochemical staining for Tβ4 in liver sections from the representative mice at 6 and 10 weeks after CCl_4 injection (×40). Magnified image from CCl_4-treated livers at 10 weeks (×63) shows Tβ4-positive HSC-looking cells. The number of Tβ4-positive cells are counted and graphed (**$p < 0.005$ vs own control group). *CON*: corn oil-treated mice, *CCl_4*: CCl_4-treated mice. *Figure modified from Kim, J., Wang, S., Hyun, J., Choi, S. S., Cha, H., Ock, M., & Jung, Y. (2015). Hepatic stellate cells express thymosin Beta 4 in chronically damaged liver.* PLoS One, 10*(3), e0122758, ©2015 Kim et al.* (See the color plate.)

expression of Tβ4 increased both in chronically damaged livers with CCl_4, a well-known chemical inducing liver fibrosis model, and during activation of HSCs (Fig. 3B and C). Primary HSCs are known to be activated during culture (De Minicis et al., 2007 Friedman, Roll, Boyles, Arenson, & Bissell, 1989), and expressed RNA levels of Tβ4 were elevated in cultured primary HSCs, when compared with quiescent HSCs that were freshly isolated and not cultured. Kim et al. revealed that both RNA and protein levels of Tβ4 were upregulated and colocalized in the α-SMA-expressing cells, activated HSCs, and CCl_4-treated livers with fibrosis (Fig. 4). Thus, these findings suggest that liver injury induces the proliferation and activation of HSCs and these cells express Tβ4 in fibrotic livers.

Given that activated HSCs express Tβ4, as supported by histological and biochemical evidence, further studies were conducted to investigate how Tβ4 influences the activation of HSCs. To test endogenous roles of Tβ4 on the activation of HSCs, both Xiao's and Kim's groups employed a small interfering RNA (siRNA) to suppress the endogenous expression of Tβ4. In Xiao's systems, depletion of Tβ4 promoted the proliferation and migration of LX-2 cells through the activation of the PI3K/AKT signaling pathway (Xiao et al., 2014). Western blot analysis showed that AKT (Ser473 and Thr308) and PI3K signaling components (p-85 regulatory subunit, PDK1 and GSK-3β) were phosphorylated in the Tβ4-inhibited LX-2 cells by Tβ4 siRNA. They also suggested that Tβ4 depletion promoted HSC activation by exhibiting increased protein levels of vimentin and α-SMA protein (Fig. 2C). However, if Tβ4 inhibited HSC activation, higher levels of

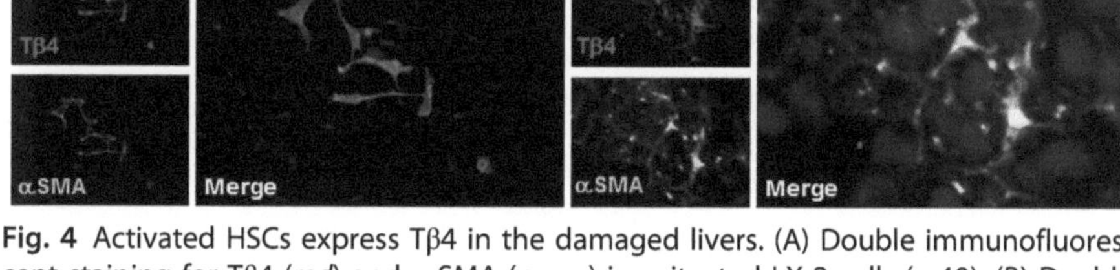

Fig. 4 Activated HSCs express Tβ4 in the damaged livers. (A) Double immunofluorescent staining for Tβ4 (*red*) and α-SMA (*green*) in acitvated LX-2 cells (×40). (B) Double immunofluorescent staining for Tβ4 (*red*) and α-SMA (green) in liver section from representative CCl_4-treated mice (×63). Coexpressing cells are shown as *yellow* to *orange*. DAPI nuclear staining is shown as *blue*. *Figure modified from Kim, J., Wang, S., Hyun, J., Choi, S. S., Cha, H., Ock, M., & Jung, Y. (2015). Hepatic stellate cells express thymosin Beta 4 in chronically damaged liver.* PLoS One, 10*(3), e0122758, ©2015 Kim et al.* (See the color plate.)

Tβ4 should have been expressed in the inactivated HSCs, not activated HSCs, and HSCs should have been activated in the absence or lower expression of Tβ4. However, there was no explanation for how the activated HSCs expressed Tβ4. In Kim's systems, the suppression of Tβ4 by siRNA induced the downregulation of profibrogenic markers, such as TGF-β and α-SMA, collagen1α1 and vimentin, and the upregulation of the inactivation markers of HSCs, PPAR-γ, and GFAP, in LX-2 cells (Kim et al., 2015). Oil Red O staining also showed cytoplasmic lipid droplets, the morphologic hallmark of inactivated HSCs, in the Tβ4-suppressed LX-2 cells (Fig. 5). Therefore, these results suggest that the endogenous expression of Tβ4 regulates HSC activation. However, Kim et al.'s experimental system did not investigate the effect of exogenous Tβ4 or the mechanism regarding how Tβ4 regulates the activation of HSCs. Thus, there is no clear explanation for the disparities between Xiao's and Kim's research. Further studies using transgenic mice to overexpress Tβ4 and investigating the possible signaling pathway associated with Tβ4 actions, might be helpful in answering those questions. Nevertheless, it is clear that Tβ4 functions as a key modulator in injured liver and HSC activation, although its fibrotic or antifibrotic effects are not elucidated. Therefore, further studies addressing the regulatory roles of Tβ4 in HSCs may aid in developing novel treatments for patients with liver fibrosis.

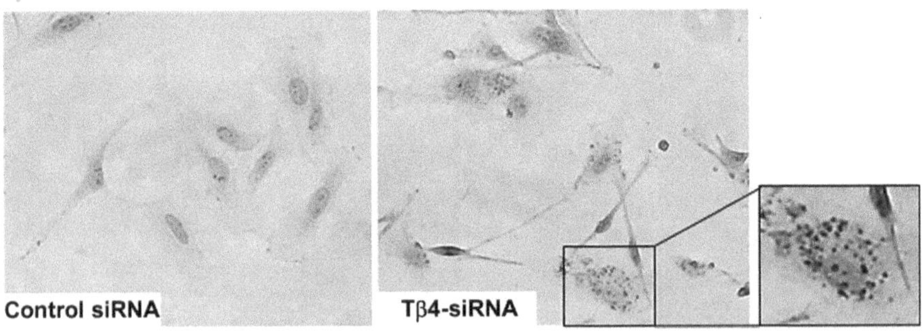

Fig. 5 Suppression of Tβ4 induces the inactivation of HSCs. Oil Red O (ORO) staining in control siRNA- or Tβ4 siRNA-treated LX-2 cells (X40). The lipid droplets appear as *red*. Magnified image (X63) from Tβ4 siRNA-treated cells show *red* lipid-like droplets around the nuclei (X63). *control siRNA*: control siRNA-treated LX-2 cells, *Tβ4 siRNA*: Tβ4 siRNA-treated LX-2 cells. *Figure modified from Kim, J., Wang, S., Hyun, J., Choi, S. S., Cha, H., Ock, M., & Jung, Y. (2015). Hepatic stellate cells express thymosin beta 4 in chronically damaged liver.* PLoS One, 10*(3), e0122758, ©2015 Kim et al.* (See the color plate.)

5.4 The Effects of the Ac-SDKP Fragment of Tβ4 in Liver Disease

Tβ4 is cleaved into N-acetyl-seryl-aspartyl-lysyl-proline (Ac-SDKP) by prolyl oligopeptidase (POP), which is a serine proteinase found in mammalian tissues (Cavasin, Rhaleb, Yang, & Carretero, 2004; Wilk, 1983). Ac-SDKP is an endogenous tetrapeptide normally present in the plasma and organs of humans and in experimental animals (Dainiak, 1992; Pradelles, Frobert, Creminon, Ivonine, & Frindel, 1991). It is also cleaved to an inactive form by the angiotensin converting enzyme (ACE) (Rousseau, Michaud, Chauvet, Lenfant, & Corvol, 1995). Emerging evidence suggests that Ac-SDKP plays a key role in inhibiting ECM accumulation in several organs under pathophysiological conditions. Decreased basal levels of endogenous Ac-SDKP caused by the downregulation of POP or the upregulation of ACE, promotes fibrosis in the heart and kidney (Cavasin, Liao, Yang, Yang, & Carretero, 2007; Zuo et al., 2013). Exogenous Ac-SDKP infusion reduces the collagen deposition in rat hearts and kidneys under hypertensive and ischemic conditions (Peng et al., 2001; Rhaleb, Peng, Harding, et al., 2001; Rhaleb, Peng, Yang, et al., 2001; Yang et al., 2004; Zhang et al., 2014). In addition, Ac-SDKP shows antifibrotic activity, whereas Tβ4 does have profibrotic effects in mice with ureteral obstruction (Zuo et al., 2013). In liver, the antifibrotic effects of Ac-SDKP were demonstrated in two different experimental models, CCl_4- or BDL-induced liver fibrosis. Endogenous Ac-SDKP was significantly alleviated in the liver of CCl_4-treated rats, but exogenous Ac-SDKP treatment reestablished the basal levels of Ac-SDKP and reduced liver fibrosis (Chen et al., 2010). In addition, Ac-SDKP suppresses the activation of HSC activation in vitro (Fig. 6). In bile duct ligation-induced liver fibrosis

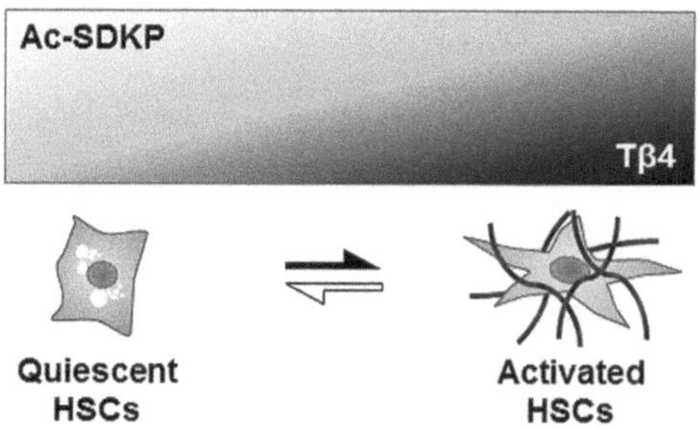

Fig. 6 Proposed model for role of Tβ4 and Ac-SDKP on activation of HSCs. Quiescent HSCs transdifferentiate into the activated HSCs following liver injury. During activation of HSCs, the endogenous level of Ac-SDKP (*white*) decreases, whereas the endogenous level of Tβ4 (*black*) increases. (See the color plate.)

models, the infusion of exogenous Ac-SDKP attenuated BDL-induced fibrosis by suppressing profibrogenic signaling, TGF-β and α-SMA, and by reducing collagen accumulation (Zhang et al., 2012). In Kim et al.'s study, POP activity decreased in CCl_4-treated mice, suggesting that this decrease in POP contributed to the accumulation of Tβ4, instead of Ac-SDKP, in the damaged liver (Fig. 7). However, the synthetic Ac-SDKP peptide was reported to inhibit the proliferation of regenerating hepatocytes in vivo by blocking the G1-S transition (Lombard, Sotty, Wdzieczak-Bakala, & Lenfant, 1990). Therefore, further studies on the action mechanisms of Ac-SDKP are needed to identify its potential role in liver.

5.5 The Tβ4 Signaling Pathway in the Liver

The PI3K/AKT signaling pathway was suggested as one of the possible targets of Tβ4 in regulating the proliferation and migration of HSCs (Xiao et al., 2014). In the cardiac repair process, Tβ4 activates ILK by stabilizing the ILK-PINCH complex, and activated ILK promotes the migration and survival of cardiac cells (Bock-Marquette et al., 2004). Tβ4 also triggers an EMT in colorectal carcinoma by upregulating ILK (Huang et al., 2007). In this EMT-promoting process, the activity of ILK is regulated in a PI3K-dependent manner (Legate, Montañez, Kudlacek, & Füssler, 2006; McDonald, Fielding, & Dedhar, 2008). Activated PI3K stimulates ILK

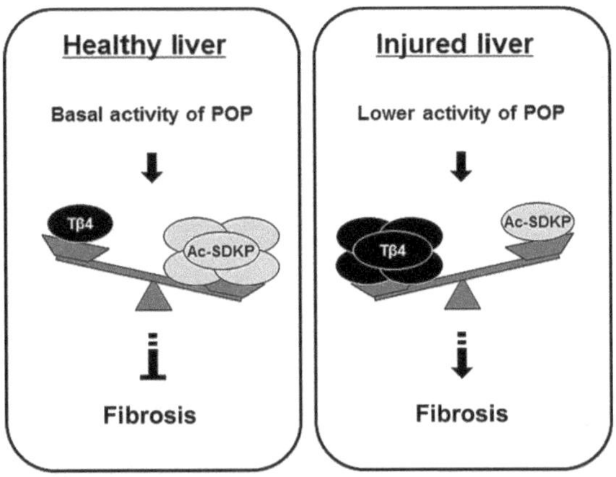

Fig. 7 Schematic illustration of the effect of balance between Tβ4 and Ac-SDKP level on liver fibrosis. It is depicted how the amount of Tβ4 and Ac-SDKP influences liver fibrosis. Tβ4 is cleaved into Ac-SDKP by prolyl oligopeptidase (POP). In healthy liver, POP with basal activity cleaves Tβ4 into Ac-SDKP normally. However, when activity of POP is reduced in damaged liver, the cleavage of Tβ4 into Ac-SDKP is suppressed. Hence, accumulation of Tβ4 with lower level of Ac-SDKP promotes liver fibrosis. (See the color plate.)

activity, which stimulates the phosphorylation of two downstream targets, AKT and GSK-3β. Although the effects of the Tβ4-PI3K/AKT pathway vary depending on the cell types, it seems to be associated with a cell response to Tβ4 stimulation. In renal fibrosis, Tβ4 is upregulated in glomerulosclerosis and is required for the angiotensin-II-induced expression of PAI-1 (plasminogen activator inhibitor-1) (Zuo et al., 2013). PAI-1, a major inhibitor of both the tissue-type plasminogen activator and urokinase-type plasminogen activator, is a key regulator of fibrinolysis by plasmin (Keijer et al., 1991). Plasmin directly or indirectly degrades ECM components through the activation of MMPs (Ghosh & Vaughan, 2012). Thus, PAI-1 influences organ fibrogenesis by impairing the plasminogen activator. Recent studies have shown that liver damage and fibrosis are ameliorated in PAI-1-deficient mice with BDL (Bergheim, Guo, Davis, Duveau, & Arteel, 2006). However, in PAI-1-knockout mice after chronic CCl_4 exposure, these injuries are dramatically enhanced (von Montfort et al., 2010). Although these studies reported different results depending on the experimental animal model used, the relationship between Tβ4 and PAI-1 might provide clues for understanding the mechanism of Tβ4 in the liver.

5.6 Tβ4 in Liver Cancers

Because the expression of Tβ4 is observed in several types of cancer, such as colorectal cancer and gliomas (Wang et al., 2004; Wirsching et al., 2014), there might be a relationship between Tβ4 and carcinogenesis. Tβ4 is overexpressed in malignant gliomas and its expressional level parallels the grade of malignancy in gliomas (Wirsching et al., 2014). In addition, Tβ4 is overexpressed in human colorectal cancer cells and it promotes cell invasion and migration (Huang et al., 2007; Nemolato et al., 2012; Wang, Chen, Hsiao, Ju, & Su, 2003; Wang et al., 2004). EMT-undergoing epithelial cells strongly express Tβ4 and this higher expression of Tβ4 is associated with a downregulation of E-cadherin (Nemolato et al., 2012; Wang et al., 2012), implying that Tβ4 influences carcinogenesis and metastasis. Several studies have been conducted to investigate the actions of Tβ4 in liver cancer. Theunissen et al. showed that Tβ4 was expressed by tumor cells localized in the parenchyma area in hepatocellular carcinoma (Theunissen et al., 2014). In addition, the association between Tβ4 and metastasis was reported in hepatoblastoma. Hepatoblastoma is the most common type of malignant liver tumors during childhood (Raney, 1997). Fu et al. showed that the upregulation of Tβ4 in hepatoblastoma cells contributed to metastasis by promoting EMT (Fu et al., 2015). In these studies, expression of Tβ4 protein levels was higher in HepG2 cells (human hepatocellular carcinoma cell line) than in L02 cells (human healthy liver cell line). Suppression of the Tβ4 gene expression resulted in decreased migration of HepG2 cells and it blocked TGF-β1-induced EMT in HepG2 cells. Tβ4 was also greatly expressed in tumor regions with or without metastasis compared with the adjacent non-tumor regions of hepatoblastoma liver tissues, suggesting that activation of the Tβ4 pathway might induce metastasis by promoting EMT. Furthermore, recent evidence suggests that EMT occurs during liver fibrogenesis. During liver injury, hepatic cells (eg, hepatocytes, cholangiocytes, and HSCs) undergo EMT, acquire MF phenotypes, and contribute to fibrogenesis (Choi et al., 2009; Nitta, Kim, Mohuczy, & Behrns, 2008; Omenetti et al., 2008; Zeisberg et al., 2007). Therefore, it is important to reveal the mechanisms underlying the involvement of Tβ4 in the EMT process, to understand the pathophysiology in the progression of chronic liver disease, including cancer.

6. CONCLUSIONS AND FUTURE DIRECTIONS

Liver fibrosis is the main characteristic of most chronic liver diseases, and HSCs are known to be the major source of fibrous matrix production (Bataller & Brenner, 2001). Therefore, studying the effective regulator that influences the activation of HSCs will provide important clues for developing therapeutics to inhibit liver fibrosis. As discussed earlier, activation of HSCs is initiated by persistent inflammation and various factors, including inflammatory cytokines, chemokines, growth factors, and products of oxidative stress. Tβ4 accelerates wound healing and modulates inflammatory mediators in various tissue injuries, such as cornea, skin, kidney, and heart (Malinda et al., 1999; Sosne et al., 2007, 2002), indicating that Tβ4 is involved in fibrogenesis for several organs. Although its effect seems to be dependent on the cellular and pathophysiological conditions, its beneficial and healing ability attracted the attention of liver researchers, leading to research investigating how Tβ4 is involved in liver pathophysiology. The exogenous Tβ4 peptide was reported to attenuate liver fibrosis by suppressing activation of HSCs and inflammatory infiltration (Barnaeva et al., 2007; Reyes-Gordillo et al., 2012, 2011). Meanwhile, increased levels of endogenous Tβ4 are associated with chronic liver disease and are involved in liver fibrosis by regulating the proliferation and activation of HSCs

(Kim et al., 2015; Xiao et al., 2014). Although it remains unclear whether Tβ4 promotes or inhibits the activation of HSCs, it is obvious that Tβ4 influences the pathogenesis of liver fibrosis. Hence, further studies are necessary to elucidate the role of endogenous Tβ4 and its interactions with other partners to develop safe and effective treatments for liver diseases. The results obtained from these studies should aid in the development of safer, more effective treatments for patients with liver fibrosis.

ACKNOWLEDGMENTS

This work was supported by the National Research Foundation of Korea (NRF) grant funded by the Korean government (MSIP: Ministry of Science, ICT and Future Planning) (No. 2014M2B2A9030333).

Competing financial interests: The authors indicate no potential conflict of interest.

REFERENCES

Barnaeva, E., Nadezhda, A., Hannappel, E., Sjogren, M. H., & Rojkind, M. (2007). Thymosin β4 upregulates the expression of hepatocyte growth factor and downregulates the expression of PDGF-β receptor in human hepatic stellate cells. *Annals of the New York Academy of Sciences*, *1112*(1), 154–160.

Bataller, R., & Brenner, D. A. (2001). Hepatic stellate cells as a target for the treatment of liver fibrosis. *Seminars in Liver Disease*, *21*(3), 437–451.

Bataller, R., & Brenner, D. A. (2005). Liver fibrosis. *Journal of Clinical Investigation*, *115*(2), 209.

Bergheim, I., Guo, L., Davis, M. A., Duveau, I., & Arteel, G. E. (2006). Critical role of plasminogen activator inhibitor-1 in cholestatic liver injury and fibrosis. *Journal of Pharmacology and Experimental Therapeutics*, *316*(2), 592–600.

Bi, W.-R., Yang, C.-Q., & Shi, Q. (2012). Transforming growth factor-β1 induced epithelial–mesenchymal transition in hepatic fibrosis. *Hepato-Gastroenterology*, *59*(118), 1960–1963.

Bica, I., McGovern, B., Dhar, R., Stone, D., McGowan, K., Scheib, R., & Snydman, D. R. (2001). Increasing mortality due to end-stage liver disease in patients with human immunodeficiency virus infection. *Clinical Infectious Diseases*, *32*(3), 492–497.

Bilzer, M., Roggel, F., & Gerbes, A. L. (2006). Role of Kupffer cells in host defense and liver disease. *Liver International*, *26*(10), 1175–1186.

Bock-Marquette, I., Saxena, A., White, M. D., DiMaio, J. M., & Srivastava, D. (2004). Thymosin β4 activates integrin-linked kinase and promotes cardiac cell migration, survival and cardiac repair. *Nature*, *432*(7016), 466–472.

Bouwens, L., De Bleser, P., Vanderkerken, K., Geerts, B., & Wisse, E. (1992). Liver cell heterogeneity: Functions of non-parenchymal cells. *Enzyme*, *46*(1–3), 155–168.

Boyer, T. D., Wright, T. L., & Manns, M. P. (2011). *Zakim and Boyer's hepatology: A textbook of liver disease*. Philadelphia: Elsevier Saunders, (Chapter 1).

Cavasin, M. A., Liao, T.-D., Yang, X.-P., Yang, J. J., & Carretero, O. A. (2007). Decreased endogenous levels of Ac-SDKP promote organ fibrosis. *Hypertension*, *50*(1), 130–136.

Cavasin, M. A., Rhaleb, N. E., Yang, X. P., & Carretero, O. A. (2004). Prolyl oligopeptidase is involved in release of the antifibrotic peptide Ac-SDKP. *Hypertension*, *43*(5), 1140–1145.

Cha, H.-J., Jeong, M.-J., & Kleinman, H. K. (2003). Role of thymosin β4 in tumor metastasis and angiogenesis. *Journal of the National Cancer Institute*, *95*(22), 1674–1680.

Cha, H., Philp, D., Lee, S., Moon, H., Kleinman, H., & Nakamura, T. (2010). Overexpression of thymosin beta 4 promotes abnormal tooth development and stimulation of hair growth. *The International Journal of Developmental Biology*, *54*(1), 135.

Chen, S. W., Chen, Y. X., Zhang, X. R., Qian, H., Chen, W. Z., & Xie, W. F. (2008). Targeted inhibition of platelet-derived growth factor receptor-beta subunit in hepatic stellate cells ameliorates hepatic fibrosis in rats. *Gene Therapy*, *15*(21), 1424–1435.

Chen, Y.-W., Liu, B.-W., Zhang, Y.-J., Chen, Y.-W., Dong, G.-F., Ding, X.-D., … Li, D.-G. (2010). Preservation of basal AcSDKP attenuates carbon tetrachloride-induced fibrosis in the rat liver. *Journal of Hepatology*, *53*(3), 528–536.

Cheng, J. H., She, H., Han, Y.-P., Wang, J., Xiong, S., Asahina, K., & Tsukamoto, H. (2008). Wnt antagonism inhibits hepatic stellate cell activation and liver fibrosis. *American Journal of Physiology. Gastrointestinal and Liver Physiology*, *294*(1), G39–G49.

Choi, S. S., Omenetti, A., Witek, R. P., Moylan, C. A., Syn, W.-K., Jung, Y., … Michelotti, G. A. (2009). Hedgehog pathway activation and epithelial-to-mesenchymal transitions during myofibroblastic transformation of rat hepatic cells in culture and cirrhosis. *American Journal of Physiology Gastrointestinal and Liver Physiology*, *297*(6), G1093–G1106.

Clark, E. A., Golub, T. R., Lander, E. S., & Hynes, R. O. (2000). Genomic analysis of metastasis reveals an essential role for RhoC. *Nature*, *406*(6795), 532–535.

Conte, E., Genovese, T., Gili, E., Esposito, E., Iemmolo, M., Fruciano, M., … Vancheri, C. (2012). Protective effects of thymosin beta4 in a mouse model of lung fibrosis. *Annals of the New York Academy of Sciences*, *1269*, 69–73.

Conte, E., Iemmolo, M., Fagone, E., Gili, E., Fruciano, M., Genovese, T., … Vancheri, C. (2014). Thymosin beta4 reduces IL-17-producing cells and IL-17 expression, and protects lungs from damage in bleomycin-treated mice. *Immunobiology*, *219*(6), 425–431.

Dainiak, N. (1992). Negative regulators of hematopoietic stem cells and progenitors. *Experimental Hematology*, *20*(10), 1154–1155.

De Minicis, S., Seki, E., Uchinami, H., Kluwe, J., Zhang, Y., Brenner, D. A., & Schwabe, R. F. (2007). Gene expression profiles during hepatic stellate cell activation in culture and in vivo. *Gastroenterology*, *132*(5), 1937–1946.

De Santis, M., Inzitari, R., Bosello, S. L., Peluso, G., Fanali, C., Iavarone, F., … Messana, I. (2011). β-Thymosins and interstitial lung disease: Study of a scleroderma cohort with a one-year follow-up. *Respiratory Research*, *12*, 22.

Elsharkawy, A., Oakley, F., & Mann, D. (2005). The role and regulation of hepatic stellate cell apoptosis in reversal of liver fibrosis. *Apoptosis*, *10*(5), 927–939.

Faouzi, S., Burckhardt, B. E., Hanson, J. C., Campe, C. B., Schrum, L. W., Rippe, R. A., & Maher, J. J. (2001). Anti-Fas induces hepatic chemokines and promotes inflammation by an NF-kB-independent, caspase-3-dependent pathway. *Journal of Biological Chemistry*, *276*(52), 49077–49082.

Fattovich, G., Giustina, G., Degos, F., Tremolada, F., Diodati, G., Almasio, P., … Brouwer, J. (1997). Morbidity and mortality in compensated cirrhosis type C: A retrospective follow-up study of 384 patients. *Gastroenterology*, *112*(2), 463–472.

Friedman, S. L. (1998). Cytokines and fibrogenesis. *Seminars in Liver Disease*, *19*(2), 129–140.

Friedman, S. L. (2000). Molecular regulation of hepatic fibrosis, an integrated cellular response to tissue injury. *Journal of Biological Chemistry*, *275*(4), 2247–2250.

Friedman, S. L. (2008a). Hepatic stellate cells: Protean, multifunctional, and enigmatic cells of the liver. *Physiological Reviews*, *88*(1), 125–172.

Friedman, S. L. (2008b). Mechanisms of hepatic fibrogenesis. *Gastroenterology*, *134*(6), 1655–1669.

Friedman, S., Roll, F. J., Boyles, J., Arenson, D., & Bissell, D. (1989). Maintenance of differentiated phenotype of cultured rat hepatic lipocytes by basement membrane matrix. *Journal of Biological Chemistry*, *264*(18), 10756–10762.

Friedman, S. L., Roll, F. J., Boyles, J., & Bissell, D. M. (1985). Hepatic lipocytes: The principal collagen-producing cells of normal rat liver. *Proceedings of the National Academy of Sciences*, *82*(24), 8681–8685.

Fu, X., Cui, P., Chen, F., Xu, J., Gong, L., Jiang, L., … Xiao, Y. (2015). Thymosin β4 promotes hepatoblastoma metastasis via the induction of epithelial-mesenchymal transition. *Molecular Medicine Reports*, *12*(1), 127–132.

Gard, A. L., White, F. P., & Dutton, G. R. (1985). Extra-neural glial fibrillary acidic protein (GFAP) immunoreactivity in perisinusoidal stellate cells of rat liver. *Journal of Neuroimmunology*, *8*, 359–375.

Ghosh, A. K., & Vaughan, D. E. (2012). PAI-1 in tissue fibrosis. *Journal of Cellular Physiology*, *227*(2), 493–507.

Giebeler, A., Boekschoten, M. V., Klein, C., Borowiak, M., Birchmeier, C., Gassler, N., … Streetz, K. L. (2009). c-Met confers protection against chronic liver tissue damage and fibrosis progression after bile duct ligation in mice. *Gastroenterology*, *137*(1), 297–308. 308.e291-294.

Goldstein, A. L., Hannappel, E., & Kleinman, H. K. (2005). Thymosin β4: Actin-sequestering protein moonlights to repair injured tissues. *Trends in Molecular Medicine*, *11*(9), 421–429.

Goldstein, A. L., Slater, F. D., & White, A. (1966). Preparation, assay, and partial purification of a thymic lymphocytopoietic factor (thymosin). *Proceedings of the National Academy of Sciences*, *56*(3), 1010.

Gomez-Marquez, J., Franco del Amo, F., Carpintero, P., & Anadon, R. (1996). High levels of mouse thymosin beta4 mRNA in differentiating P19 embryonic cells and during development of cardiovascular tissues. *Biochimica et Biophysica Acta*, *1306*(2–3), 187–193.

Goodall, G. J., Hempstead, J. L., & Morgan, J. I. (1983). Production and characterization of antibodies to thymosin beta 4. *The Journal of Immunology*, *131*(2), 821–825.

Grant, D. S., Kinsella, J. L., Kibbey, M. C., LaFlamme, S., Burbelo, P. D., Goldstein, A. L., & Kleinman, H. K. (1995). Matrigel induces thymosin beta 4 gene in differentiating endothelial cells. *Journal of Cell Science*, *108*(Pt. 12), 3685–3694.

Grant, D. S., Rose, W., Yaen, C., Goldstein, A., Martinez, J., & Kleinman, H. (1999). Thymosin β4 enhances endothelial cell differentiation and angiogenesis. *Angiogenesis*, *3*(2), 125–135.

Gressner, A. M. (1995). Cytokines and cellular crosstalk involved in the activation of fat-storing cells. *Journal of Hepatology*, *22*(2 Suppl.), 28–36.

Gressner, A. M., Weiskirchen, R., Breitkopf, K., & Dooley, S. (2002). Roles of TGF-beta in hepatic fibrosis. *Frontiers in Bioscience*, 7(1), d793–d807.

Grisham, J. W. (1962). A morphologic study of deoxyribonucleic acid synthesis and cell proliferation in regenerating rat liver; autoradiography with thymidine-H3. *Cancer Research*, *22*, 842–849.

Gupta, S., Kumar, S., Sopko, N., Qin, Y., Wei, C., & Kim, I. K. (2012). Thymosin β4 and cardiac protection: Implication in inflammation and fibrosis. *Annals of the New York Academy of Sciences*, *1269*(1), 84–91.

Hannappel, E., & Leibold, W. (1985). Biosynthesis rates and content of thymosin beta 4 in cell lines. *Archives of Biochemistry and Biophysics*, *240*(1), 236–241.

Hannappel, E., & van Kampen, M. (1987). Determination of thymosin beta 4 in human blood cells and serum. *Journal of Chromatography*, *397*, 279–285.

Hannappel, E., Xu, G. J., Morgan, J., Hempstead, J., & Horecker, B. L. (1982). Thymosin beta 4: A ubiquitous peptide in rat and mouse tissues. *Proceedings of the National Academy of Sciences*, *79*(7), 2172–2175.

Henkel, C., Schwamborn, K., Zimmermann, H. W., Tacke, F., Kuhnen, E., Odenthal, M., … Weiskirchen, R. (2011). From proteomic multimarker profiling to interesting proteins: Thymosin-beta(4) and kininogen-1 as new potential biomarkers for inflammatory hepatic lesions. *Journal of Cellular and Molecular Medicine, 15*(10), 2176–2188.

Higgins, G. M. (1931). Experimental pathology of the liver. *Archives of Pathology, 12*, 186–202.

Hu, S.-K., Low, T. L., & Goldstein, A. L. (1981). Modulation of terminal deoxynucleotidyl transferase activity by thymosin. *Molecular and Cellular Biochemistry, 41*(1), 49–58.

Huang, H., Hu, C., Tang, M., Wang, W., Chen, P., & Su, Y. (2007). Thymosin β4 triggers an epithelial–mesenchymal transition in colorectal carcinoma by upregulating integrin-linked kinase. *Oncogene, 26*(19), 2781–2790.

Huff, T., Muller, C. S., Otto, A. M., Netzker, R., & Hannappel, E. (2001). beta-Thymosins, small acidic peptides with multiple functions. *International Journal of Biochemistry & Cell Biology, 33*(3), 205–220.

Huff, T., Otto, A. M., Muller, C. S., Meier, M., & Hannappel, E. (2002). Thymosin beta4 is released from human blood platelets and attached by factor XIIIa (transglutaminase) to fibrin and collagen. *FASEB Journal, 16*(7), 691–696.

Imamura, M., Ogawa, T., Sasaguri, Y., Chayama, K., & Ueno, H. (2005). Suppression of macrophage infiltration inhibits activation of hepatic stellate cells and liver fibrogenesis in rats. *Gastroenterology, 128*(1), 138.

Inagaki, Y., & Okazaki, I. (2007). Emerging insights into transforming growth factor beta Smad signal in hepatic fibrogenesis. *Gut, 56*(2), 284–292.

Jaeschke, H. (2002). Inflammation in response to hepatocellular apoptosis. *Hepatology, 35*(4), 964–966.

Ji, Y.-I., Lee, B.-Y., Kang, Y.-J., Jo, J.-O., Lee, S. H., Kim, H. Y., … Kim, A. (2013). Expression patterns of thymosin β4 and cancer stem cell marker CD133 in ovarian cancers. *Pathology & Oncology Research, 19*(2), 237–245.

Jiang, F., Parsons, C. J., & Stefanovic, B. (2006). Gene expression profile of quiescent and activated rat hepatic stellate cells implicates Wnt signaling pathway in activation. *Journal of Hepatology, 45*(3), 401–409.

Keijer, J., Linders, M., van Zonneveld, A. J., Ehrlich, H. J., de Boer, J. P., & Pannekoek, H. (1991). The interaction of plasminogen activator inhibitor 1 with plasminogen activators (tissue-type and urokinase-type) and fibrin: Localization of interaction sites and physiologic relevance. *Blood, 78*(2), 401–409.

Kim, N. S., Kang, Y.-J., Jo, J.-O., Kim, H. Y., Oh, Y. R., Kim, Y.-O., … Cha, H.-J. (2011). Elevated expression of thymosin β4, vascular endothelial growth factor (VEGF), and hypoxia inducible factor (HIF)-1α in early-stage cervical cancers. *Pathology & Oncology Research, 17*(3), 493–502.

Kim, J., Wang, S., Hyun, J., Choi, S. S., Cha, H., Ock, M., & Jung, Y. (2015). Hepatic stellate cells express thymosin Beta 4 in chronically damaged liver. *PLoS One, 10*(3), e0122758.

Kmiec, Z. (2001). *Cooperation of liver cells in health and disease*, Vol. 161, Berlin, Germany: Springer Berlin Heidelberg.

Knittel, T., Kobold, D., Saile, B., Grundmann, A., Neubauer, K., Piscaglia, F., & Ramadori, G. (1999). Rat liver myofibroblasts and hepatic stellate cells: Different cell populations of the fibroblast lineage with fibrogenic potential. *Gastroenterology, 117*(5), 1205–1221.

Kobayashi, T., Okada, F., Fujii, N., Tomita, N., Ito, S., Tazawa, H., … Fujita, H. (2002). Thymosin-β4 regulates motility and metastasis of malignant mouse fibrosarcoma cells. *American Journal of Pathology, 160*(3), 869–882.

Kong, D., Zhang, F., Zhang, Z., Lu, Y., & Zheng, S. (2013). Clearance of activated stellate cells for hepatic fibrosis regression: Molecular basis and translational potential. *Biomedicine & Pharmacotherapy, 67*(3), 246–250.

Kountouras, J., Billing, B. H., & Scheuer, P. J. (1984). Prolonged bile duct obstruction: A new experimental model for cirrhosis in the rat. *British Journal of Experimental Pathology*, *65*(3), 305–311.
Kumar, S., & Gupta, S. (2011). Thymosin beta 4 prevents oxidative stress by targeting antioxidant and anti-apoptotic genes in cardiac fibroblasts. *PLoS One*, *6*(10), e26912.
Larsson, L.-I., & Holck, S. (2007). Occurrence of thymosin β4 in human breast cancer cells and in other cell types of the tumor microenvironment. *Human Pathology*, *38*(1), 114–119.
Legate, K. R., Montañez, E., Kudlacek, O., & Füssler, R. (2006). ILK, PINCH and parvin: The tIPP of integrin signalling. *Nature Reviews. Molecular Cell Biology*, 7(1), 20–31.
Li, D., & Friedman, S. (1999). Liver fibrogenesis and the role of hepatic stellate cells: New insights and prospects for therapy. *Journal of Gastroenterology and Hepatology*, *14*(7), 618–633.
Lombard, M. N., Sotty, D., Wdzieczak-Bakala, J., & Lenfant, M. (1990). In vivo effect of the tetrapeptide, N-acetyl-Ser-Asp-Lys-Pro, on the G1-S transition of rat hepatocytes. *Cell and Tissue Kinetics*, *23*(2), 99–103.
Low, T., Hu, S.-K., & Goldstein, A. L. (1981). Complete amino acid sequence of bovine thymosin beta 4: A thymic hormone that induces terminal deoxynucleotidyl transferase activity in thymocyte populations. *Proceedings of the National Academy of Sciences*, *78*(2), 1162–1166.
Lugo, D. I., Chen, S.-C., Hall, A. K., Ziai, R., Hempstead, J. L., & Morgan, J. I. (1991). Developmental regulation of β-thymosins in the rat central nervous system. *Journal of Neurochemistry*, *56*(2), 457–461.
Luk, J. M., Zhang, Q. S., Lee, N. P., Wo, J. Y., Leung, P. P., Liu, L. X., ... Fan, S. T. (2007). Hepatic stellate cell-targeted delivery of M6P-HSA-glycyrrhetinic acid attenuates hepatic fibrogenesis in a bile duct ligation rat model. *Liver International*, *27*(4), 548–557.
Malinda, K. M., Sidhu, G. S., Mani, H., Banaudha, K., Maheshwari, R. K., Goldstein, A. L., & Kleinman, H. K. (1999). Thymosin β4 accelerates wound healing. *Journal of Investigative Dermatology*, *113*(3), 364–368.
Mann, J., Oakley, F., Akiboye, F., Elsharkawy, A., Thorne, A. W., & Mann, D. A. (2007). Regulation of myofibroblast transdifferentiation by DNA methylation and MeCP2: Implications for wound healing and fibrogenesis. *Cell Death and Differentiation*, *14*(2), 275–285.
Matsuoka, M., & Tsukamoto, H. (1990). Stimulation of hepatic lipocyte collagen production by Kupffer cell-derived transforming growth factor β: Implication for a pathogenetic role in alcoholic liver fibrogenesis. *Hepatology*, *11*(4), 599–605.
McDonald, P. C., Fielding, A. B., & Dedhar, S. (2008). Integrin-linked kinase–essential roles in physiology and cancer biology. *Journal of Cell Science*, *121*(19), 3121–3132.
Michalopoulos, G. K., & DeFrances, M. C. (1997). Liver regeneration. *Science*, *276*(5309), 60–66.
Mora, C. A., Baumann, C. A., Paino, J. E., Goldstein, A. L., & Badamchian, M. (1997). Biodistribution of synthetic thymosin β4 in the serum, urine, and major organs of mice. *International Journal of Immunopharmacology*, *19*(1), 1–8.
Nemolato, S., Ekstrom, J., Cabras, T., Gerosa, C., Fanni, D., Di Felice, E., ... Faa, G. (2013). Immunoreactivity for thymosin beta 4 and thymosin beta 10 in the adult rat oro-gastro-intestinal tract. *European Journal of Histochemistry*, *57*(2), e17.
Nemolato, S., Restivo, A., Cabras, T., Coni, P., Zorcolo, L., Orru, G., ... Faa, G. (2012). Thymosin beta 4 in colorectal cancer is localized predominantly at the invasion front in tumor cells undergoing epithelial mesenchymal transition. *Cancer Biology & Therapy*, *13*(4), 191–197.

Nemolato, S., Van Eyken, P., Cabras, T., Cau, F., Fanari, M., Locci, A., ... Castagnola, M. (2011). Expression pattern of thymosin beta 4 in the adult human liver. *European Journal of Histochemistry*, *55*(3), e25.

Nieto, N. (2006). Oxidative-stress and IL-6 mediate the fibrogenic effects of rodent Kupffer cells on stellate cells. *Hepatology*, *44*(6), 1487–1501.

Nieto, N., & Cederbaum, A. I. (2003). Increased Sp1-dependent transactivation of the LAMgamma 1 promoter in hepatic stellate cells co-cultured with HepG2 cells overexpressing cytochrome P450 2E1. *Journal of Biological Chemistry*, *278*(17), 15360–15372.

Nieto, N., Friedman, S. L., & Cederbaum, A. I. (2002a). Cytochrome P450 2E1-derived reactive oxygen species mediate paracrine stimulation of collagen I protein synthesis by hepatic stellate cells. *Journal of Biological Chemistry*, *277*(12), 9853–9864.

Nieto, N., Friedman, S. L., & Cederbaum, A. I. (2002b). Stimulation and proliferation of primary rat hepatic stellate cells by cytochrome P450 2E1-derived reactive oxygen species. *Hepatology*, *35*(1), 62–73.

Nieto, N., Friedman, S. L., Greenwel, P., & Cederbaum, A. I. (1999). CYP2E1-mediated oxidative stress induces collagen type I expression in rat hepatic stellate cells. *Hepatology*, *30*(4), 987–996.

Nieto, N., Greenwel, P., Friedman, S. L., Zhang, F., Dannenberg, A. J., & Cederbaum, A. I. (2000). Ethanol and arachidonic acid increase alpha 2(I) collagen expression in rat hepatic stellate cells overexpressing cytochrome P450 2E1. Role of H2O2 and cyclooxygenase-2. *Journal of Biological Chemistry*, *275*(26), 20136–20145.

Nitta, T., Kim, J. S., Mohuczy, D., & Behrns, K. E. (2008). Murine cirrhosis induces hepatocyte epithelial mesenchymal transition and alterations in survival signaling pathways. *Hepatology*, *48*(3), 909–919.

Omenetti, A., Porrello, A., Jung, Y., Yang, L., Popov, Y., Choi, S. S., ... Diehl, A. M. (2008). Hedgehog signaling regulates epithelial–mesenchymal transition during biliary fibrosis in rodents and humans. *Journal of Clinical Investigation*, *118*(10), 3331–3342.

Paulussen, M., Landuyt, B., Schoofs, L., Luyten, W., & Arckens, L. (2009). Thymosin beta 4 mRNA and peptide expression in phagocytic cells of different mouse tissues. *Peptides*, *30*(10), 1822–1832.

Peng, H., Carretero, O. A., Raij, L., Yang, F., Kapke, A., & Rhaleb, N. E. (2001). Antifibrotic effects of N-acetyl-seryl-aspartyl-Lysyl-proline on the heart and kidney in aldosterone-salt hypertensive rats. *Hypertension*, *37*(2 Pt. 2), 794–800.

Philp, D., Badamchian, M., Scheremeta, B., Nguyen, M., Goldstein, A. L., & Kleinman, H. K. (2003). Thymosin β4 and a synthetic peptide containing its actin-binding domain promote dermal wound repair in db/db diabetic mice and in aged mice. *Wound Repair and Regeneration*, *11*(1), 19–24.

Philp, D., Goldstein, A., & Kleinman, H. (2004). Thymosin β4 promotes angiogenesis, wound healing, and hair follicle development. *Mechanisms of Ageing and Development*, *125*(2), 113–115.

Pinzani, M. (2002). PDGF and signal transduction in hepatic stellate cells. *Frontiers in Bioscience*, 7, d1720–d1726.

Pinzani, M., Gentilini, A., Caligiuri, A., De Franco, R., Pellegrini, G., Milani, S., ... Gentilini, P. (1995). Transforming growth factor-β1 regulates platelet-derived growth factor receptor β subunit in human liver fat-storing cells. *Hepatology*, *21*(1), 232–239.

Pinzani, M., & Rombouts, K. (2004). Liver fibrosis: From the bench to clinical targets. *Digestive and Liver Disease*, *36*(4), 231–242.

Pradelles, P., Frobert, Y., Creminon, C., Ivonine, H., & Frindel, E. (1991). Distribution of a negative regulator of haematopoietic stem cell proliferation (AcSDKP) and thymosin beta 4 in mouse tissues. *FEBS Letters*, *289*(2), 171–175.

Raney, B. (1997). Hepatoblastoma in children: A review. *Journal of Pediatric Hematology/Oncology*, *19*(5), 418–422.

Reeves, H. L., & Friedman, S. L. (2002). Activation of hepatic stellate cells—A key issue in liver fibrosis. *Frontiers in Bioscience*, 7(4), 808–826.
Reif, S., Lang, A., Lindquist, J. N., Yata, Y., Gäbele, E., Scanga, A., ... Rippe, R. A. (2003). The role of focal adhesion kinase-phosphatidylinositol 3-kinase-akt signaling in hepatic stellate cell proliferation and type I collagen expression. *Journal of Biological Chemistry*, *278*(10), 8083–8090.
Reyes-Gordillo, K., Shah, R., Arellanes-Robledo, J., Rojkind, M., & Lakshman, M. R. (2012). Protective effects of thymosin β4 on carbon tetrachloride-induced acute hepatotoxicity in rats. *Annals of the New York Academy of Sciences*, *1269*(1), 61–68.
Reyes-Gordillo, K., Shah, R., Popratiloff, A., Fu, S., Hindle, A., Brody, F., & Rojkind, M. (2011). Thymosin-β4 (Tβ4) blunts PDGF-dependent phosphorylation and binding of AKT to actin in hepatic stellate cells. *American Journal of Pathology*, *178*(5), 2100–2108.
Rhaleb, N.-E., Peng, H., Harding, P., Tayeh, M., LaPointe, M. C., & Carretero, O. A. (2001a). Effect of N-acetyl-seryl-aspartyl-lysyl-proline on DNA and collagen synthesis in rat cardiac fibroblasts. *Hypertension*, *37*(3), 827–832.
Rhaleb, N. E., Peng, H., Yang, X. P., Liu, Y. H., Mehta, D., Ezan, E., & Carretero, O. A. (2001b). Long-term effect of N-acetyl-seryl-aspartyl-lysyl-proline on left ventricular collagen deposition in rats with 2-kidney, 1-clip hypertension. *Circulation*, *103*(25), 3136–3141.
Rockey, D. C., Boyles, J., Gabbiani, G., & Friedman, S. (1992). Rat hepatic lipocytes express smooth muscle actin upon activation in vivo and in culture. *Journal of Submicroscopic Cytology and Pathology*, *24*(2), 193–203.
Rousseau, A., Michaud, A., Chauvet, M. T., Lenfant, M., & Corvol, P. (1995). The hemoregulatory peptide N-acetyl-Ser-Asp-Lys-Pro is a natural and specific substrate of the N-terminal active site of human angiotensin-converting enzyme. *Journal of Biological Chemistry*, *270*(8), 3656–3661.
Safer, D., Elzinga, M., & Nachmias, V. (1991). Thymosin beta 4 and Fx, an actin-sequestering peptide, are indistinguishable. *Journal of Biological Chemistry*, *266*(7), 4029–4032.
Sanders, M. C., Goldstein, A. L., & Wang, Y.-L. (1992). Thymosin beta 4 (Fx peptide) is a potent regulator of actin polymerization in living cells. *Proceedings of the National Academy of Sciences*, *89*(10), 4678–4682.
Schirmacher, P., Geerts, A., Pietrangelo, A., Dienes, H. P., & Rogler, C. E. (1992). Hepatocyte growth factor/hepatopoietin A is expressed in fat-storing cells from rat liver but not myofibroblast-like cells derived from fat-storing cells. *Hepatology*, *15*(1), 5–11.
Schuppan, D., & Afdhal, N. H. (2008). Liver cirrhosis. *The Lancet*, *371*(9615), 838–851.
Sicklick, J. K., Li, Y.-X., Choi, S. S., Qi, Y., Chen, W., Bustamante, M., ... Torbenson, M. S. (2005). Role for hedgehog signaling in hepatic stellate cell activation and viability. *Laboratory Investigation*, *85*(11), 1368–1380.
Siegmund, S. V., & Brenner, D. A. (2005). Molecular pathogenesis of alcohol-induced hepatic fibrosis. *Alcoholism, Clinical and Experimental Research*, *29*(s2), 102S–109S.
Sopko, N., Qin, Y., Finan, A., Dadabayev, A., Chigurupati, S., Qin, J., ... Gupta, S. (2011). Significance of thymosin beta4 and implication of PINCH-1-ILK-alpha-parvin (PIP) complex in human dilated cardiomyopathy. *PLoS One*, *6*(5), e20184.
Sosne, G., Qiu, P., & Kurpakus-Wheater, M. (2007). Thymosin beta 4: A novel corneal wound healing and anti-inflammatory agent. *Clinical Ophthalmology (Auckland, NZ)*, *1*(3), 201.
Sosne, G., Szliter, E. A., Barrett, R., Kernacki, K. A., Kleinman, H., & Hazlett, L. D. (2002). Thymosin beta 4 promotes corneal wound healing and decreases inflammation in vivo following alkali injury. *Experimental Eye Research*, *74*(2), 293–299.
Sosne, G., Xu, L., Prach, L., Mrock, L. K., Kleinman, H. K., Letterio, J. J., ... Kurpakus-Wheater, M. (2004). Thymosin beta 4 stimulates laminin-5 production independent of TGF-beta. *Experimental Cell Research*, *293*(1), 175–183.

Stocker, E., & Heine, W. D. (1971). Regeneration of liver parenchyma under normal and pathological conditions. *Beiträge zur Pathologie*, *144*(4), 400–408.

Sung, J. Y., Costerton, J. W., & Shaffer, E. A. (1992). Defense system in the biliary tract against bacterial infection. *Digestive Diseases and Sciences*, *37*(5), 689–696.

Svegliati Baroni, G., D'Ambrosio, L., Ferretti, G., Casini, A., Di Sario, A., Salzano, R., ... Benedetti, A. (1998). Fibrogenic effect of oxidative stress on rat hepatic stellate cells. *Hepatology*, *27*(3), 720–726.

Taub, R. (2004). Liver regeneration: From myth to mechanism. *Nature Reviews. Molecular Cell Biology*, *5*(10), 836–847.

Theunissen, W., Fanni, D., Nemolato, S., Di Felice, E., Cabras, T., Gerosa, C., ... Faa, G. (2014). Thymosin beta 4 and thymosin beta 10 expression in hepatocellular carcinoma. *European Journal of Histochemistry*, *58*(1), 2242.

Tsukamoto, H. (1999). Cytokine regulation of hepatic stellate cells in liver fibrosis. *Alcoholism, Clinical and Experimental Research*, *23*(5), 911–916.

Tsukamoto, H. (2002). Redox regulation of cytokine expression in Kupffer cells. *Antioxidants & Redox Signaling*, *4*(5), 741–748.

Vartiainen, N., Pyykönen, I., Hökfelt, T., & Koistinaho, J. (1996). Induction of thymosin [beta] 4 mRNA following focal brain ischemia. *Neuroreport*, *7*(10), 1613–1616.

von Montfort, C., Beier, J. I., Kaiser, J. P., Guo, L., Joshi-Barve, S., Pritchard, M. T., & Arteel, G. E. (2010). PAI-1 plays a protective role in CCl_4-induced hepatic fibrosis in mice: Role of hepatocyte division. *American Journal of Physiology. Gastrointestinal and Liver Physiology*, *298*(5), G657–G666.

Wang, W.-S., Chen, P.-M., Hsiao, H.-L., Ju, S.-Y., & Su, Y. (2003). Overexpression of the thymosin beta-4 gene is associated with malignant progression of SW480 colon cancer cells. *Oncogene*, *22*(21), 3297–3306.

Wang, W.-S., Chen, P.-M., Hsiao, H.-L., Wang, H.-S., Liang, W.-Y., & Su, Y. (2004). Overexpression of the thymosin β-4 gene is associated with increased invasion of SW480 colon carcinoma cells and the distant metastasis of human colorectal carcinoma. *Oncogene*, *23*(39), 6666–6671.

Wang, Y., Jiang, X.-Y., Liu, L., & Jiang, H.-Q. (2008). Phosphatidylinositol 3-kinase/Akt pathway regulates hepatic stellate cell apoptosis. *World Journal of Gastroenterology*, *14*(33), 5186.

Wang, Z. Y., Zeng, F. Q., Zhu, Z. H., Jiang, G. S., Lv, L., Wan, F., ... Xing, S. A. (2012). Evaluation of thymosin β4 in the regulation of epithelial–mesenchymal transformation in urothelial carcinoma. *Urologic Oncology: Seminars and Original Investigations*, *30*(2), 167–176 Elsevier.

Wilk, S. (1983). Prolyl endopeptidase. *Life Sciences*, *33*(22), 2149–2157.

Wirsching, H. G., Krishnan, S., Florea, A. M., Frei, K., Krayenbuhl, N., Hasenbach, K., ... Tabatabai, G. (2014). Thymosin beta 4 gene silencing decreases stemness and invasiveness in glioblastoma. *Brain*, *137*(Pt. 2), 433–448.

Wisse, E., Braet, F., Luo, D., De Zanger, R., Jans, D., Crabbe, E., & Vermoesen, A. (1996). Structure and function of sinusoidal lining cells in the liver. *Toxicologic Pathology*, *24*(1), 100–111.

Wong, J. B., McQuillan, G. M., McHutchison, J. G., & Poynard, T. (2000). Estimating future hepatitis C morbidity, mortality, and costs in the United States. *American Journal of Public Health*, *90*(10), 1562.

Xiao, Y., Qu, C., Ge, W., Wang, B., Wu, J., Xu, L., & Chen, Y. (2014). Depletion of thymosin β4 promotes the proliferation, migration, and activation of human hepatic stellate cells. *Cellular Physiology and Biochemistry: International Journal of Experimental Cellular Physiology, Biochemistry, and Pharmacology*, *34*(2), 356–367.

Xu, G. J., Hannappel, E., Morgan, J., Hempstead, J., & Horecker, B. L. (1982). Synthesis of thymosin beta 4 by peritoneal macrophages and adherent spleen cells. *Proceedings of the National Academy of Sciences*, *79*(13), 4006–4009.

Yamamoto, M., Yamagishi, T., Yaginuma, H., Murakami, K., & Ueno, N. (1994). Localization of thymosin beta 4 to the neural tissues during the development of Xenopus laevis, as studied by in situ hybridization and immunohistochemistry. *Brain Ressearch. Developmental Brain Research*, *79*(2), 177–185.

Yang, F., Yang, X.-P., Liu, Y.-H., Xu, J., Cingolani, O., Rhaleb, N.-E., & Carretero, O. A. (2004). Ac-SDKP reverses inflammation and fibrosis in rats with heart failure after myocardial infarction. *Hypertension*, *43*(2), 229–236.

Zeisberg, M., Yang, C., Martino, M., Duncan, M. B., Rieder, F., Tanjore, H., & Kalluri, R. (2007). Fibroblasts derive from hepatocytes in liver fibrosis via epithelial to mesenchymal transition. *Journal of Biological Chemistry*, *282*(32), 23337–23347.

Zhang, L., Chopp, M., Teng, H., Ding, G., Jiang, Q., Yang, X. P., … Zhang, Z. G. (2014). Combination treatment with N-acetyl-seryl-aspartyl-lysyl-proline and tissue plasminogen activator provides potent neuroprotection in rats after stroke. *Stroke*, *45*(4), 1108–1114.

Zhang, L., Xu, L. M., Chen, Y. W., Ni, Q. W., Zhou, M., Qu, C. Y., & Zhang, Y. (2012). Antifibrotic effect of N-acetyl-seryl-aspartyl-lysyl-proline on bile duct ligation induced liver fibrosis in rats. *World Journal of Gastroenterology*, *18*(37), 5283–5288.

Zuo, Y., Chun, B., Potthoff, S. A., Kazi, N., Brolin, T. J., Orhan, D., … Myöhänen, T. (2013). Thymosin β4 and its degradation product, Ac-SDKP, are novel reparative factors in renal fibrosis. *Kidney International*, *84*(6), 1166–1175.

CHAPTER SEVEN

Immune Modulation with Thymosin Alpha 1 Treatment

R. King[1], C. Tuthill
SciClone Pharmaceuticals, Inc., Foster City, CA, United States
[1]Corresponding author: e-mail address: rking@sciclone.com

Contents

Abstract

Thymosin alpha 1 (Ta1) is a peptide originally isolated from thymic tissue as the compound responsible for restoring immune function to thymectomized mice. Ta1 has a pleiotropic mechanism of action, affecting multiple immune cell subsets that are involved in immune suppression. Ta1 acts through Toll-like receptors in both myeloid and plasmacytoid dendritic cells, leading to activation and stimulation of signaling pathways and initiation of production of immune-related cytokines.

Due to the immune stimulating effects of Ta1, the compound would be expected to show utility for treatment of immune suppression, whether related to aging or to diseases such as infection or cancer. Extensive studies in both the preclinical and clinical setting will be summarized in the subsequent sections. These studies have demonstrated improvements in immune system cell subsets and the potential of Ta1 for the treatment of a range of diseases.

Vitamins and Hormones, Volume 102
ISSN 0083-6729
http://dx.doi.org/10.1016/bs.vh.2016.04.003

1. INTRODUCTION

Thymosin alpha 1 (INN: thymalfasin; commonly abbreviated Ta1) is currently sold in over 30 countries worldwide under the brand name ZADAXIN®. Ta1, initially selected for its activity in restoring immune function to thymectomized mice, was the first peptide to be isolated from thymic tissue (Goldstein, Low, McAdoo, et al., 1977; Low & Goldstein, 1979; Low, Thurman, McAdoo, et al., 1979). Ta1 is an N-terminal acetylated acidic peptide of 28 amino acids with a molecular weight of 3108 Da (Low & Goldstein, 1985). Circulating Ta1 is the amino terminal proteolytic cleavage product of the chromatin-remodeling protein prothymosin (Haritos, Goodall, & Horecker, 1984), and is derived from cleavage of prothymosin by the lysosomal asparaginyl endopeptidase legumain (Sarandeses, Covelo, Diaz-Jullien, & Freire, 2003). Ta1 is a highly conserved peptide, and therefore of biological significance, as its amino acid sequence is homologous in bovine, porcine, ovine, and human species (Schulof, 1985), and similar peptides have even been found in crustaceans (Oates & Erdos, 1989).

Endogenous Ta1 serum levels measured in healthy adults by immunoassay are in the 0.1–1.0 ng/mL range (Weller, Shah, Cummings, Chretien, & Mutchnick, 1992), although the circulating concentration tends to be lower in diseased individuals and higher during pregnancy (Jevremovic et al., 1997; Sherman, Jones, Goldstein, & Naylor, 1991; Welch, Lee, Sokol, & Mutchnick, 1988; Welch, Mutchnick, Weller, & Sokol, 1987). While the highest concentrations of Ta1 are found in the thymus, the peptide has also been found in spleen, lung, kidney, brain, blood, and a number of other tissues.

A chemically synthesized version of Ta1 shows activity similar to the native peptide (Wang, Makofske, Bach, & Merrifield, 1980); this compound has been used for preclinical and clinical evaluation leading to regulatory approval.

2. MECHANISM OF IMMUNE RECONSTITUTION WITH TA1

In keeping with Ta1's original isolation as a compound responsible for reconstitution of immune function in preclinical models of immune-compromised animals, investigation of the mechanism of action at the

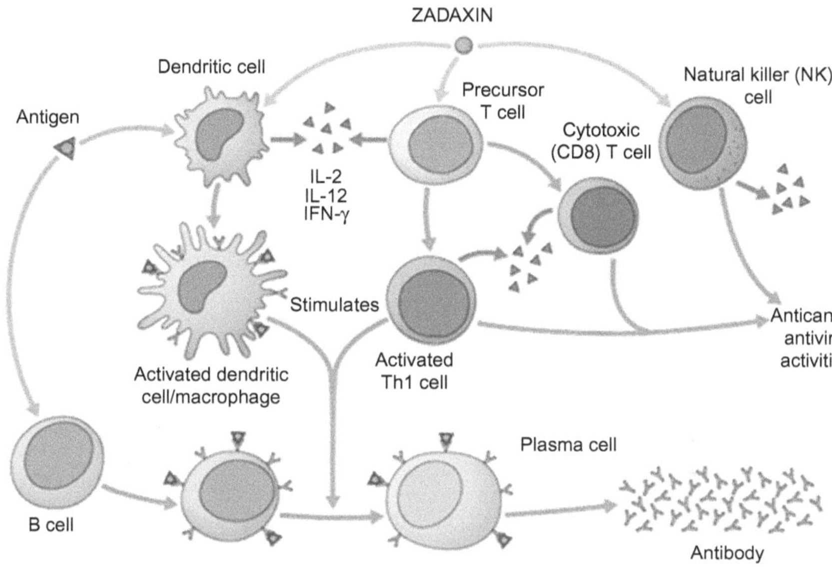

Fig. 1 Immune-stimulating mechanism of action of Ta1. (See the color plate.)

cellular level has implicated a number of intracellular cell-signaling pathways that are associated with stimulation of the immune system (Fig. 1).

Ta1 has been shown to be a Toll-like receptor (TLR)-9 and TLR-2 agonist (Romani, Bistoni, Montagnoli, et al., 2007; Romani, Bistoni, Perruccio, et al., 2006; Romani, Moretti, Fallarino, Bozza, et al., 2012). The TLRs are a family of proteins that mediate innate immunity; stimulation of one or more TLRs by a TLR agonist can enhance the adaptive immune response which is critical for fighting viral, bacterial, and fungal infections and cancers, as well as stimulation of humoral immunity for vaccine effectiveness. Ta1 affects TLRs in both myeloid and plasmacytoid dendritic cells (DCs), the professional antigen-presenting cells, leading to activation and stimulation of signaling pathways and initiation of production of immune-related cytokines that fight infections (Romani, Bistoni, Gaziano, et al., 2004; Serafino, Pierimarchi, Pica, et al., 2012).

The effects of Ta1 on precursor T cells leads to an increase in the number of activated T helper (Th) cells (CD4 T cells) (Bozza, Gaziano, Bonifzi, et al., 2007; Peng, Chen, Yu, et al., 2008) and a shift toward the Th1 subclass (Cursaro, Margotti, Favarelli, et al., 1998; Gramenzi et al., 2005). This shift leads to increased expression of Th1-type cytokines such as interleukin-2 (IL-2) (Serrate, Schulof, Leondaridis, Goldstein, & Sztein, 1987; Sztein & Serrate, 1989; Sztein, Serrate, & Goldstein, 1986; Yao, Doan, Zhang, et al., 2007), and interferon (IFN)-alpha (Hsia, Sarin, Oliver, & Goldstein, 1989; Leichtling, Serrate, & Sztein, 1990; Serrate et al., 1987; Svedersky, Hui, May, McKay, & Stebbing, 1982; Sztein & Serrate, 1989; Sztein et al., 1986; Yao et al., 2007). The activated DCs and Th1 cells then act in concert to kill bacterial, fungal, or viral infections or tumor cells and lead to the stimulation of differentiation of specific B cells to antibody-producing plasma cells and an improvement in response to vaccines by stimulation of antibody production (Effros, Casillas, & Walford, 1988; Ershler, Hebert, Blow, Granter, & Lynch, 1985; Ershler, Moore, & Socinski, 1984). Further, Tα1 acting through TLR9 leads to stimulation of the NFkappaB and p38 MAPK pathways (Peng et al., 2008; Romani et al., 2004; Zhang, Chan, Dragoi, et al., 2005), both of which play critical roles in the maturation of DCs (Arrighi, Rebsamen, Rousset, Kindler, & Hauser, 2001; Iijima, Yanagawa, & Onoe, 2003) and in antigen presentation by DCs (Yoshimura, Bondeson, Brennan, Foxwell, & Feldmann, 2001). Tα1 leads to increased expression of the thymopoetic cytokines IFN-alpha, IL-7, and IL-15 (Knutsen, Freeman, Mueller, Roodman, & Bouhasin, 1999; Weller et al., 1992).

The wide-ranging effects of Tα1 include a reduction in apoptosis of immune cells, as shown in mouse (Baumann, Badamchian, & Goldstein, 1997; Osheroff, 1981) and human (Ho, Stehle, Dietz, Hunstein, & Hoffbrand, 1985; Knutsen et al., 1999) thymocytes, and stem cell expansion or differentiation in immunosuppressed mice (Hu, Low, & Goldstein, 1981; Ohta, Sueki, Yoneyama, Tezuka, & Yagi, 1983; Ohta, Tezuka, Tamura, & Yagi, 1985). Tα1 treatment also leads to an increase in intracellular glutathione (GSH) (Palamara, Bue, Savini, et al., 1998), important for antiviral effects, and to directly inhibit the in vitro growth of certain cancer cells (Moody, Badamchian, & Goldstein, 1998; Moody, Fagarasan, Zia, Cesnjaj, & Goldstein, 1993).

In addition to its effects on DCs and T helper cells, Tα1 also stimulates innate immunity, including natural killer (NK) cells and macrophages, additionally supporting its antiviral and anticancer activities. NK-cell activity has been shown to be increased by Tα1 in a variety of model systems, including infections (mice with herpes simplex virus (HSV) (Shiau, Wu, & Huang, 1988) or influenza (D'Agostini, Palamara, Favalli, et al., 1996), as well as various cancers in mice and rats and polymorphonuclear blood cells from human subjects (Ni et al., 2015; Serrate et al., 1987; Umeda, Sakamoto, Nakamura, Ishitsuka, & Yagi, 1983). In human monocyte-derived macrophages, Tα1 helps implement pathogen internalization and phagocytosis (Serafino et al., 2014).

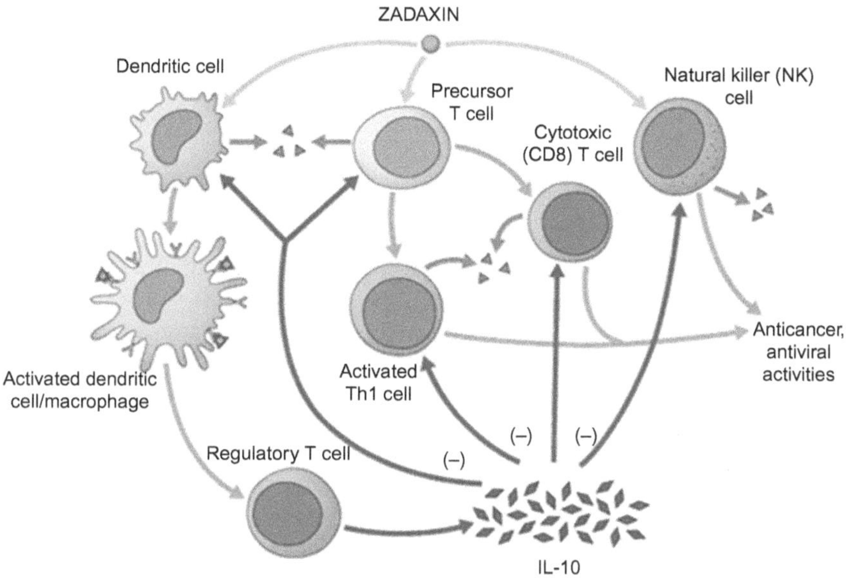

Fig. 2 Immune-modulating mechanism of action of Ta1. (See the color plate.)

Importantly, it has also been shown that Ta1 stimulates activity of indoleamine-2,3-dioxygenase (IDO) in plasmacytoid DCs (Romani, Moretti, et al., 2012; Romani et al., 2007). Stimulation of IDO leads to an increase in FoxP3^{+} IL-10 producing regulatory T cells, and this increase leads to feedback inhibition of cytokine production, hence dampening immune response to prevent a proinflammatory cytokine storm (Fig. 2).

In summary, due to its pleiotropic mechanisms on affected immune cells, Ta1 could be useful in treatment of subjects who are immunosuppressed, whether the suppression is a result of infection, cancer, renal disease, or age. The utility of treatment with Ta1 for these indications has been investigated in animal model systems and clinical studies, as outlined later.

3. PRECLINICAL STUDIES OF TA1 IN IMMUNE SUPPRESSED ANIMALS

3.1 Animal Models of Infectious Disease

Microbial infections lead to multiple pathways of immune suppression, so that infections can lead to chronic, debilitating, and often fatal conditions. Ta1 treatment has been investigated in animal models of both acute and chronic viral, bacterial, and fungal infections.

Improvements in in vitro immune responses to mitogens by mouse lymphocytes were seen after treatment of aged mice immune suppressed by hydrocortisone treatment (Hadden, Saha, Sosa, & Hadden, 1995). As expected from these effects, Tα1 treatment provided statistically significant protection against a variety of acute lethal bacterial and fungal infections in animals immunosuppressed with 5-flurouracil (5-FU), including *Serratia marcescens*, *Pseudomonas aeruginosa*, *Listeria monocytogenes*, and *Candida albicans* (Ishitsuka, Umeda, Nakamura, & Yagi, 1983). Tα1 also prolonged survival of mice, and prevented the increased susceptibility to infection, in animals infected with *C. albicans* and immune suppressed with either cyclophosphamide or morphine treatment (Bistoni, Marconi, Frati, Bonmassar, & Garaci, 1982; Di Francesco et al., 1994). Recent studies have investigated the efficacy of Tα1 against *Aspergillus fumigatus* in more detail, including the evaluation of knockout mice to determine the cellular pathways involved in the increased survival, decreased infectious agent burden, and increased T cells and NK-cell activity that are seen (Romani et al., 2004). Tα1 has shown benefit in treatment of generalized sepsis in several mouse models, both as a monotherapy and in combination with dexamethasone (Xiang, Li, Zhao, Li, & Li, 2014).

With respect to acute viral infections, Tα1 treatment has also led to increased survival, in mice infected with HSV (Shiau et al., 1988), influenza (D'Agostini et al., 1996; Effros et al., 1988), and cytomegalovirus (Bozza et al., 2007). Studies with mice infected with influenza, for example, showed that Tα1 in combination with IFN increased survival, reduced viral titer, and stimulated NK cells; Tα1 in combination with both IFN and the antiviral agent amantadine led to statistically significant increases in survival, NK-cell activity, CD4 and CD8 counts, and cytotoxic activity in the lung, while viral titers in the lung were reduced (D'Agostini et al., 1996).

Chronic infections have also been shown to respond positively to treatment with Tα1, with improved immune parameters and decreased viral loads, in models for hepatitis B virus (HBV) (Gerin, Korba, Cote, & Tennant, 1992; Tennant et al., 1993) hepatitis C virus (HCV) (Cursaro et al., 1998), and human immunodeficiency virus (HIV) (Garaci, Rocchi, Perroni, et al., 1994).

3.2 Animal Models of Cancer

Tα1 has been shown to have beneficial effects in several experimental models of cancer. When given in combination with chemotherapy and either IL-2 or IFN, Tα1 has shown not only to increase cytotoxic responses of T cells and NK-cell activity, but also to reduce tumor size and, importantly, to increase survival in models of colorectal cancer (Rasi, Silecchia, Sinibaldi-Vallebona, et al., 1994; Silecchia, Guarino, Sinibaldi-Vallebona, et al., 1999), melanoma (King & Tuthill, 2015; Pica, Fraschetti, Matteucci, Tuthill, & Rasi, 1998), non-small cell lung cancer (NSCLC) (Moody et al., 1993), lung carcinoma (Garaci, Mastino, Pica, & Favalli, 1990; Mastino, Favalli, Grelli, et al., 1992), erythroleukemia (Garaci et al., 1993), leukemia (Umeda et al., 1983), and fibrosarcoma (Tomazic, Sacasa, Loftus, Suter, & Elias, 1988).

In colorectal cancer, for example, Tα1 treatment in combination with 5-FU and IL-2 led to significantly improved median survival time compared to the 5-FU control (70.0 ± 8.2 days vs 54.6 ± 5.3 days, $p < 0.0001$) (Rasi et al., 1994). In addition to these positive effects on survival, there were also greatly reduced growth of liver metastases, reduced liver invasion (20% vs 60% in control), and reduced extra-hepatic spread. In addition, the improvement of survival time in these rats from the addition of Tα1 treatment allowed for a second cycle of treatment to be provided (Silecchia et al., 1999), which led to a further significant increase in survival time (81% survival at 100 days vs 39% with 5-FU alone or 44% for 5-FU plus IL-2).

Studies with a mouse model of melanoma showed that combination of increasing doses of Tα1 with chemotherapy and IFN chemotherapy led to significantly increased time to relapse, decreased the tumor growth rate, and improved survival in a Tα1 dose-dependent fashion (Pica et al., 1998). In fact, the addition of a single cycle of Tα1 treatment led to a cure in 24% of the mice (5 of 21 animals were alive and disease free 1 year after treatment). As in other studies investigating the effects of Tα1 treatment in animal models of cancer, immune parameters were also improved. Splenocytes from treated mice showed markedly increased cytotoxic activities, and the tumor-induced reduction in percentages of CD3 and CD4 cells was reversed to nontumor levels (Pica et al., 1998).

The beneficial effects of Tα1 have also been seen when the compound is used as a sole therapy: Tα1 monotherapy has shown to prevent lung carcinogenesis in mice injected with a chemical carcinogen (Moody et al., 1993, 1998, 2000) and breast cancer in rats (Moody, Tuthill, Badamchian, & Goldstein, 2002). In addition, recent studies have shown that Tα1 monotherapy and/or in combination with anti-PD-1 antibody was able to significantly reduce lung metastases, and also to significantly reduce tumor growth of the highly metastatic melanoma clone B16F10;

widely used in studying the mechanisms of metastasis and in evaluating cancer therapeutics (King & Tuthill, 2015).

3.3 Animal Models of Improvement in Vaccine Response

As expected from the effects of Tα1 on antigen-presenting DCs, administration of Tα1 increases antibody response to vaccination. For example, while older mice (23 months old) have a significantly lower response to vaccination than young mice (2–3 months), when mice were treated with Tα1 not only was their antibody response to tetanus toxoid significantly increased ($p < 0.05$), but the response of older mice was restored to levels seen in young animals (Ershler et al., 1985). In a similar study comparing response to influenza virus infection in old mice (>24 months), Tα1 treatment was shown to increase IgM titer along with virus-specific cytotoxic T-cell responses (Effros et al., 1988).

The ability of Tα1 to enhance response to a pandemic influenza vaccine was tested with the 2008 seasonal trivalent vaccine. Studies in both older mice and ferrets showed an improvement in antibody titers when Tα1 was given along with the vaccine (Tuthill et al., 2010). Administration of Tα1 twice, 7 days prior to vaccination and again on the day of vaccination, was found to be the optimal dosing regimen, leading to improvement in response to all three strains of influenza tested. The titer of antibody was greater when determined 21 days after vaccination, and persisted when evaluated 42 days after vaccination. The improvement in titer was seen whether animals were given a vaccine booster or not.

4. CLINICAL STUDIES OF TA1 IN IMMUNE SUPPRESSION

As anticipated from the immune-modulating mechanism of action of Tα1 and the positive effects seen in animal models as described above, Tα1 has proven to be a useful therapeutic in a wide range of clinical indications. Over 4400 subjects have been enrolled in US, European and Chinese clinical trials investigating the use of Tα1, including primary treatment for subjects with acute infections, such as seen in severe sepsis, and for chronic infections including chronic hepatitis B (CHB), chronic hepatitis C (CHC), and HIV; as an adjunct treatment for cancers, including melanoma, hepatocellular carcinoma (HCC), and NSCLC; and as an enhancement to both hepatitis B and influenza vaccines in immune-depressed individuals. Results from some of these trials will be discussed later; based on these clinical studies Tα1 (trade name ZADAXIN) was first licensed in

Italy in 1993 for use as a vaccine enhancer, and has since obtained market approvals in over 30 countries in Asia, Latin and South America and Eastern Europe as a mono or combination therapy for CHB and CHC, as an adjuvant therapy for chemotherapy-induced immune suppression, and as a vaccine-enhancement agent.

To date, over 17 million doses of ZADAXIN (estimated as >350,000 individuals exposed) have been administered postmarket. ZADAXIN is generally well tolerated and has an excellent safety profile. Over the past 20 years, adverse experiences have been infrequent and mild.

4.1 Clinical Studies in Primary Immune Deficiency

The first clinical evaluation of Tα1 was a physician-sponsored study of 10 individuals with congenital immune deficiencies including hypogammaglobulinemia, ataxia telangectasia, Down's syndrome, and chronic mucocutaneous candidiasis (unpublished data). The trial was open-label and uncontrolled; the age of 7 subjects ranged from infancy to 16 years. Data were analyzed for eight subjects who had completed 6 months of treatment, and improvement in some parameters of T-lymphocyte function were observed in five of these subjects. No adverse experiences associated with Tα1 were reported.

A single child with DiGeorge anomaly was also treated with Tα1 (Gupta, Aggarwal, & Nguyen, 1997). Blood cells were taken from this 13-month-old infant before and after 3 months of treatment with Tα1 and examined for evidence of lymphocyte apoptosis compared to an age-matched healthy control. Prior to treatment with Tα1, the subject showed increased apoptosis (increased Fas and FasL, decreased Bcl-2 in both CD4 and CD8 cells, increased DNA fragmentation); after treatment with Tα1 the proportion of lymphocytes undergoing apoptosis decreased. The T-cell responses (response to mitogen) and B-cell function (specific antibody formation) also improved after treatment, and the subject showed a marked clinical improvement evidenced by a significant decrease in infection. No adverse experiences associated with Tα1 were reported.

4.2 Clinical Studies in Infectious Disease

4.2.1 Acute Infections

Subsequent to these early physician-sponsored studies, Tα1 showed promise in larger clinical studies of acute infections, including severe sepsis (Chen, 2007; Li, Zhou, Qiang, et al., 2009; Wu, Zhou, Liu, et al., 2013; Zhao,

Cao, Fei, et al., 2007), severe acute pancreatitis (Wang, Li, et al., 2011), acute respiratory distress syndrome (ARDS) (Ji, Li, Sun, et al., 2007; Sun, Liu, Chen, et al., 2006), infectious episodes in chronic obstructive pulmonary disease (Li, Wang, et al., 2007; Zheng et al., 2008), and lung infections (Huang, Yang, Pen, et al., 2006; Li, Xu, Zhang, et al., 2007).

Severe sepsis, a syndrome defined as the presence of two or more features of systemic inflammation (such as fever or hypothermia, leucocytosis or leukopenia, tachycardia, tachyponea or supranormal minute ventilation), has historically been thought to be a manifestation of over-stimulation of the immune system in response to infection and thus was originally treated with immune suppressive drugs. Severe sepsis begins with a bacterial or fungal infection, which triggers immune signaling and subsequent release of noxious mediators, including proinflammatory cytokines and reactive oxygen free radicals that cause direct tissue injury. It is a remarkably diverse and heterogeneous condition, ranging in severity from mild systemic inflammation without significant clinical consequences to multisystem failure in septic shock with a high mortality rate. Also occurring very early on, however, is a compensatory antiinflammatory response leading to lymphocyte apoptosis, monocyte hyporesponsiveness and a state of immune suppression including a change to the T helper cell type 2 antiinflammatory cytokine phenotype (LaRosa, 2002). The pathophysiologic response in a given patient is often determined by the infectious insult, as well as the patient's age, comorbid illnesses and genome (Hotchkiss & Karl, 2003). In fact, only a small minority of previously well young patients die from an overwhelming proinflammatory cytokine storm in severe sepsis, and thus the use of immune-stimulatory therapies, in contrast to previous use of immune suppressive treatments, has recently been recommended, particularly in the elderly or with patients with serious coexisting illnesses in whom immunosuppression is identified or predicted (Angus & van der Poll, 2013).

Supporting this idea, several small clinical studies demonstrated the benefit of Tα1 in the treatment of severe sepsis, showing higher 28-day cumulative survival rates and improvements in other secondary endpoints (Gupta et al., 1997; Li et al., 2009; Zhao et al., 2007). A large (361 subjects), prospective, multicenter, single-blind, randomized, placebo-controlled trial was therefore conducted to further evaluate the concept (Wu et al., 2013). Twenty-eight-day mortality from any cause was 26% in the Tα1-treated group, vs 35% in control group ($p=0.062$ nonstratified analysis, $p=0.049$ log-rank; Fig. 3). The relative risk of death in the Tα1 group compared to control was 0.74 (95% confidence interval [CI] 0.54–1.02). There

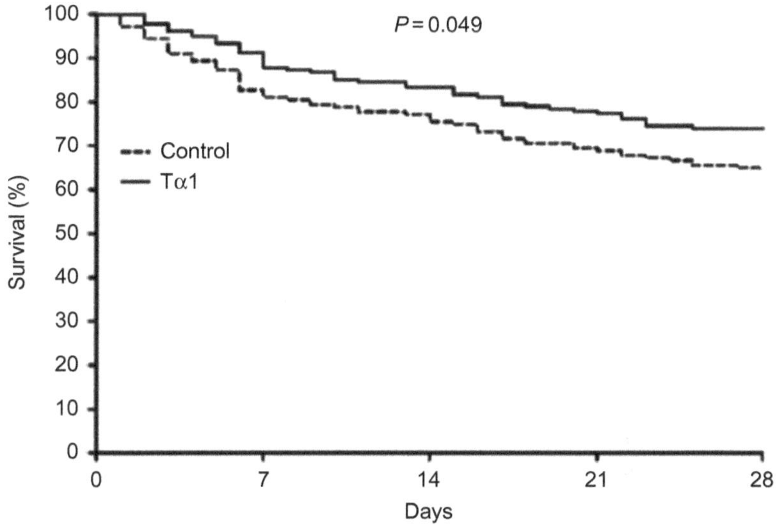

Numbers at risk

Day	0	7	14	21	28
Control	180	149	139	125	117
Tα1	181	165	151	141	134

Fig. 3 Ta1 improves survival in severe sepsis.

was a 9.0% (95% CI −0.5% to 18.5%) absolute reduction in mortality in the Ta1-treated subjects. As a measure of immune response, the levels of HLA-DR were evaluated and greater improvements were seen in subjects treated with Ta1 ($p=0.037$).

Another study in acute infection is also of particular interest. Renal transplant patients with ARDS due to CMV infection were given Ta1 treatment at the time that they presented with low CD4 counts, and were treated until the CD4 levels returned to normal (Ji et al., 2007). Mortality was decreased from 50% (7/14) in control subjects, to 22% (7/32) in subjects treated with Ta1. All subjects were being treated with immunosuppressive drugs on admission; those treated with Ta1 were able to have the doses of immune suppressants decreased. Furthermore, in the survivors, acute cellular rejection of the renal graft was seen in only 12% of the Ta1 group vs 28.5% in the control group; renal transplant success rate was significantly higher (78% vs 50%) in the Ta1 treatment group. Finally, Ta1 treatment also significantly increased the number of CD4 and CD8 lymphocytes ($p<0.05$), suggesting that repairing cellular immunity with Ta1 reinforces resistance to CMV.

A recent phase 2 clinical study was conducted in yet another acute infection setting: evaluating the use of Tα1 after HLA matched or haploidentical stem cell transplant in 75 subjects (Perruccio, Bonifazi, Topini, et al., 2010; Romani, Aversa, Garaci, & Velardi, 2012). This study found quite promising results. The cumulative incidence of transplant-related mortality (TRM; the majority of which was CMV or *Aspergillus* infection related) was 32% in control subjects but only 7% in subjects treated with Tα1 ($p=0.02$). Event-free survival was increased in Tα1 (40% vs 20% in controls; $p=0.02$). Tα1 treatment was a significant independent factor predicting a lower incidence of TRM ($p=0.04$), which tended to provide better survival ($p=0.09$). And, as expected from its mechanism of action, Tα1 administration was associated with increased T-cell counts and earlier appearance of functional pathogen-specific T-cells responses (against *Aspergillus*, *Candida*, CMV, varicella-zoster virus, HSV, and *Toxoplasma*).

4.2.2 Chronic Infections

Chronic infections resulting from infection with HBV, HCV, and HIV are considered hallmarks of immune suppression, resulting from the myriad pathways of immune system evasion that the viruses have evolved. During the course of infection with HBV, for example, the virus disrupts innate immunity pathways (including TLR signaling, DCs, NK, and NKT cells) early during infection, which then also compromises the quality of the adaptive immune response, including a decrease in the number and function of HBV-specific T lymphocytes (Busca & Kumar, 2014). Although most adults are able to clear the acute infection, those individuals who have impaired cellular immune mechanisms, including the young, do not effectively clear HBV-infected hepatocytes and chronic infection results, correlated with a greatly increased risk for developing cirrhosis, liver failure, and HCC. About one-third of the world's population will be infected by HBV at one point in their lives; CHB has caused epidemics in many parts of Asia and Africa and is considered an endemic in China, affecting up to 240 million persons worldwide (WHO, 2015).

Interest in using Tα1 for treatment of CHB was based on its immunomodulating effects, primarily the improved maturation of lymphocytes and augmentation of T-cell function. Clinical studies with Tα1 have resulted in disease remission in 26–41% of the subjects treated. An independent meta-analysis of 435 subjects entered into randomized, controlled studies of Tα1 monotherapy for CHB demonstrated a statistically significant benefit in favor of Tα1 therapy, inducing a sustained virological response over placebo (odds ratio, OR = 2.87; 95% CI 1.58–5.22; $p = 0.0005$) (Chan, Tang, & Sung, 2001). The same study also demonstrated a trend in favor of Tα1 treatment compared with IFN for sustained virological response (OR = 2.62; 95% CI 0.80–8.56). Other trials have shown that the addition of Tα1 to other agents, such as IFN and nucleoside analogs, can lead to improved responses (Lau et al., 1999; Lim, Wai, Lee, et al., 2006; Saruc, Yuceyar, Kucukmetin, Demir, & Kandiloglu, 2002; Saruc, Ozden, et al., 2002).

4.3 Clinical Studies in Cancer

Most cancer patients have clearly depressed cellular immunity, and progression of cancer appears to be related to impaired suppression of the tumors by the immune system. Immune-based therapies are emerging as efficacious treatments for cancer, including antibodies or compounds that manipulate cellular signals to induce better recognition of cancer cells by the immune system, such as anti-CTLA-4, which can enable T-cell activation via stimulation of CD80 and CD86 (Acharya & Jeter, 2013); anti-PD-L1, which can reverse immunosuppression (John, Westwood, Darcy, & Kershaw, 2013); or anti-CD47, which enhances phagocytosis of cancer cells (Weissman & Majeti, 2012). General immune modulation (eg, with IL-2) has also shown promising results for treatment of human cancers (Balch et al., 1997; Richards, Gale, Mahta, & Lestingi, 1999; Rosenberg, Mul, Spiess, Reichert, & Schwarz, 1985). A robust host immune response appears to be useful or necessary for the complete clearance of tumor cells, and the establishment of effective antitumor immune responses may be essential to the success of treatment (Burkholder et al., 1845).

Tα1 treatment has been shown to provide several benefits to cancer patients, including an increase in effectiveness of chemotherapy, a decrease in treatment side effects, and an overall improvement in quality of life (QOL), without the types of adverse events seen with IL-2 or IFN.

4.3.1 Melanoma

Melanoma is a cancer of the skin caused by a malignant tumor of melanocytes found predominantly in adults, responsible for more than 75% of deaths caused by skin cancers, one of the most common types of cancer. If diagnosed early, patients with localized melanoma may be cured with surgery; however, there is no current cure for malignant melanoma after the cancer has metastasized. The average survival of patients with melanoma that has spread outside the local area is approximately 7.5 months, with a 90% mortality rate after 5 months (American Cancer Society, 2009).

Pilot studies showed that Tα1 treatment resulted in an improvement in response, time to progression, and overall survival compared to historical controls in subjects with metastatic melanoma being treated with dacarbazine (DTIC) chemotherapy plus IL-2 (Lopez et al., 1994) or IFN (Rasi, Terzoli, Izzo, et al., 2000). A significant protection against chemotherapy-induced reductions in NK activity and CD4 T cells ($p < 0.001$) was also seen.

A large randomized study was subsequently conducted in 488 subjects with metastatic melanoma (Fig. 4) (Maio et al., 2010). In this study, addition of various doses of Tα1 to DTIC, with or without IFN, led to an increase in median overall survival (9.4 months for Tα1-treated subjects vs 6.6 months for control; $p=0.08$), progression-free survival (16 months for Tα1 vs 9 months, $p=0.06$; significant for the subset of subjects given 3.2 mg Tα1 plus DTIC, $p=0.04$), a significant improvement in tumor response (10.3 or 12.1% in subjects treated with 3.2 mg Tα1 with DTIC with or without IFN, vs 4% in the control group; $p<0.05$), and an increase in duration of

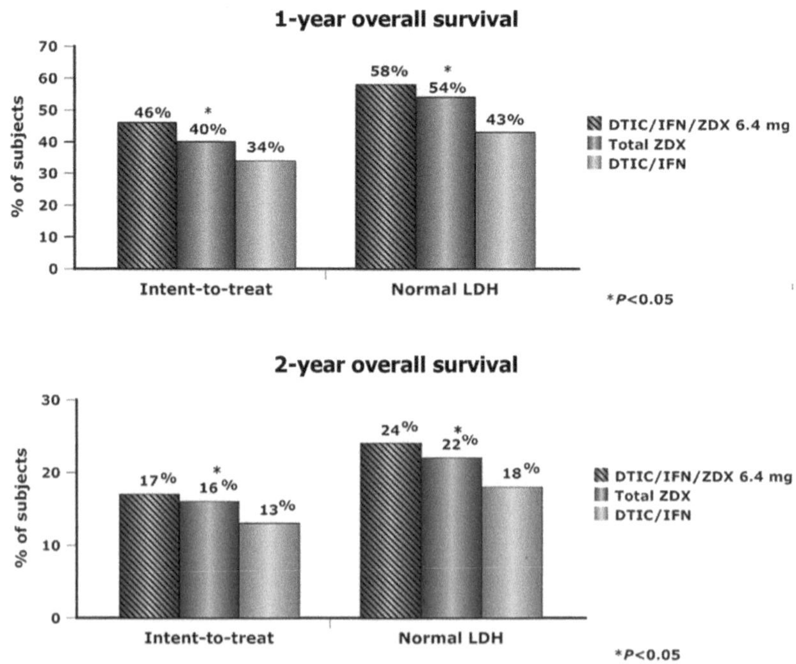

Fig. 4 Tα1 improves survival in Stage IV melanoma. (See the color plate.)

response (from a range of only 4.4–8.4 months in the control, to a range of 1.9–23.2 for Ta1-treated subjects). The clinical benefit rates were significantly higher for subjects given 3.2 mg Ta1 plus DTIC compared to control ($p=0.009$), and no additional toxicity was seen with the addition of Ta1.

4.3.2 Hepatocellular Carcinoma

HCC is a primary malignancy of the liver affecting over 500,000 people worldwide. It is the third leading cause of cancer deaths worldwide with the highest incidence in Asia and Africa due to the high prevalence of HBV and HCV (El-Serag & Rudolph, 2007). Although surgical resection or liver transplant may benefit some patients, due to the advanced stage of cancer and liver disease at initial diagnosis, surgery may not be a viable option. Standard treatments for patients who cannot receive a liver transplant involve radiation and chemotherapies. Systemic chemotherapy results are at best dismal, although a number of reports have provided encouraging perspectives for regional chemotherapy. Transcatheter arterial chemoembolization (TACE) is a combination of regional chemotherapy and some form of hepatic artery occlusion. Consistently higher response rates have been reported for TACE when compared with systemic chemotherapy. For patients who have either failed TACE or who present with more advanced HCC, sorafenib (a multikinase inhibitor with activity against Raf-1, B-Raf, vascular endothelial growth factor receptor 2, platelet-derived growth factor receptor, and c-Kit receptors, among other kinases) is now considered first-line treatment and has shown a clinically relevant improvement in time to progression and in survival.

Based on the immune-modulating mechanism of action of Ta1, it is expected to be of benefit when used in combination with either sorafenib or TACE in treatment of HCC, with no addition of toxicity. Pilot studies were conducted to evaluate the benefit of Ta1 after TACE, which demonstrated significant improvement in survival compared to historical controls at 6, 9, and 12 months after TACE with or without Ta1 (91% compared to 77%, 88% compared to 58%, and 78% compared to 46%, respectively; $p<0.05$) (Li, 2001; Stefanini, Foschi, Castelli, et al., 1998; Zhang, 2000). Significantly increased cytotoxic T cells, CD3 cells, CD4/CD8 ratio, and NK cells were also seen.

Further controlled studies confirmed these results, some showing only a trend toward survival (Cheng et al., 2005b; Gish, Gordon, Nelson, Rustgi, & Rios, 2009). But one study showed significantly improved survival (10 vs 7 months; $p=0.002$) and delayed time to tumor recurrence

($p=0.039$) compared to TACE treatment controls (Cheng et al., 2004a, 2004b).

Subjects with HCC and concurrent HBV infection were evaluated for the addition of Tα1 treatment to lamivudine after tumor resection or partial hepatectomy (Cheng et al., 2005a, 2006; Cheng, Ding, et al., 2005). In this study, the addition of Tα1/lamivudine combination treatment resulted in statistically significant increases in time to tumor recurrence (10.0 vs 6.5 months; $p=0.0032$), median survival time (12.5 vs 6.0 months; $p=0.0023$), and HBV-DNA clearance (100% vs 4%; $p=0.0000$) and seroconversion to HBeAg (73% vs 7.5%; $p<0.05$).

4.3.3 Non-Small Cell Lung Cancer

Lung cancer is the most common form of cancer worldwide and has the highest mortality rate (World Cancer Research Fund International, 2012). NSCLC is any type of epithelial lung cancer excluding small cell lung carcinoma. NSCLC accounts for approximately 85% of all lung cancers with more than half of the cases having already metastasized at time of diagnosis. NSCLC is typically treated by surgical resection due to the low efficacy of chemotherapy. Since NSCLC cannot be cured via current treatment modalities, palliative care is an important part of the standard treatment regimen.

Clinical studies have shown notable improvements after treatment with Tα1 in combination with chemotherapy or radiation, including statistically significantly improvement in survival time and survival rates compared to chemotherapy or radiation alone, even in subjects with stages III or IV cancers (Garaci et al., 1995; Ma, Lin, Mei, & Ping, 2006; Salvati, Rasi, Portalone, Antilli, & Garaci, 1996; Schulof, Lloyd, Cleary, et al., 1985). As expected, these studies also demonstrated increases in immune cell numbers and function, and decreased toxicity from chemotherapy.

4.3.4 Quality of Life in Cancer

A large number of studies conducted with Tα1 in various cancers have shown improvements in QOL, as expected from an agent that counteracts the immune suppression that results from the cancers themselves, the chemotherapy utilized, and other confounding factors such as age.

In NSCLC, the Karnovsky performance score (KPS) measurement of QOL was increased from about 30–60% ($p<0.05$) (Hou, 2007; Liang, Zhou, & Yang, 2010), and immune parameters were significantly improved (Shi, Ding, & Yang, 2007; Sun, Gao, Liu, & Li, 2009). The significant improvements in side effects evaluated included an abrogation of decreases in white blood cell counts and platelets, and decreases in the number of subjects experiencing nausea and vomiting.

Subjects with lung, gastric, or breast cancer were given Tα1 treatment during one of two cycles of chemotherapy, and QOL was evaluated. Tα1 provided a significant improvement in all measures of QOL (appetite, sleep, fatigue, daily activity, overall feeling, and depression), lowered occurrences of infections during chemotherapy, increased immune parameters (Chen et al., 2000; Luna et al., 2000), and decreased neurotoxicity (An, Liu, Fang, & Wu, 2004; Liu & An, 2003). Elderly subjects with advanced malignant cancer (lung, liver, stomach, colon, and rectal) also showed significant improvement in KPS and immune parameters (Yang, Lu, & Huang, 2003), and a second study in the same patient population also showed a significant increase in survival (median survival improved from 6 to 24 months; $p < 0.01$); the effect was so pronounced that at 24 months, when there was a 43.3% survival in the Tα1-treated group, the control group of subjects in the study had no survivors at all (Wang, Zhen, et al., 2010).

Improvements in QOL have also been observed in treatment of subjects with pancreatic cancer (Ni et al., 2001) and postmenopausal women undergoing treatment for breast cancer with aromatase inhibitors who had joint symptoms and pain related to the treatment (Zhang, Tang, & Zhao, 2010). In the latter study, subjects treated with Tα1 showed significant improvement in the Western Ontario and McMaster Universities Osteoarthritis Index ($p < 0.001$) and Functional Assessment of Cancer Therapy-General physical well-being ($p < 0.001$), and reported statistically significant improvement in worst pain scores (decreasing from a score of 5.7 to 3.4; $p < 0.001$), pain severity (from 3.9 to 2.9; $p < 0.01$), and pain-related functional interference (4.2 to 1.8; $p < 0.001$) using the brief pain inventory scale.

4.4 Clinical Studies in Vaccine Enhancement

Immune senescence, a normal aging process, has been related to a gradual decline in thymus function and thymic hormone production. The lack of thymic hormones may contribute to the decline in immune function, particularly the T-cell component (Makinodan & Peterson, 1962; Rosenberg et al., 1985; Wade & Szewczuk, 1984). In the elderly, for example, analysis of a specific antibody response after vaccination has been shown to be impaired when compared with response in young subjects (Bramwell, Tsakiris, Briggs, et al., 1985).

Decreased antibody response to T-cell-dependent antigens, particularly in the elderly, may be one factor that accounts for insufficient efficacy of certain vaccination programs (eg, influenza). Diminished antibody responses have also been reported in patients with end-stage renal disease (ESRD). The evidence for impairment of cell-mediated immunity in hemodialysis patients has been attributed to incompetence in T-cell-mediated immune responses (Dammin, Couch, & Murray, 1957; Lawrence, 1965; Revie, Shen, Ordonez, et al., 1985; Sanders, Luby, Sanford, & Hull, 1971). Several studies have reported poor antibody response after hepatitis B vaccination in hemodialysis patients (Crosnier, Jungers, Courouc, et al., 1981; Grob, Binswanger, Zaruba, et al., 1983; Stevens, Alter, Taylor, Zang, et al., 1984).

Tα1 had been shown to improve T-cell-dependent specific antibody production in animal models (Effros et al., 1988; Ershler et al., 1985); in vitro studies also demonstrated that influenza antibody synthesis was augmented in cells from vaccinated elderly individuals (Ershler et al., 1984).

Clinical studies have evaluated the benefit of Tα1 as an adjuvant for influenza and hepatitis B antiviral vaccines in subjects immunocompromised due to age or hemodialysis. When compared with vaccine plus placebo, administration of Tα1 in conjunction with vaccine increased and sustained the specific antibody response, increased protection against illness, and overcame previous lack of specific antibody response and age-associated decline in specific antibody response (Carraro et al., 2012; Gravenstein, Duthie, Miller, et al., 1989; Gravenstein, Ershler, Durmaskin, Schwab, & Weksler, 1986; McConnell et al., 1989; Shen, Corteza, Josselson, et al., 1990; Shen, Josselson, McRoy, Sadler, & Chretien, 1987a, 1987b). These studies also confirm that Tα1 is safe for administration to immunocompromised subjects, as no serious adverse effects were observed in any of the studies.

The most recent study evaluated the addition of Tα1 as an enhancer of the immunogenicity of the 2009 H1N1 monovalent vaccine (Focetria®, Novartis) in adults with ESRD on chronic dialysis (Carraro et al., 2012). Tα1 was given twice, 1 week before and on the day of vaccination, and was administered at two different doses (40 subjects received 3.2 mg and 42 subjects received 6.4 mg). Results show that subjects who were treated with either dose of Tα1 achieved a marked and significant increase in their antibody titers compared to placebo; the percent of subjects who seroconverted by 21 days after vaccination was 88% and 89% for the Tα1-treated subjects, respectively, but only 53% for the placebo group ($p < 0.01$). The percent of subjects who were seroprotected by 21 days after vaccination were 94% and 93%, vs 77%, respectively, for these treatment groups. The positive effects from Tα1 were maintained at 42 days after vaccination, although by 84 days the differences were smaller and only the higher dose of Tα1 showed a continued improvement over vaccine alone.

5. CONCLUSIONS AND FUTURE DIRECTIONS

Immune suppression occurs during aging, and is a hallmark of infectious disease and cancer. The cell-signaling pathways involved are particular targets of Tα1's mechanism of action, and can explain the utility of Tα1 shown for treatment of immune-sensitive diseases, including both acute and chronic viral, bacterial, and fungal infections, multiple types of cancers, and vaccine enhancement. For example, Tα1 has shown promise in treatment of severe sepsis, and in clinical studies monotherapy with Tα1 was as effective as or better than IFN in treatment of HBV, especially in populations with low response rates to IFN (eg, vertically transmitted HBV; previous nonresponders to other therapy; immune tolerant patients with high HBV-DNA levels). For treatment of cancer, Tα1 has shown utility in clinical trials in melanoma, HCC, and NSCLC, in which the addition of Tα1 to treatment regimen has led to significantly increased survival. Tα1 can decrease toxicity from chemotherapy and significantly improve QOL in cancer patients. Tα1 improves response to vaccines, as shown in elderly persons and subjects with end-stage renal disease. Patients receiving Tα1 report few serious drug-related toxicities during treatment, even in combination with other agents, making Tα1 particularly attractive for evaluation in immune-depressed patients from all causes.

ACKNOWLEDGMENTS

Robert King is an employee, and Cynthia Tuthill is a consultant, for SciClone Pharmaceuticals.

REFERENCES

Acharya, U. H., & Jeter, J. M. (2013). Use of ipilimumab in the treatment of melanoma. *Clinical Pharmacology*, *5*, 21–27.

American Cancer Society. (2009). *Cancer facts and figures 2009*. Atlanta, GA: American Cancer Society.

An, T.-T., Liu, X.-Y., Fang, J., & Wu, M.-N. (2004). Primary assessment of treatment effect of thymosin alpha 1 on chemotherapy-induced neurotoxicity. *Chinese Journal of Cancer*, *23*(11 Suppl.), 1428–1430.

Angus, D. C., & van der Poll, T. (2013). Severe sepsis and septic shock. *New England Journal of Medicine, 369*(9), 840–851.
Arrighi, J.-F., Rebsamen, M., Rousset, F., Kindler, V., & Hauser, C. (2001). A critical role for p38 mitogen-activated protein kinase in the maturation of human blood-derived dendritic cells induced by lipopolysaccharide, TNF-q, and contact sensitizers. *Journal of Immunology, 166*, 3837–3845.
Balch, C. M., Reintgen, D. S., Kirkwood, J. M., Houghton, A., Peters, L., & Aug, K. K. (1997). Cutaneous melanoma. In V. DeVita, S. Hellman, & S. Rosenberg (Eds.), *Cancer: Principles and practice of oncology* (5th ed., pp. 1947–1994). Philadelphia, PA: Lippincott Williams & Wilkins.
Baumann, C. A., Badamchian, M., & Goldstein, A. L. (1997). Thymosin alpha 1 antagonizes dexamethasone and CD3-induced apoptosis of CD4+CD8+ thymocytes through the activation of cAMP and protein kinase C dependent second messenger pathways. *Mechanics of Ageing and Development, 94*, 85–101.
Bistoni, F., Marconi, P., Frati, L., Bonmassar, E., & Garaci, E. (1982). Increase of mouse resistance to Candida albicans infection by thymosin alpha 1. *Infection and Immunity, 36*, 609–614.
Bozza, S., Gaziano, R., Bonifzi, P., et al. (2007). Thymosin alpha 1 activates the TLR9/MyD88/IRF7-dependent murine cytomegalovirus sensing for induction of anti-viral responses in vivo. *International Immunology, 19*, 1261–1270.
Bramwell, S. P., Tsakiris, D. J., Briggs, J. D., et al. (1985). Dinitrochlorobenzene skin testing predicts response to hepatitis B vaccine in dialysis patients. *Lancet, 1*, 1412–1415.
Burkholder, B., Huang, R.-Y., Burgess, R., Luo, S., Jones, V. S., Zhang, W., et al. (1845). Tumor-induced perturbations of cytokines and immune cell networks. *Biochimica et Biophysica Acta, 2014*, 182–201.
Busca, A., & Kumar, A. (2014). Innate immune responses in hepatitis B infection. *Virology Journal, 11*, 22.
Carraro, G., Naso, A., Montomoli, E., Gasparini, R., Camerini, R., Panatoo, D., et al. (2012). Thymosin alpha 1 (ZADAXIN) enhances the immunogenicity of an adjuvanted pandemic H1N1v influenza vaccine (Focetria) in hemodialyzed patients: A pilot study. *Vaccine, 30*, 1170–1180.
Chan, H. L., Tang, J. L., & Sung, J. Y. (2001). Thymosin a1 for the treatment of chronic hepatitis B virus (HBV) infection: A meta-analysis. In *Paper presented at the digestive disease week 2001; May 20–23, Atlanta, GA*.
Chen, J. (2007). Effects of thymosin α1 on cell immunity function in patients with septic shock. *Zhongguo Wei Zhong Bing Ji Jiu Yi Xue, 19*, 153–155.
Chen, J., Huang, F. L., Zheng, X. L., Chen, L. X., Qin, S. K., Li, G. F., et al. (2000). Thymosin alpha 1 positively alters quality of life in chemotherapy treated patients. In *Annual ASCO proceedings abstract 2450*.
Cheng, S., Ding, G., Shi, J., Guo, W., Zhao, Y., Sheng, L., et al. (2005). Role of antivirus therapy in treatment of hepatocellular carcinoma with chronic hepatitis B infection. *Chinese-German Journal of Clinical Oncology, 4*(6), 330–333.
Cheng, S., Wu, M., Chen, H., Chen, H., Shen, F., Yang, J., et al. (2006). Antiviral therapy using lamivudine and thymosin alpha 1 for hepatocellular carcinoma coexisting with chronic hepatitis B infection. *Hepato-Gastroenterology, 53*, 249–252.
Cheng, S., Wu, M., Chen, H., Shen, F., Yang, J., Cong, W., et al. (2004a). Combination transcatheter hepatic arterial chemoembolization with thymosin alpha 1 on recurrence prevention of hepatocellular carcinoma. *Hepato-Gastroenterology, 51*, 1445–1447.
Cheng, S., Wu, M., Chen, H., Shen, F., Yang, J., Cong, W., et al. (2004b). Transcatheter hepatic arterial chemoembolization and thymosin in postoperative treatment of hepatocellular carcinoma. *Chinese Journal of Oncology, 26*(5), 305–307.

Cheng, S., Wu, M., Chen, H., Shen, F., Yang, J., Cong, W., et al. (2005a). Anti-viral therapy using lamivudine and thymosin is helpful to prevent recurrence in hepatocellular carcinoma with coexisting active hepatitis B. *Chinese Journal of Oncology*, *27*(2), 114–116.

Cheng, S., Wu, M., Chen, H., Shen, F., Yang, J., Cong, W., et al. (2005b). Hepatocellular carcinoma with tumor thrombus in the portal vein: a comparison of therapeutic effects by different treatments. *Chinese Journal of Oncology*, *27*(3), 183–185.

Crosnier, J., Jungers, P., Courouc, A. M., et al. (1981). Randomised placebo-controlled trial of hepatitis B surface antigen vaccine in French haemodialysis units: I, medical staff. *Lancet*, *1*, 455–459.

Cursaro, C., Margotti, M., Favarelli, L., et al. (1998). Thymosin alpha 1 (Ta1) plus interferon a (IFNa) enhance the immune and antiviral response of patients with hepatitis C virus infection. *Hepatology*, *28*(4), 361A.

D'Agostini, C., Palamara, A. T., Favalli, C., et al. (1996). Efficacy of combination therapy with amantadine, thymosin alpha 1 and alpha/beta interferon in mice infected with influenza A virus. *International Journal of Immunopharmacology*, *18*, 95–102.

Dammin, G., Couch, N., & Murray, J. (1957). Prolonged survival of skin homografts in uremic patients. In *Paper presented at the second tissue homotransplantation conference, New York, NY.*

Di Francesco, P., Gaziano, R., Casalinuovo, I. A., Belogi, L., Palamara, A. T., & Favalli, C. (1994). Combined effect of fluconazole and thymosin alpha 1 on systemic candidiasis in mice immunosuppressed by morphine treatments. *Clinical and Experimental Immunology*, *97*, 347–352.

Effros, R. B., Casillas, A., & Walford, R. L. (1988). The effect of thymosin alpha 1 on immunity to influenza in aged mice. In *Aging: Immunology and infectious disease: Vol. 1.* (pp. 31–40). New Rochelle, NY: Mary Ann Liebert, Inc.

El-Serag, H. B., & Rudolph, K. L. (2007). Hepatocelluar carcinoma: Epidemiology and molecular carcinogenesis. *Gastroenterology*, *132*(7), 2557–2576.

Ershler, W., Hebert, J., Blow, A., Granter, S., & Lynch, J. (1985). Effect of thymosin alpha one on specific antibody response and susceptibility to infection in young and aged mice. *International Journal of Immunopharmacology*, 7, 465–471.

Ershler, W., Moore, A., & Socinski, M. (1984). Influenza and aging: Age-related changes and the effects of thymosin on the antibody response to influenza vaccine. *Journal of Clinical Immunology*, *4*, 445–454.

Garaci, E., Lopez, M., Bonsignore, G., Della Giulia, M., D'Aprile, M., Favalli, C., et al. (1995). Sequential chemoimmunotherapy for advanced non-small cell lung cancer using cisplatin, etoposide, thymosin alpha 1 and interferon alfa2a. *European Journal of Cancer*, *31A*(13/14), 2403–2405.

Garaci, E., Mastino, A., Pica, F., & Favalli, C. (1990). Combination treatment using thymosin alpha 1 and interferon after cyclophosphamide is able to cure Lewis lung carcinoma in mice. *Cancer Immunology, Immunotherapy*, *32*, 154–160.

Garaci, E., Pica, F., Mastino, A., Palamara, A. T., Belardelli, F., & Favalli, C. (1993). Antitumor effect of thymosin alpha-1 interleukin-2 or thymosin a-1/interferon ab following cyclophosphamide in mice injected with highly metastatic Friend erythroleukemia cells. *Journal of Immunotherapy*, *13*, 7–17.

Garaci, E., Rocchi, G., Perroni, L., et al. (1994). Combination treatment with zidovudine, thymosin alpha 1 and interferon-alpha in human immunodeficiency virus infection. *International Journal of Clinical & Laboratory Research*, *24*(1), 23–28.

Gerin, J. L., Korba, B. E., Cote, P. J., & Tennant, B. C. (1992). A preliminary report of a controlled study of thymosin alpha 1 in the woodchuck model of hepadnavirus infection. In T. Block (Ed.), *Innovations in antiviral development and the detection of virus infection* (pp. 121–123). Philadelphia, PA: Jefferson Medical.

Gish, R. G., Gordon, S. C., Nelson, D., Rustgi, V., & Rios, I. (2009). A randomized controlled trial of thymalfasin plus transarterial chemoembolization for unresectable hepatocellular carcinoma. *Hepatology International*, *3*(3), 480–489.

Goldstein, A. L., Low, T. L., McAdoo, M., et al. (1977). Thymosin alpha1: Isolation and sequence analysis of an immunologically active thymic polypeptide. *Proceedings of the National Academy of Sciences of the United States of America*, *74*(2), 725–729.

Gramenzi, et al. (2005). *In vitro effect of thymosin alpha 1 and interferon alpha on Th1 and Th2 cytokine synthesis in patients with HBEAg-negative chronic hepatitis B*. Alexandria, VA: American Association for the Study of Liver Disease.

Gravenstein, S., Duthie, E. H., Miller, B. A., et al. (1989). Augmentation of influenza antibody response in elderly men by thymosin alpha one. A double-blind placebo-controlled clinical study. *Journal of American Geriatrics Society*, *37*, 1–8.

Gravenstein, S., Ershler, W. B., Drumaskin, S., Schwab, R., & Weksler, M. E. (1986). Anti-influenza antibody response: Augmentation in elderly-non-responders by thymosin alpha 1. *Gerontologist*, *26*, 150A.

Grob, P., Binswanger, U., Zaruba, K., et al. (1983). Immunogenicity of a hepatitis B subunit vaccine in hemodialysis and in renal transplant recipients. *Antiviral Research*, *3*, 43–52.

Gupta, S., Aggarwal, S., & Nguyen, T. (1997). Accelerated spontaneous programmed cell death in lymphocytes in DiGeorge syndrome. *The Journal of Allergy and Clinical Immunology*, *99*(1), S3(#11).

Hadden, J. W., Saha, A., Sosa, M., & Hadden, E. M. (1995). Immunotherapy with natural interleukins and/or thymosin alpha 1 potently augments T-lymphocyte responses of hydrocortisonetreated aged mice. *International Journal of Immunopharmacology*, *17*, 821–828.

Haritos, A. A., Goodall, G. J., & Horecker, B. L. (1984). Prothymosin alpha: Isolation and properties of the major immunoreactive form of thymosin alpha 1 in rat thymus. *Proceedings of the National Academy of Sciences of the United States of America*, *81*, 1008–1011.

Ho, A. D., Stehle, B., Dietz, G., Hunstein, W., & Hoffbrand, A. V. (1985). Terminal differentiation of cord blood lymphocytes induced by thymosin fraction 5 and thymosin alpha 1. *Scandinavian Journal of Immunology*, *21*(3), 221–225.

Hotchkiss, R. S., & Karl, I. E. (2003). The pathophysiology and treatment of sepsis. *New England Journal of Medicine*, *348*, 138–150.

Hou, X. (2007). Clinical effects of advanced patients with non-small cell lung cancer treatment by thymosin alpha 1 combined with NP regimens. *Journal of Bethune Military Medical College*, *5*(5), 272–274.

Hsia, J., Sarin, N., Oliver, J. H., & Goldstein, A. L. (1989). Aspirin and thymosin increase interleukin-2 and interferon-gamma production by human peripheral blood lymphocytes. *Immunopharmacology*, *17*(3), 167–173.

Hu, S., Low, T., & Goldstein, A. (1981). Modulation of terminal deoxynucleotidyl transferase activity by thymosin. *Molecular and Cellular Biochemistry*, *41*, 49–58.

Huang, D.-P., Yang, M., Pen, W.-P., et al. (2006). Prevention and management of lung infections with thymosin alpha 1 in critical patients with tracheotomy. *Journal of Southern Medical University*, *26*(11), 128–129.

Iijima, N., Yanagawa, Y., & Onoe, K. (2003). Role of early- or late-phase activation of p38 mitogen-activated protein kinase induced by tumor necrosis factor-a or 2,4-dinitrochlorobenzene during maturation of murine dendritic cells. *Immunology*, *110*, 322–328.

Ishitsuka, H., Umeda, Y., Nakamura, J., & Yagi, Y. (1983). Protective activity of thymosin against opportunistic infections in animal models. *Cancer Immunology, Immunotherapy*, *14*, 145–150.

Jevremovic, M., Kartaljevic, G., Jelusic, V., Vodnik, T., Pesic, M., & Filipovic, S. (1997). Determination of thymosin alpha 1 with enzyme-immunoassay in colorectal cancer patients. *Archive of Oncology*, *5*, 193–194.

Ji, S.-M., Li, L.-S., Sun, Q.-Q., et al. (2007). Immunoregulation of thymosin alpha 1 treatment of cytomegalovirus infection accompanied with acute respiratory distress syndrome after renal transplantation. *Transplantation Proceedings, 39,* 115–119.

John, L. B., Westwood, J. A., Darcy, P. K., & Kershaw, M. H. (2013). Immune modulation of the tumor microenvironment for enhancing cancer immunotherapy. *Oncoimmunology, 2,* e25961.

King, R. S., & Tuthill, C. (2015). Evaluation of thymosin alpha 1 (Ta1) in nonclinical models of the immune-suppressing indications melanoma and sepsis. *Expert Opinion on Biological Therapy, 15*(Suppl. 1), S41–S49.

Knutsen, A. P., Freeman, J. J., Mueller, K. R., Roodman, S. T., & Bouhasin, J. D. (1999). Thymosin-alpha1 stimulates maturation of CD34+ stem cells into CD3+4+ cells in an in vitro thymic epithelia organ coculture model. *International Journal of Immunopharmacology, 21*(1), 15–26.

LaRosa, S. P. (2002). Sepsis: Menu of new approaches replaces one therapy for all. *Cleveland Clinic Journal of Medicine, 69*(1), 65–73.

Lau, G., Yuen, S., Kwok, A., Lai, S., Lim, W., & Lam, S. (1999). Six-months follow-up on a 26 week trial of thymosin alpha 1 plus famciclovir in the treatment of Chinese immune tolerant adult patients with chronic hepatitis B. *Gastroenterology, 116*(4), Abstract no. L0251.

Lawrence, H. S. (1965). Uremia: Nature's immunosuppressive device. *Annals of Internal Medicine, 62,* 166–170.

Leichtling, K. D., Serrate, S. A., & Sztein, M. B. (1990). Thymosin alpha 1 modulates the expression of high affinity interleukin-2 receptors on normal human lymphocytes. *International Journal of Immunopharmacology, 12,* 19–29.

Li, Z. S. (2001). HCC patients treated with TACE combined with thymalfasin: One year follow up. In *Paper presented at the Shanghai International Oncology Conference.*

Li, C., Wang, C.-H., Meng, Q.-H., Ye, S.-L., Wang, X.-J., & Jiang, C. (2007). Effect of the thymosin alpha 1 on immune function in aged chronic obstructive pulmonary disease during acute period. *Chinese Journal of Hospital Pharmacy, 27*(5), 637–639.

Li, P., Xu, L.-H., Zhang, Q., et al. (2007). Treatment of drug-resistant Pseudomonas auroginosa pneumonia in elderly patients by using thymosin alpha 1 with sulperazone. *Chinese Journal of Nosocomial, 17,* 1271–1273.

Li, Y. N., Zhou, L. X., Qiang, X. H., et al. (2009). Effect of continuous blood purification and thymosin alpha 1 on the cellular immunity in patients with severe sepsis: A prospective, randomized, controlled clinical trial. *Zhongguo Wei Zhong Bing Ji Jiu Yi Xue, 21*(3), 139–142.

Liang, G., Zhou, Z., & Yang, J. (2010). Combination of thymosin alpha 1 with low-dose gemcitabine for advanced on-small cell lung cancer in elderly patients. *Journal of Practical Oncology, 25*(4), 470–473.

Lim, S. G., Wai, C. T., Lee, Y. M., et al. (2006). A randomized, placebo-controlled trial of thymosin-alpha1 and lymphoblastoid interferon for HBeAg-positive chronic hepatitis B. *Antiviral Therapy, 11*(2), 245–253.

Liu, X., & An, T. (2003). Thymalfasin for the protection of chemotherapy-induced neurotoxicity: A pilot study. *Proceedings of ASCO, 22,* 3153.

Lopez, M., Carpano, S., Cavaliere, R., DiLauro, L., Ameglio, F., & Vitelli, G. (1994). Biochemotherapy with thymosin alpha 1, interleukin-2 and dacarbazine in patients with metastatic melanoma: Clinical and immunological effects. *Annals of Oncology, 5,* 741–746.

Low, T. L. K., & Goldstein, A. L. (1979). The chemistry and biology of thymosin II. Amino acid sequence analysis of thymosin alpha 1 and polypeptide beta 1. *Journal of Biological Chemistry, 254,* 987–995.

Low, T. L. K., & Goldstein, A. L. (1985). Thymosin alpha 1 and polypeptide beta 1. *Methods in Enzymology, 116,* 233–248.

Low, T. L., Thurman, G. B., McAdoo, M., et al. (1979). The chemistry and biology of thymosin. I. Isolation, characterization, and biological activities of thymosin alpha1 and polypeptide beta1 from calf thymus. *Journal of Biological Chemistry*, *254*(3), 981–986.
Luna, G. C., Naval, G. R., Gorospe, A. D., Daep, M., Jurilla, M., Batac, E. N., et al. (2000). The effect of thymosin alpha on the cellular immune profile of cancer patients receiving chemotherapy. *Journal of Immunotherapy*, *23*(5), 588.
Ma, D., Lin, D., Mei, L., & Ping, Y. (2006). The use of thymosin alpha 1 in advanced non-small cell lung cancer treatment. *Shangdong Pharmaceutical*, *46*(14), 77.
Maio, M., Mackiewicz, A., Testori, A., Trefzer, U., Ferraresi, V., Jassem, J., et al. (2010). Large randomized study of thymosin alpha 1, interferon alfa, or both in combination with dacarbazine in patients with metastatic melanoma. *Journal of Clinical Oncology*, *28*, 1780–1787.
Makinodan, T., & Peterson, W. (1962). Relative antibody-forming capacity of spleen cells as a function of age. *Proceedings of the National Academy of Sciences of the United States of America*, *48*, 234–238.
Mastino, A., Favalli, C., Grelli, S., et al. (1992). Combination therapy with thymosin alpha 1 potentiates the antitumor activity of interleukin-2 with cyclophosphamide in the treatment of the Lewis lung carcinoma in mice. *Cancer Research*, *50*, 493–499.
McConnell, L., Gravenstein, S., Roecker, E., Spencer, S., Simon, G., & Erschler, W. (1989). Augmentation of influenza antibody levels and reduction in attack rates in elderly subjects by thymosin alpha 1. *The Gerentologist*, *29*, 188A.
Moody, T., Badamchian, M., & Goldstein, A. (1998). Thymosin alpha 1 prevents lung carcinogenesis. *FASEB Journal*, *12*, A1457.
Moody, T. W., Fagarasan, M., Zia, F., Cesnjaj, M., & Goldstein, A. L. (1993). Thymosin alpha 1 down-regulates the growth of human non-small cell lung cancer cells in vitro and in vivo. *Cancer Research*, *53*, 5214–5218.
Moody, T. W., Leyton, J., Zia, F., Tuthill, C., Badamchian, M., & Goldstein, A. L. (2000). Thymosin alpha 1 is chemopreventive for lung adenoma formation in A/J mice. *Cancer Letters*, *155*, 121–127.
Moody, T., Tuthill, C., Badamchian, M., & Goldstein, A. (2002). Thymosin alpha 1 inhibits mammary carcinogenesis in fisher rats. *Peptides*, *23*, 1011–1014.
Ni, Q., Fu, D., Yu, X., Xu, J., Hua, Y., Chen, W., et al. (2001). Efficacy of thymalfasin on cellular immune function and chemotherapy induced toxicity in pancreatic cancer. In *37th annual ASCO proceedings*: Vol. 20(2), (p. 2630).
Ni, Chao, Wu, Pin, Wu, Xianguo, Zhang, Ting, Zhang, Tao, Wang, Zhen, et al. (2015). Thymosin alpha1 enhanced cytotoxicity of iNKT cells against colon cancer via upregulating CD1d expression. *Cancer Letters*, *356*(2 Pt B), 579–588.
Oates, K. K., & Erdos, M. (1989). Biochemical identification of thymosin alpha 1: Its phylogenetic distribution and evolutionary implications. *Comparative Biochemistry and Physiology. B*, *94*(4), 759–763.
Ohta, Y., Sueki, K., Yoneyama, Y., Tezuka, E., & Yagi, Y. (1983). Immunomodulating activity of thymosin fraction 5 and thymosin alpha 1 in immunosuppressed mice. *Cancer Immunology, Immunotherapy*, *15*, 108–113.
Ohta, Y., Tezuka, E., Tamura, S., & Yagi, Y. (1985). Thymosin alpha 1 exerts protective effect against the 5-FU induced bone marrow toxicity. *International Journal of Immunopharmacology*, *7*, 761–768.
Osheroff, P. L. (1981). The effect of thymosin on glucocorticoid recepters in lymphoid cells. *Cellular Immunology*, *60*, 376–385.
Palamara, A., Bue, M., Savini, P., et al. (1998). Thymosin alpha 1 inhibits Sendai virus replication: Involvement of intracellular redox state. In *Paper presented at the 6th international expert forum of immunotherapy and gene therapy*.
Peng, Y., Chen, Z., Yu, W., et al. (2008). Effects of thymic polypeptides on the thymopoiesis of mouse embryonic stem cells. *Cell Biology International*, *32*, 1265–1271.

Perruccio, K., Bonifazi, P., Topini, F., et al. (2010). Thymosin alpha 1 to harness immunity to pathogens after haploidentical hematopoietic transplantation. *Annals of the New York Academy of Sciences*, *1194*, 153–161.
Pica, F., Fraschetti, M., Matteucci, C., Tuthill, C., & Rasi, G. (1998). High doses of thymosin alpha 1 enhance the anti-tumor efficacy of combination chemo-immunotherapy for murine B16 melanoma. *Anticancer Research*, *18*, 3571–3578.
Rasi, G., Silecchia, G., Sinibaldi-Vallebona, P., et al. (1994). Anti-tumor effect of combined treatment with thymosin alpha 1 and interleukin-2 after 5-fluorouracil in liver metastases from colorectal cancer in rats. *International Journal of Cancer*, *57*, 701–705.
Rasi, G., Terzoli, E., Izzo, F., et al. (2000). Combined treatment with thymosin alpha 1 and low-dose interferon alpha after dacarbazine in advanced melanoma. *Melanoma Research*, *10*, 189–192.
Revie, D., Shen, S., Ordonez, J., et al. (1985). T-cell subsets and status of hepatitis B surface antigen and antibody in end-stage renal disease patients. *Kidney International*, *27*, 150.
Richards, J., Gale, D., Mahta, N., & Lestingi, T. (1999). Combination of chemotherapy with interleukin-2 and interferon alfa for treatment of metastatic melanoma. *Journal of Clinical Oncology*, *17*, 651–657.
Romani, L., Aversa, F., Garaci, E., & Velardi, A. (2012). Thymosin alpha 1 improves immune reconstitution in hematopoietic transplantation. *Annals of the New York Academy of Sciences*, *1269–1270*, 83–84.
Romani, L., Bistoni, F., Gaziano, R., et al. (2004). Thymosin alpha 1 activates dendritic cells for antifungal Th1 resistance through toll-like receptor signaling. *Blood*, *103*(11), 4232–4239.
Romani, L., Bistoni, F., Montagnoli, C., et al. (2007). Thymsoin alpha 1: An endogenous regulator of inflammation, immunity, and tolerance. *Annals of the New York Academy of Sciences*, *1112*, 326–338.
Romani, L., Bistoni, F., Perruccio, K., et al. (2006). Thymosin alpha1 activates dendritic cell tryptophan catabolism and establishes a regulatory environment for balance of inflammation and tolerance. *Blood*, *108*(7), 2265–2274.
Romani, L., Moretti, S., Fallarino, F., Bozza, S., et al. (2012). Jack of all trades: Thymosin alpha 1 and its pleiotropy. *Annals of the New York Academy of Sciences*, *1269–1270*, 5–10.
Rosenberg, S. A., Mul, J. J., Spiess, P. J., Reichert, C. M., & Schwarz, S. L. (1985). Regression of established pulmonary metastases and subcutaneous tumor mediated by the systemic administration of high dose recombinant interleukin 2. *Journal of Experimental Medicine*, *161*, 1169–1188.
Salvati, F., Rasi, G., Portalone, L., Antilli, A., & Garaci, E. (1996). Combined treatment with thymosin alpha 1 and low-dose interferon-alpha after ifosfamide in non-small cell lung cancer: A phase II controlled trial. *Anticancer Research*, *16*, 1001–1004.
Sanders, C. V., Jr., Luby, J. P., Sanford, J. P., & Hull, A. R. (1971). Suppression of interferon response in lymphocytes from patients with uremia. *Journal of Laboratory and Clinical Medicine*, *77*, 768–776.
Sarandeses, C. S., Covelo, G., Diaz-Jullien, C., & Freire, M. (2003). Prothymosin alpha is processed to thymosin alpha 1 and thymosin alpha 11 by a lysosomal asparaginyl endopeptidase. *Journal of Biological Chemistry*, *278*, 13286–13293.
Saruc, M., Ozden, N., Turkel, N., Ayhan, S., Hock, L., Tuzcouglu, I., et al. (2002). Long-term outcomes of thymosin alpha 1 and interferon alpha-2b combination therapy in patients with hepatitis B e antigen (HBeAg) negative chronic hepatitis B. *Journal of Pharmaceutical Sciences*, *92*(7), 1386–1395.
Saruc, M., Yuceyar, H., Kucukmetin, N., Demir, M. A., & Kandiloglu, A. R. (2002). Combination thymosin-alpha 1 and interferon-alpha 2b in the treatment of anti-HBe-positive chronic hepatitis B in Turkey. *Hepato-Gastroenterology*, *49*(45), 798–802.
Schulof, R. (1985). Thymic peptide hormones: Basic properties and clinical applications in cancer. *Critical Reviews in Oncology/Hematology*, *3*, 309–376.

Schulof, R. S., Lloyd, M. J., Cleary, P. A., et al. (1985). A randomized trial to evaluate the immunorestorative properties of synthetic thymosin alpha 1 in patients with lung cancer. *Journal of Biological Response Modifiers*, *4*, 147–158.

Serafino, A., Pica, F., Andreola, F., Gaziano, R., Moroni, N., Moroni, G., et al. (2014). Thymosin α1 activates complement receptor-mediated phagocytosis in human monocyte-derived macrophages. *Journal of Innate Immunity*, *6*(1), 72–88.

Serafino, A., Pierimarchi, P., Pica, F., et al. (2012). Thymosin alpha 1 as a stimulatory agent of innate cell-mediated immune response. *Annals of the New York Academy of Sciences*, *1269–1270*, 43–50.

Serrate, S., Schulof, R., Leondaridis, L., Goldstein, A. L., & Sztein, M. B. (1987). Modulation of human natural killer cell cytotoxic activity, lymphokine production, and interleukin 2 receptor expression by thymic hormones. *Journal of Immunology*, *139*, 2338–2343.

Shen, S. Y., Corteza, Q. B., Josselson, J., et al. (1990). Age-dependent enhancement of influenza vaccine responses by thymosin in chronic hemodialysis patients. In A. L. Goldstein (Ed.), *Biomedical advances in aging* (pp. 523–530). New York, NY: Plenum Press.

Shen, S., Josselson, J., McRoy, C., Sadler, J., & Chretien, P. (1987a). Effect of thymosin alpha 1 on heptavax-B vaccination among hemodialysis patients. *Kidney International*, *1987*, 217.

Shen, S., Josselson, J., McRoy, C., Sadler, J., & Chretien, P. (1987b). Effects of thymosin alpha 1 on peripheral T-cell and Heptavax-B vaccination in previously non-responsive hemodialysis patients. *Hepatology*, 7, 1120.

Sherman, K., Jones, C., Goldstein, A., & Naylor, P. (1991). Low thymosin alpha-1 concentrations in patients chronically infected with the hepatitis B virus. *Viral Immunology*, *4*, 195–199.

Shi, X., Ding, Q., & Yang, Q. (2007). Effects of domestic thymosin alpha 1 in combination with chemotherapy in treatment of advanced non-small cell lung cancer. *Practical Journal of Cardiac, Cerebral, Pneumal, and Vascular Diseases*, *15*(10), 764–766.

Shiau, A. L., Wu, C. L., & Huang, K. Y. (1988). The effect of thymosin on experimental herpes simplex virus infections. *Journal of the Formosan Medical Association*, *87*, 34–42.

Silecchia, G., Guarino, E., Sinibaldi-Vallebona, P., et al. (1999). Efficacy of repeated cycles of chemo-immunotherapy with thymosin alpha 1 and interleukin-2 after intraperitoneal 5-fluorouracil delivery. *Cancer Immunology, Immunotherapy*, *48*, 172–178.

Stefanini, G. F., Foschi, F. G., Castelli, E., et al. (1998). Alpha-1-thymosin and transcatheter arterial chemoembolization in hepatocellular carcinoma patients: A preliminary experience. *Hepato-Gastroenterology*, *45*, 209–215.

Stevens, C. E., Alter, H. J., Taylor, P. E., Zang, E. A., et al. (1984). Hepatitis B vaccine in patients receiving hemodialysis: Immunogenicity and efficacy. *New England Journal of Medicine*, *311*, 496–501.

Sun, X., Gao, F., Liu, X., & Li, G. (2009). The influence of thymosin alpha 1 on cellular immune function during chemotherapy for non-small cell lung cancer patients. *Shangdong Medical Journal*, *49*(31), 83.

Sun, Q., Liu, Z.-H., Chen, J., et al. (2006). An aggressive systematic strategy for acute respiratory distress syndrome caused by severe pneumonia after renal transplantation. *Transplant International: Official Journal of the European Society for Organ Transplantation*, *19*, 110–116.

Svedersky, L., Hui, A., May, L., McKay, P., & Stebbing, N. (1982). Induction and augmentation of mitogen-induced immune interferon production in human peripheral blood lymphocytes by Na-desacetylthymosin alpha 1. *European Journal of Immunology*, *12*, 244–247.

Sztein, M., & Serrate, S. (1989). Characterization of the immunoregulatory properties of thymosin alpha 1 on interleukin-2 production and interleukin-2 receptor expression in normal human lymphocytes. *International Journal of Immunopharmacology*, *11*, 789–800.

Sztein, M., Serrate, S., & Goldstein, A. (1986). Modulation of interleukin-2 receptor expression on normal human lymphocytes by thymic hormones. *Proceedings of the National Academy of Sciences of the United States of America*, *83*, 6107–6111.

Tennant, B. C., Korba, B. E., Baldwin, B. H., Goddard, L. A., Hornbuckle, W. E., & Cote, P. J. (1993). Treatment of chronic woodchuck hepatitis virus infection with thymosin alpha-1 (TA1). *Antiviral Research*, *20*(Suppl. 1), 163.

Tomazic, V., Sacasa, C., Loftus, A., Suter, C., & Elias, G. (1988). Thymic factor-induced reduction of pulmonary metastases in mice with FSA-1 fibrosarcoma. *Clinical & Experimental Metastasis*, *6*, 17–25.

Tuthill, C., DeRosa, A., Camerini, R., Rios, I., DeMarco, M. A., Donatelli, I., et al. (2010). The immunomodulatory peptide thymosin alpha 1 enhances response to influenza vaccine. In *National foundation for infectious diseases 13th annual meeting, Bethesda, MD.*

Umeda, Y., Sakamoto, A., Nakamura, J., Ishitsuka, H., & Yagi, Y. (1983). Thymosin alpha 1 restores NK-cell activity and prevents tumor progression in mice immunosuppressed by cytostatics or X-rays. *Cancer Immunology, Immunotherapy*, *15*(2), 78–83.

Wade, A. W., & Szewczuk, M. R. (1984). Aging, idiotype repertoire shifts, and compartmentalization of the mucosal-associated lymphoid system. *Advances in Immunology*, *36*, 143–188.

Wang, X., Li, W., Niu, C., Pan, L., Li, N., & Li, J. (2011). Thymosin alpha 1 is associated with improved cellular immunity and reduced infection rate in severe acute pancreatitis patients in a double-blind randomized control study. *Inflammation*, *34*(3), 198–202.

Wang, S. S., Makofske, R., Bach, A., & Merrifield, R. B. (1980). Automated solid phase synthesis of thymosin alpha 1. *International Journal of Peptide and Protein Research*, *15*, 1–4.

Wang, Y., Zhen, S., Qin, J., Zhang, Y., Yang, W., Fu, Y., et al. (2010). Curative effect analysis of thymosin alpha 1 in treating senile patients with advanced malignant cancer. *Journal of Military Surgeon in Southwest China*, *12*(5), 821–822.

Weissman, I. L., & Majeti, R. (2012). The CD47-SIRPalpha pathway in cancer immune evasion and potential therapeutic implications. *Current Opinion in Immunology*, *24*, 225–232.

Welch, R., Lee, H., Sokol, R., & Mutchnick, M. (1988). Amniotic fluid thymosin alpha 1 levels increase during gestation. *American Journal of Reproductive Immunology and Microbiology*, *17*, 96–97.

Welch, R., Mutchnick, M., Weller, F., & Sokol, R. (1987). Maternal and fetal circulating levels of thymosin alpha 1 during parturition. *American Journal of Reproductive Immunology and Microbiology*, *13*, 125–127.

Weller, F. E., Shah, U., Cummings, G. D., Chretien, P. B., & Mutchnick, M. G. (1992). Serum levels of immunoreactive thymosin alpha 1 and thymosin beta 4 in large cohorts of healthy adults. *Thymus*, *19*, 45–52.

World Cancer Research Fund International. Worldwide data, 2012.

World Health Organization, March 2015. Hepatitis B fact sheet number 204.

Wu, J., Zhou, L., Liu, J., et al. (2013). The efficacy of thymosin alpha 1 for severe sepsis (ETASS): A multicenter, single-blind, randomized and controlled trial. *Critical Care*, *17*, R8. http://dx.doi.org/10.1186/cc11932.

Xiang, X. S., Li, N., Zhao, Y. Z., Li, Q. R., & Li, J. S. (2014). Combination therapy with thymosin alpha1 and dexamethasone helps mice survive sepsis. *Inflammation*, *37*(2), 402–416.

Yang, Y.-M., Lu, X.-Y., & Huang, W.-D. (2003). Effect of thymosin alpha 1 on cellular immune function in elderly patients with malignant tumor. *Journal of Zhejiang University Medical Sciences*, *32*(4), 339–341.

Yao, Q., Doan, L. X., Zhang, R., et al. (2007). Thymosin-a1 modulated dendritic cell differentiation and functional maturation from peripheral blood CD14+ monocytes. *Immunology Letters*, *110*, 110–120.

Yoshimura, S., Bondeson, J., Brennan, F. M., Foxwell, B. M., & Feldmann, M. (2001). Role of NFkappaB in antigen presentation and development of regulatory T cells elucidated by treatment of dendritic cells with proteasome inhibitor PSI. *European Journal of Immunology*, *31*, 1883–1893.

Zhang, W. J. (2000). Transcatheter arterial chemotherapy and embolization plus thymosin alpha 1 for treatment of hepatocellular carcinoma. *Journal of Gastroenterology and Hepatology*, *15*, A401.

Zhang, P., Chan, J., Dragoi, A. M., et al. (2005). Activation of IKK by thymosin alpha1 requires the TRAF6 signalling pathway. *EMBO Reports*, *6*(6), 531–537.

Zhang, Q., Tang, D., & Zhao, H. (2010). Immunological therapies can relieve aromatase inhibitor-related joint symptoms in breast cancer survivors. *American Journal of Clinical Oncology*, *33*(6), 557–560.

Zhao, M.-Y., Cao, Y., Fei, D., et al. (2007). Influence of thymosin α1 on the cellular immune function in patients with severe sepsis. *Chinese Journal of Critical Care Medicine*, *27*(3), 206–208.

Zheng, B.-X., Cheng, D.-Y., Xu, G., Fan, L.-L., Yang, Y., & Yang, W. (2008). The prophylactic effect of thymosin alpha 1 on the acute exacerbation of chronic obstructive pulmonary disease. *Journal of Sichuan University*, *39*(4), 588–590.

CHAPTER EIGHT

Prothymosin Alpha and Immune Responses: Are We Close to Potential Clinical Applications?

P. Samara*,[2], K. Ioannou*,†,[2], O.E. Tsitsilonis*,[1]

*Faculty of Biology, National and Kapodistrian University of Athens, Panepistimiopolis, Athens, Greece
†King's College London, Rayne Institute, London, United Kingdom
[1]Corresponding author: e-mail address: rtsitsil@biol.uoa.gr

Contents

Abstract

The thymus gland produces soluble molecules, which mediate significant immune functions. The first biologically active thymic extract was thymosin fraction V, the

[2] P. Samara and K. Ioannou contributed equally to this work.

Vitamins and Hormones, Volume 102
ISSN 0083-6729
http://dx.doi.org/10.1016/bs.vh.2016.04.008

fractionation of which led to the isolation of a series of immunoactive polypeptides, including prothymosin alpha (proTα).

ProTα displays a dual role, intracellularly as a survival and proliferation mediator and extracellularly as a biological response modifier. Accordingly, inside the cell, proTα is implicated in crucial intracellular circuits and may serve as a surrogate tumor biomarker, but when found outside the cell, it could be used as a therapeutic agent for treating immune system deficiencies. In fact, proTα possesses pleiotropic adjuvant activity and a series of immunomodulatory effects (eg, anticancer, antiviral, neuroprotective, cardioprotective). Moreover, several reports suggest that the variable activity of proTα might be exerted through different parts of the molecule. We first reported that the main immunoactive region of proTα is the carboxy-terminal decapeptide proTα (100–109). In conjunction with data from others, we also revealed that proTα and proTα(100–109) signal through Toll-like receptor 4. Although their precise molecular mechanism of action is yet not fully elucidated, proTα and proTα(100–109) are viewed as candidate adjuvants for cancer immunotherapy.

Here, we present a historical overview on the discovery and isolation of thymosins with emphasis on proTα and data on some immune-related new activities of the polypeptide and smaller immunostimulatory peptides thereof. Finally, we propose a compiled scenario on proTα's mode of action, which could eventually contribute to its clinical application.

1. INTRODUCTION

Biologic response modifiers (BRMs) are endogenous (ie, naturally produced in the body) or exogenous (administered together with a drug) agents that modulate immunity. BRMs regulate, among others, the type, duration, and intensity of immune responses and are characterized by pleiotropy and redundancy. The thymic polypeptide prothymosin alpha (proTα) has been incorporated in the large family of BRMs, mainly because of its modulating effects on several properties of immune effectors. Its wide distribution in cells, tissues, and organs, its broad phylogenetic dissemination and the lack of a mechanism supporting its secretion, questioned the initial characterization of proTα as "thymic hormone." Now, it is widely acknowledged that proTα possesses an essential intracellular role related to cell survival and growth, and at the same time, extracellularly it enhances the functionalities of diverse subpopulations of the immune system. Several novel functions, beyond immunomodulation, have also been ascribed to proTα.

Accumulated data suggest that its immunopotentiating activity could be therapeutically exploited in various clinical conditions associated with immunodeficiency, immunosenescence, cancer, and autoimmune diseases.

Herein, we present the most prominent effects of the polypeptide, as reported by various research teams for over 30 years, propose a compiled scenario on its mode of action, and provide means, which eventually could lead to its incorporation in clinical trials as an immunostimulant/adjuvant.

2. HISTORICAL OVERVIEW ON ProTα ISOLATION AND PROPERTIES

2.1 Thymosin Fraction V: The First Immunoactive Thymic Extract

The thymus had remained an enigmatic organ for centuries, as its clinical significance and true function were much disputed. Although its lymphopoietic role had been repeatedly demonstrated (Miller, 2002), the long-prevailing opinion claimed that the thymus was a redundant organ with no immunological function. It was not until the early 1960s that Miller, Good, and their colleagues tipped the balance in favor of a crucial and central role for thymus in establishing a competent immune system (Good et al., 1962; Miller, 1961). They were the first to show that neonatally thymectomized mice exhibited a marked deficiency of lymphocytes and impaired immune responses that were further associated with their inability to resist infections and reject histoincompatible transplants. However, even these studies could not fully elucidate the exact mechanism underlying the continued thymic control over lymphocytic function. A year later, it was unequivocally proven that the thymus is actually a gland mediating its function, at least partially, through humoral messages and not merely via cellular interactions (Aisenberg & Wilkes, 1965; Levey, Trainin, & Law, 1963).

In the years to follow, Goldstein and his colleagues isolated and identified a series of hormonal-like agents with immune-restorative capacity (Goldstein, 2007). These initial efforts led to the isolation of a partially purified thymic extract that could enhance in vivo the incorporation of tritiated thymidine into DNA of mouse lymph nodes, indicative of increased lymphocytic proliferation (Klein, Goldstein, & White, 1965). With the invaluable aid of the newly developed in vitro proliferation assay, this soluble factor inducing lymphocytopoiesis was purified, characterized, and named "*thymosin*" (Goldstein, Slater, & White, 1966). However, the major breakthrough was the isolation of *thymosin fraction V* (*TFV*), a further purified preparation with immunoregulatory activity (Goldstein et al., 1981). Using a novel and complex process that enabled its purification in large amounts (Hooper et al., 1975), TFV's properties and activity were explored in several in vitro and in vivo studies. In vitro, TFV was shown to stimulate lymphocytes deriving from immunosuppressed humans and restore their functions to normal levels (Wara & Ammann, 1975). Additionally, TFV enhanced murine spleen cell responses in mixed lymphocyte reactions (MLRs) and in mixed lymphocyte-tumor cultures (MLTCs; Talmadge, Uithoven, Lenz, & Chirigos, 1984). In vivo, TFV increased the survival of neonatally thymectomized mice (Spangelo, Hall, & Goldstein, 1987), restored graft-vs-host reactivity (Spangelo et al., 1987), and "corrected" T-cell abnormalities in mice with autoimmune disease (Dauphinee, Tala, Goldstein, & White, 1974). Interestingly, TFV also possessed direct antiproliferative properties against malignant cells, as demonstrated in vitro against murine pituitary adenocarcinoma and glioma (Spangelo, Farrimond, Pompilius, & Bowman, 2000), and against human acute T lymphoblastic (Ho, Ma, Price, Hunstein, & HoVbrand, 1983) and promyelocytic leukemia cell lines (Spangelo et al., 2007). Finally, in vivo treatment with TFV conferred resistance to Dunning (Khaw & Rule, 1973) and murine lymphoblastic leukemias (Petro & Watson, 1982).

2.2 Dissecting TFV: Isolation of the First Immunoactive Thymosins

The immune-stimulating and restorative activities attributed to TFV triggered the need to further identify its several components. Thorough fractionation and analysis revealed that TFV consisted of at least 40 different molecules, of which 10–15 were principal and 20 or more were secondary components, and the majority of these molecules were acidic and of varying molecular weights (between 1000 and 15,000 Da). A nomenclature system based upon the isoelectric point (p*I*) of the peptides has resulted in TFV's subdivision into three different groups, each one identified by a Greek letter, ie, alpha (α), beta (β), and gamma (γ). Thus, peptides with p*I*s less than 5.0 were named α-peptides, these with p*I*s of 5.0–7.0 were termed β-peptides, and peptides with p*I*s greater than 7.0 were considered γ-peptides. A subscript number was used to identify the peptides from each group, indicative of the order in which they were isolated from TFV, eg, α_1, α_2, α_3, etc. (Goldstein, 2007).

The first two peptides identified and fully characterized were *thymosin* α_1 (Tα_1; Goldstein et al., 1966) and *thymosin* β_4 (Tβ_4; Low, Hu, & Goldstein, 1981), and their immunopotentiating properties were extensively studied in the years to follow (Spangelo et al., 1987). Subsequently, a number of α- and β-peptides were isolated and sequenced from thymus and other tissues (Hannappel & Huff, 2003). Utilizing a newly developed method that reduced the effect of proteolysis during peptide isolation, Hannappel and colleagues demonstrated that Tβ_4 was the major component of calf thymus extract and that Tα_1 was absent or present in only trace amounts (Hannappel, Davoust, & Horecker, 1982a). The latter observation led the investigators to hypothesize that the presence of Tα_1 in TFV was the result of proteolytic cleavage of a precursor peptide. This speculation was further fueled by evidence showing that a previously isolated peptide, Tβ_8, was actually the proteolytic product of a longer molecule, Tβ_9 (Hannappel, Davoust, & Horecker, 1982b). In addition, the same group isolated two peptides from TFV, whose sequence strongly resembled that of Tα_1. When compared to Tα_1, one peptide lacked four amino acids at its carboxy (C)-terminus and it was therefore named des-(25–28)-Tα_1, and the second had an additional seven amino acids at its C-terminus, and was given the name Tα_{11} (Caldarella et al., 1983). Consequently, the search for a larger α-thymosin precursor molecule began.

In an effort to isolate the native polypeptide, Haritos and colleagues developed a radioimmunoassay based on an antibody raised against synthetic Tα_1 (Haritos & Horecker, 1985). By coupling this assay with an isolation procedure designed to minimize any proteolytic activity in cell extracts, they eventually isolated, initially from rat thymus, a polypeptide of 112 amino acids long, which contained the Tα_1 sequence (amino acid residues 1–28) at its amino (N)-terminus (Haritos, Goodall, & Horecker, 1984). They named this polypeptide *proTα* to indicate that it was the source of Tα_1 and Tα_1-related peptide fragments present in TFV preparations.

2.3 ProTα: Major Structural Characteristics and Properties

Human proTα is 109 amino acids long and is encoded by the *PTMA* gene located on chromosome 2 (Szabo et al., 1993). Nearly half of the total residues in proTα are accounted by glutamic and aspartic acid and, as a result, the molecule acquires a particularly low p*I* 3.55. Although it was initially referred to as a "thymic hormone," proTα is not solely expressed in the thymus, but detected in all tissues (Haritos, Tsolas, & Horecker, 1984). It is a noticeably conserved polypeptide characterized by a high degree of sequence homology among mammals (Hannappel & Huff, 2003) and over-expressed in cells with increased physiological (eg, young thymus) or abnormal (eg, malignant) proliferative capacity (Haritos, 1987).

As proTα's acidic residues are found primarily within the central segment of the polypeptide chain, the molecule has no specific secondary structure and, eventually, adopts a random coil conformation (Gast et al., 1995). However, the possibility that proTα acquires a secondary structure upon interaction with other proteins has not been ruled out (Piñeiro, Cordero, & Nogueira, 2000). In support of this, it has been demonstrated that under specific conditions (low pH and high concentration) the polypeptide forms amyloid fibrils (Pavlov, Cherny, Heim, Jovin, & Subramaniam, 2002). Although proTα is released by necrotic neurons via a unique non-classical pathway (Halder, Matsunaga, & Ueda, 2012; Matsunaga & Ueda, 2010), the peptide lacks a signal peptide sequence required for secretion. Instead, it bears a bipartite nuclear localization signal which consists of two blocks of basic residues (^{87}KR^{89} and ^{100}TKKQKT^{105}), and is both necessary and sufficient for import of the protein to the nucleus, where it is predominantly located (Manrow, Sburlati, Hanover, & Berger, 1991; Rubtsov et al., 1997). In contradiction, many studies have also supported proTα's cytoplasmic localization (Sburlati, Manrow, & Berger, 1990; Tsitsiloni, Yialouris, Sekeri-Pataryas, & Haritos, 1989). In an effort to address this antiphasis, Enkemann and coworkers demonstrated that proTα is in principle detected at active transcription sites in the nucleus, while a smaller fraction remains in the cytoplasm (Enkemann, Wang, Trumbore, & Berger, 2000). It was additionally suggested that, owing to its small size and its negative charge, proTα may facilitate the movement of other positively charged molecules (eg, of histones) into and within the nucleus, particularly in highly charged environments where there is a need to overcome electrostatic interactions. In terms of proTα's biological role, a vast number of studies have attributed, among others, various and diverse immune-related properties to the polypeptide, the most important of which are analyzed later.

3. THE MULTIFACETED IMMUNE ACTIVITIES OF ProTα

3.1 The Thoroughly Studied Anticancer Activity of ProTα

The potent anticancer activity of proTα was studied in the early 1990s, when it was first reported that it regulates MHC class II expression on human and mouse antigen-presenting cells (APCs; Baxevanis et al., 1992). The significance of this observation was verified in three serial studies in a leukemic in vivo animal model, where mice inoculated with L1210 cells and therapeutically treated with proTα, survived for over 2 months (Papanastasiou, Baxevanis, & Papamichail, 1992). The immunological modifications caused by proTα included the in vivo generation of MHC-restricted tumor-specific CD8+ cytotoxic T lymphocytes (CTLs;

Baxevanis, Gritzapis, Spanakos, Tsitsilonis, & Papamichail, 1995), concomitantly with the enhancement of tumor-reactive NK cell-mediated cytotoxicity (Baxevanis et al., 1994). Both effectors efficiently lysed the syngeneic L1210 tumor cells. The immunoenhancing effect of proTα was exerted upstream lymphocyte activation via an interleukin (IL)-2-dependent manner. Most importantly, proTα shifted antitumor-reactive immune responses toward the stimulation of the most suitable effectors, CTLs or NK cells, depending on the presence or absence of tumor-specific antigenic peptides, respectively.

Animal studies were followed by in vitro human studies in lymphocytes from cancer patients. Cancer induces severe immune dysfunctions, which are further intensified by anticancer therapies administered to patients. ProTα was shown to restore the deficiencies of peripheral blood lymphocytes deriving from patients with advanced solid tumors by enhancing: (1) the allogeneic cell-mediated lympholysis; (2) antigen presentation, as confirmed by the increased values recorded in MLR; and (3) the reduced NK and T-cell cytotoxic activity, by regulating the levels of prostaglandin E2 (PGE2) and IL-2 (Baxevanis, Reclos, & Papamichail, 1993). Two consecutive studies in melanoma and colon cancer patients showed that proTα could act beneficially at early-stage cancers, and when combined with low-dose interferon (IFN)-γ or IL-2, significantly enhanced monocyte, NK and LAK cell tumoristatic and tumorilytic activities (Eckert et al., 1995; Garbin, Eckert, Büttner, Garbe, & Maurer, 1994). ProTα in conjunction with either IFN-γ or IL-2 increased the adhesion of monocytes to tumor targets and the expression of characteristic NK cell markers (eg, CD56, CD16), enhanced the lytic activity of LAK cells, in particular of CD16+CD2− cells, inhibited the secretion of TGF-β and PGE2, and induced the production and secretion of high levels of the proinflammatory cytokines IL-1β and tumor necrosis factor (TNF)-α (Cordero, Sarandeses, López-Rodríguez, & Nogueira, 1995; Eckert, Grünberg, Garbin, & Maurer, 1997; Eckert, Grünberg, Immenschuh, et al., 1997; Garbin, Eckert, Immenschuh, Kreuser, & Maurer, 1997; López-Rodríguez, Cordero, Sarandeses, Viñuela, & Nogueira, 1994). Based on the aforementioned data, the first means of action of proTα was shaped, indicating that the polypeptide restores cancer patient lymphocyte deficiencies by selectively controlling the functions of monocytes. Monocytes, in turn, produce cytokines that generate a favorable cytokine milieu, facilitating lymphocyte activation. To optimize this effect, synergy between proTα and low concentrations of other BRMs was assessed. For example, proTα was combined with a monoclonal antibody to CD3 (anti-CD3), and this synergy further increased tumor cell-lysis by both MHC- and non-MHC-restricted PBMC effectors (Baxevanis et al., 1999).

All these results were complemented by the elegant study of Voutsas et al. (2000), where, using MLTCs, proTα in synergy with low-dose IL-2 led to expansion of tumor-reactive CD4+ T cells and the subsequent generation of MHC class I-restricted autologous tumor-specific CTLs. This was the first study showing that in order for proTα to fully exert its beneficial effect on CTLs, the concomitant presence of autologous CD4+ T cells and monocytes was required.

Combining all previous data, in an attempt to fill in the gaps, our research team used proteomics to elucidate the mechanism of action of proTα on healthy donor- and cancer patient-derived PBMCs. Based on proteins up- and downregulated upon PBMC treatment with proTα, we were the first to analytically describe the scenario underlying its immunoenhancing activity. We suggested that proTα ligates innate immunity receptor(s) on APCs, leading to formation of strong APC-T-cell synapses, proinflammatory cytokine secretion and thus, the indirect increase in the cytotoxicity of CTLs, NK cells, and other effectors (Skopeliti et al., 2009). We and others further confirmed the realism of this scenario, showing that proTα ligates Toll-like receptor 4 (TLR4) (Mosoian et al., 2010; Omotuyi, Matsunaga, & Ueda, 2015) and matures monocyte-derived dendritic cells (DCs), which express costimulatory molecules (Skopeliti et al., 2009) and secrete proinflammatory cytokines. Moreover, proTα-matured DCs loaded with tumor antigens induced the polarization of T_H1-type tumor-reactive immune responses, resulting in the generation of polyfunctional highly lytic tumor antigen-specific CTLs (Ioannou et al., 2013). Besides DCs, proTα could also activate other immune cell types expressing TLR4, eg, neutrophils to secrete $\cdot O_2^-$ and kill tumor cells in vitro (Samara, Ioannou, et al., 2013), and macrophages (Mosoian et al., 2010). Ex vivo experiments showed that proTα significantly restored the reduced cytotoxicity of immunosuppressed ascites-derived tumor-associated lymphocytes against ovarian tumor cells and inhibited ovarian tumor growth in SCID mice inoculated with human tumors (Voutsas et al., 2013).

3.2 Antiviral Activities of ProTα

Control of viral replication in humans requires the involvement of robust CTL responses, which in turn demand the presence of type I IFNs, that further support CD8+ T-cell development and function (Welsh, Bahl, Marshall, & Urban, 2012). Interestingly, proTα has been shown to be a regulator and enhancer of type I IFN secretion and as such, its antiviral properties were investigated. Qiu and colleagues provided the first evidence of proTα's capacity to enhance IFN-α secretion by murine macrophages (Qiu et al., 2002). Later, proteomic analysis of human mononuclear cells treated with proTα revealed that the polypeptide enhanced the expression of mixovirus-resistance protein 2, an IFN-α-inducible protein that possesses significant antiviral activity (Skopeliti et al., 2007). Moreover, proTα was identified as a potent suppressor of human immunodeficiency virus (HIV)-1 replication in primary macrophages, when released by virus-infected CD8+ T cells. It was also shown that this HIV-1 inhibition occurs following viral integration, it is not virus specific (Mosoian et al., 2006), and is rather mediated by the upregulation of type I IFNs, following activation of TLR4 by proTα (Mosoian et al., 2010; Teixeira et al., 2015). Similarly, release of proTα upon infection of guinea pigs with an attenuated strain of *Pichinde* virus led to induction of potent antiviral immunity and consequent viral clearance (Bowick et al., 2010). In terms of in vivo applications, proTα has been used as a potent adjuvant for hepatitis B virus DNA vaccines, where it successfully enhanced both humoral and cellular immune responses (Jin et al., 2005). Impressively, a most recent report showed that proTα displays antiviral activity in other classes than mammals, specifically in fish (tongue sole) infected with megalocytivirus (Zhang & Sun, 2015).

3.3 ProTα Exerts Neuroprotective Functions

Neuronal death, induced by stroke or trauma, occurs via both necrosis and apoptosis. While apoptosis is a more regulated process that limits the damage in the brain, necrosis tends to expand and is therefore being targeted for stroke treatment (Ueda, 2009). Based on this premise, in an effort to identify soluble molecules that could inhibit necrosis, Ueda and colleagues isolated and identified proTα as a unique antineuronal necrosis factor in the conditioned medium of cortical neurons (Ueda, Fujita, Yoshida, Matsunaga, & Ueda, 2007). In more detail, proTα could reverse the rapid decrease in survival of cortical neurons, abolish the typical necrosis features in these cells, and switch the cell death mode from necrosis to apoptosis (Ueda et al., 2007). Later on, proTα's potent neuroprotective functions were also demonstrated in in vivo experimentally induced cerebral and retinal ischemia (Fujita & Ueda, 2007; Fujita, Ueda, Fujiwara, & Ueda, 2009). As for the mechanism underlying proTα's neuroprotective role, it has been shown that the polypeptide is released via a nonclassical manner under ischemic stress conditions and its extracellular release is facilitated by its interaction with a cargo molecule, namely S100A13 (Halder et al., 2012; Matsunaga & Ueda, 2010). Upon release, proTα activates the TLR4–TRIF-signaling pathway, inducing the expression of neuroprotective factors (Halder, Matsunaga, Ishii, & Ueda, 2015).

3.4 Miscellaneous Functions Reported for ProTα

Given its neuroprotective properties, the effect of proTα on cardiomyocytes during ischemic injury was also investigated (Cannavo et al., 2013). In vitro treatment of cardiomyocytes with recombinant proTα during simulated ischemia significantly decreased the apoptotic response and enhanced cell survival. Consistent with the in vitro findings, in vivo administration of proTα following myocardial infarction successfully reduced the infarct size in mice, when compared to untreated controls, an effect that was mediated by an Akt-dependent mechanism.

Li and colleagues had initially reported that proTα transgenic mice exhibited the polycystic kidney disease phenotype, as well as emphysema-like changes in the lung (Li, Shiau, Chiou, Yo, & Wu, 2005). Further studies by the same research group showed that proTα enhanced the acetylation of histones and nuclear factor-kappa B (NF-κB), or inhibited the TGF-β/Smad signaling, and thus contributed to the development of emphysema (Su et al., 2013; Su et al., 2016). Furthermore, high levels of proTα were positively correlated with the severity of emphysema, both in transgenic mice and in emphysema patients (Su et al., 2013).

Finally, the polypeptide's involvement in the induction of insulin resistance was investigated, as proTα reportedly regulated some inflammatory responses and oxidative stress, features that are associated with diabetes (Su et al., 2015). Patients with type 2 diabetes had significantly higher levels of serum proTα compared to normal individuals and proTα transgenic mice exhibited an insulin-resistant phenotype. Exploitation of the underlying mechanism revealed that proTα induces insulin resistance through a TLR4–NF-κB-dependent pathway.

4. THE IMMUNOSTIMULATORY ACTIVITY OF ProTα-DERIVED PEPTIDES

Since $T\alpha_1$ was the first thymosin isolated and identified from TFV and proTα was the natural precursor of $T\alpha_1$ and other smaller α-thymosins

(Haritos, Goodall, et al., 1984), $T\alpha_1$ was initially considered as the major immunoactive fragment of the polypeptide. However, data accrued in particular over the last years revealed that proTα exerts its immunomodulating role through diverse fragments, such as the C-terminal fragment spanning residues 100–109 (Skopeliti et al., 2009) and the central, negatively charged region 50–89 (Mosoian et al., 2010). Two points that we consider very important are: (1) when compared to its fragments, intact proTα seems to perform better at least in most immune-based assays; and (2) different areas of the molecule seem to be responsible for proTα's diverse activities (Table 1).

Table 1 Prothymosin Alpha Fragments with Distinct Activities

ProTα Fragment	Type of Activity	Reported in:
ProTα(1–28), referred to as "$T\alpha_1$"	Immunomodulatory; DC activation; anticancer; antiviral; antifungal; vaccine enhancement	Camerini and Garaci (2015)
ProTα(1–35), referred to as "$T\alpha_{11}$"	Antifungal	Hannappel and Huff (2003)
ProTα(49–78) (P30) ProTα(52–60) (P9)	Neuroprotective	Halder, Sugimoto, Matsunaga, and Ueda (2013)
ProTα(50–89), referred to as "Mosoian domain"	Anti-HIV-1	Mosoian et al. (2010)
ProTα variants (p7 [proTα(32–49)] and isoB [proTα(32–41), proTα(51–55), proTα (56–61), proTα(65–71)])	Anti-HIV-1	Teixeira et al. (2015)
ProTα(100–109), referred to as "Skopelitian domain"	Immunomodulatory; anticancer; DC maturation; enhancement of phagocytosis, respiratory burst, and cytotoxicity of human neutrophils	Skopeliti et al. (2009), Voutsas et al. (2013), Ioannou et al. (2013), and Samara, Ioannou, et al. (2013)
ProTα(1–100)	Cardioprotective	Cannavo et al. (2013)

4.1 The Amino-Terminal Peptide Tα_1 Has Shown Some Immune Activity

The 28-amino acid-long N-terminal proTα fragment Tα_1 was reported to enhance human cell-mediated immunity and stimulate endothelial cell migration, angiogenesis, and wound healing (Malinda et al., 1998). Being the only α-thymosin tested in man, Tα_1 has been used in a broad range of clinical applications, both as a single agent and in combination with other standard treatments (eg, chemotherapeutics), showing an excellent safety profile (Romani et al., 2012). Among other clinical cases, Tα_1 has been used for treating infectious diseases including viral (chronic hepatitis B and C, AIDS), fungal (aspergillosis in bone marrow-transplanted patients) and bacterial (*Pseudomonas aeruginosa*) sepsis, severe lung pathologies (eg, chronic obstructive pulmonary disease, acute respiratory distress syndrome, and severe acute respiratory syndrome), age-related deficiencies, cancer, and as a vaccine enhancer (Camerini & Garaci, 2015). However, despite the promising data generated so far, the clinical benefit of Tα_1 administration in these specific pathologies is still disputed and needs to be verified in larger well-designed randomized clinical trials.

4.2 The Carboxy-Terminal ProTα Peptides Exhibit Improved Immune Functions

The C-terminal decapeptide proTα(100–109) (TKKQKTDEDD) was identified by our research team as the immunoactive area of the polypeptide and a potent lymphocyte stimulator (Skopeliti et al., 2006). ProTα (100–109) has been shown to stimulate PBMC proliferation and cytotoxicity and to promote the phenotypic maturation of DCs (Skopeliti et al., 2009), and, consequently, it improved the functionality of immunogenic peptide-pulsed DCs, induced T_H1-type immune response polarization (Ioannou et al., 2013), augmented basic properties of human neutrophils (Samara, Ioannou, et al., 2013), enhanced the depressed cytotoxicity of tumor-associated lymphocytes against autologous tumor cells in vitro, and retarded tumor growth in vivo (Voutsas et al., 2013). Using as control a scrambled decapeptide with the same amino acid composition but a different primary structure, the immunoenhancing activity of proTα(100–109) was shown to be sequence-specific and comparable to that of intact proTα. Most recently, we reported that proTα(100–109) radiolabeled with (99m)Tc binds on the surface of human neutrophils via a complex involving TLR4 (Karachaliou et al., 2015) and selectively accumulates in sites of experimentally induced inflammation (C.E. Karachaliou et al., personal communication).

Based on the fact that the decapeptide proTα(100–109) is in vivo generated upon caspase cleavage of proTα during apoptosis (Enkemann et al., 2000; Evstafieva et al., 2003), we developed a highly sensitive and specific competitive ELISA for proTα(100–109), using high affinity-purified polyclonal antibodies (Samara, Kalbacher, et al., 2013). The decapeptide was quantified in the serum of healthy humans, where it was detected at very low concentrations ($\leq$0.5 ng/mL; P. Samara et al., unpublished data), whereas higher levels were recorded in the serum of mice infected with the bacterium *Streptococcus pyogenes*, suggesting a correlation between proTα(100–109) levels and the progress of bacterial infection.

Two more C-terminal immunomodulatory peptide fragments of proTα have been reported. In an earlier study, our research group identified a slightly smaller segment, namely proTα(103–109), which was also effective in vitro in restoring the immune functions of PBMCs derived from cancer patients (Skopeliti et al., 2006). Most recently, gravimetric assays and molecular dynamics simulation revealed that proTα, via its C-terminal segment (91–111) and similar to LPS, biophysically interacts in vitro with the TLR4/MD-2 complex at overlapping LPS-binding positions (Omotuyi et al., 2015).

4.3 The Immune Activity of ProTα Middle Segment Peptides

The research group of Mosoian introduced the synthetic peptide spanning amino acids 50–89 of proTα, designated proTα(50–89), in an attempt to comprehend how proTα mediates its anti-HIV-1 activity and induces type I IFN production. Incubation of human macrophages and myeloid DCs with proTα or proTα(50–89) stimulated IFN-α1 and IFN-β production, as shown by the increase in type I IFN mRNA (Mosoian et al., 2010). Furthermore, the same research team recently reported that proTα variants (p7 and isoB), spanning middle segment sequences of the polypeptide (amino acids 32–41, 32–49, 51–55, 56–61, and 65–71), also induced mRNA expression of type I and type III IFNs in human macrophages, suggesting that these peptides possess strong antiviral activities, responsible for the registered suppression of HIV-1 replication in this cell type (Teixeira et al., 2015).

In addition, Ueda and colleagues reported that a 30-amino acid middle peptide sequence of proTα [termed P30; proTα(49–78)] exerts substantial neuroprotective activity in vitro and in vivo and inhibits cerebral blood vessel damage caused by ischemic stress in retina and brain. The minimum neuroprotective sequence encompasses the 9-amino acid peptide 52–60 (P9), which was shown to comprise the full neuroprotective effect of proTα in cultured cortical neurons under ischemic conditions (Halder et al., 2013).

5. EVIDENCE SUPPORTING A DUAL—INTRACELLULAR AND EXTRACELLULAR—ROLE FOR ProTα

It has long been disputed whether a strictly intracellular, in principle nuclear molecule like proTα, could modulate immune responses and several researchers initially rejected the idea that the polypeptide could exert an extracellular role. The wide distribution of proTα in tissues and cells strongly supported the sole implication of the polypeptide in crucial intracellular processes.

5.1 The Intracellular Role of ProTα

Indeed, in normal cells, proTα was shown to be important for survival and proliferation. High levels of proTα mRNA were detected in lymphocytes (Eschenfeldt & Berger, 1986; Szabo, Ehleiter, Whittington, & Weksler, 1992) and in NIH3T3 cells stimulated to divide (Wu, Shiau, & Lin, 1997), whereas proliferation of myeloma cells was hindered in the presence of proTα antisense oligomers (Sburlati, Manrow, & Berger, 1991). Over the years, research on the intracellular activity of proTα revealed its implication in: (1) *gene transcription*; proTα was reported to localize in nuclear sites of active transcription, physically interact with the CREB-binding protein (CBP), bind to histone H1, and stimulate transcription (Karetsou, Kretsovali, Murphy, Tsolas, & Papamarcaki, 2002); (2) *DNA remodeling*; during proliferation proTα increased the accessibility of micrococcal nuclease to chromatin (Gomez-Marquez & Rodriguez, 1998); (3) *inhibition of apoptosis*; proTα was shown to bind Apaf-1 and inhibit apoptosome formation (Jiang et al., 2003); and (4) *protection against oxidative stress*; proTα interacted with INrf2 leading to its nuclear translocation, where it mediated the degradation of Nrf2 and promoted cell survival and growth (Niture, Kaspar, Shen, & Jaiswal, 2010) (Fig. 1).

In cancer cells, which are highly proliferative and divide constantly, increased proTα content was also reported. High proTα levels were detected in human colon (Mori et al., 1993), colorectal (Zhang et al., 2014), lung (Sasaki, Nonaka, et al., 2001), bladder (Tsai et al., 2014), and liver cancers (Wu, Habib, et al., 1997); and in neuroblastoma (Sasaki, Sato, et al., 2001), rhabdomyosarcoma (Carey et al., 2006), thyroid carcinomas (Kashat et al., 2010; Letsas, Vartholomatos, Tsepi, Tsatsoulis, & Frangou-Lazaridis, 2007), pituitary tumors (Pawlikowski et al., 2014), head and neck cancers (Tripathi et al., 2011), gastric adenocarcinomas (Leys et al., 2007), hepatocellular (Ha, Song, Hwang, Cho, & Park, 2015), and urinary tract transitional cell carcinomas (Jou et al., 2009). We and others reported that proTα was elevated in breast tumors (Tsitsiloni et al., 1993) and tumor proTα levels correlated with the disease outcome (Dominguez et al.,

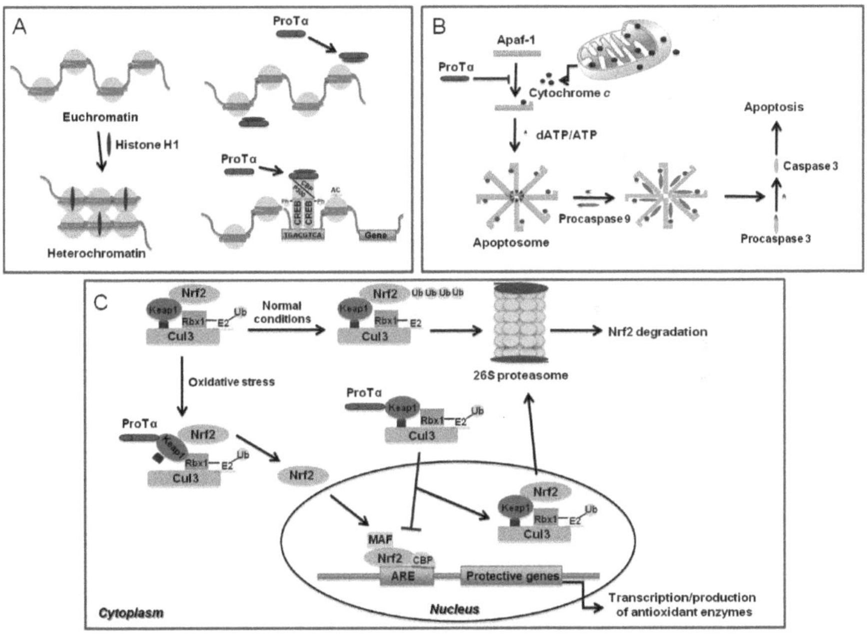

Fig. 1 The intracellular role of proTα. (A) In the absence of proTα, histone H1 binds to nucleosomes and induces condensation of euchromatin to heterochromatin. ProTα interacts with histone H1, mediates its transfer from and to chromatin, and leads to the formation of CREB–CBP–p300 complex, chromatin remodeling, and gene transcription. (B) Apoptotic stimuli induce the release of cytochrome *c*, which binds to Apaf-1 and forms the apoptosome. The subsequent activation of caspase 9 results in conversion of procaspase 3 to caspase 3, leading to apoptosis. ProTα hinders binding of cytochrome *c* to Apaf-1, and the apoptotic cascade is inhibited. (C) Normally, the Nrf2–Keap1–Cul3–Rbx1 complex is ubiquitinated and degraded by proteasomes. Under oxidative stress, proTα binds to the complex via Keap1, and Nrf2 is released, migrates to the nucleus, and promotes gene transcription and antioxidant enzyme production.

1993), the metastatic potential of the tumor (Tsitsilonis et al., 1998), and the risk of recurrence and death (Magdalena, Dominguez, Loidi, & Puente, 2000). It is worth mentioning that high proTα levels detected in prostate cancer samples (Klimentzou et al., 2008) were shown to progressively increase as prostate tissue progressed from normal epithelium, through prostatic intraepithelial neoplasia to carcinoma, and correlated with Gleason's score (Suzuki et al., 2006). Therefore, following the appropriate validation, proTα tumor content and/or its serum levels may be used as a surrogate tumor biomarker for cancer development, prognosis, and follow-up.

5.2 The Extracellular Role of ProTα

Nevertheless, evidence supporting the extracellular activity of proTα was also reported by in vivo studies in animals, which showed that proTα can protect mice against opportunistic infections, in particular with *Candida albicans* (Pan et al., 1986). Numerous in vitro studies further confirmed the extracellular role of the polypeptide. As already aforementioned, peripheral blood T cells stimulated with proTα produced high amounts of IL-2 and increased the expression of IL-2 receptor on their surface (Baxevanis et al., 1990; Cordero et al., 1991), while APCs and DCs activated with proTα upregulated MHC class II (Baxevanis et al., 1992) and costimulatory molecules (CD11b, CD80, CD83, CD86, and CD40; Skopeliti et al., 2009), respectively. NK and LAK cells cultured in the presence of proTα augmented their cytotoxicity (Cordero, Sarandeses, López, & Nogueira, 1992; López-Rodríguez et al., 1994), and proTα-activated neutrophils showed increased chemotaxis, produced high amounts of ROS, and became cytotoxic against cancer cell targets (Heidecke, Eckert, Schulze-Forster, & Maurer, 1997; Samara, Ioannou, et al., 2013). All these multiple immunological responses suggested that proTα acts pleiotropically, especially in immunosuppressed environments.

At that time, it was obvious that the cytokine-like activities of proTα would be justified if specific-binding sites (receptors?) on immune cell surface were discovered. In an initial attempt, Cordero and colleagues radiolabeled proTα and searched for binding sites on the surface of lymphoblasts. Two binding sites, of low and high affinity, were identified (Cordero, Sarandeses, & Nogueira, 1996). Five years later, the same researchers demonstrated, by affinity cross-linking and chromatography, the existence of three binding partners for proTα on the membrane of PHA-activated lymphoblasts, which were associated with lipid rafts (Piñeiro, Begoña

Bugia, Pilar Arias, Cordero, & Nogueira, 2001; Salgado, Piñeiro, Canda-Sánchez, Lojo, & Nogueira, 2005). However, a specific proTα receptor was never cloned. In 2007, based on the detailed proteomic analysis of protein changes induced by proTα on immune cells and on the upregulation of the signaling protein IRAK4, we proposed that proTα ligates TLRs (Skopeliti et al., 2007). In 2010, Mosoian and colleagues confirmed that proTα ligates TLR4 and signals through both the TRIF-dependent and the MyD88 pathways for induction of IFN-β and proinflammatory cytokines, respectively (Mosoian et al., 2010). Most recently, proTα was shown to bind to the TLR4/MD-2 complex and activate the TRIF–IRF3-signaling pathway downstream TLR4 (Halder et al., 2015; Omotuyi et al., 2015).

5.3 Evidence Suggesting That ProTα May Act as an Alarmin

Although more data need to be generated to specifically define the pathways activated in response to proTα, it is more than tempting to speculate that the polypeptide probably belongs to the vague family of molecules termed damage-associated molecular patterns (DAMPs) or else "alarmins" (Bianchi, 2007), and as such, can simultaneously and distinctively exert both an intracellular and an extracellular role.

With the term "alarmins" we practically refer to endogenous pathogen-associated molecular patterns (PAMPs), or in more detail, to multifunctional endogenous constitutively available molecules, passively released from damaged cells or rapidly secreted by stimulated leukocytes and epithelial cells following tissue destruction. Alarmins activate both innate and adaptive immune responses. Uncontrolled and excessive release of alarmins extracellularly induces concomitantly the recruitment and activation of APCs through pattern recognition receptors, such as TLRs. In the absence of such stimuli, alarmins exert their vital intracellular roles.

The alarmin family is rapidly growing and, as for now, its best-characterized members are high-mobility group protein B1 (HMGB1) and some heat-shock proteins (Bianchi, 2007). Thymosins were suggested as candidate alarmins, although, to our knowledge, direct comparison of their characteristics with known alarmins has not been performed. Herein, we compile evidence that proTα possesses several attributes to be considered an alarmin (Table 2). Among others and in comparison to HMGB1, proTα is a ubiquitously expressed, nonhistone nuclear protein with a marked intracellular physiological role in regulating transcription. It is released via a nonclassical pathway upon ischemic stress, and when localized extracellularly, proTα can recruit and activate innate immune cells, promoting inflammation and cytokine secretion, similar to HMGB1. In addition, proTα signals through TLR4, activates leukocytes and orchestrates immune responses (Ioannou et al., 2012). Finally, it is implicated in neuro- and cardio-tissue repair, suggesting the regenerative potential of the polypeptide (Cannavo et al., 2013; Halder et al., 2012).

Table 2 Characteristics Ascribed to Alarmins (HMGB1) and Comparison with Reported Properties of Prothymosin Alpha

Characteristic	HMGB1 (Chan et al., 2012; Schiller et al., 2013; Telusma et al., 2006)	ProTα (Cannavo et al., 2013; Halder et al., 2013; Ioannou, Samara, Livaniou, Derhovanessian, & Tsitsilonis, 2012)
Origin	Nonhistone nuclear protein	Nonhistone nuclear protein
Expression	Expressed in all cells	Expressed in all cells
Physiological intracellular role	DNA organization, transcriptional regulator	DNA organization, transcriptional regulator, antiapoptotic and oxidative stress regulator
Extracellular role	Cytokine/inflammatory mediator	Cytokine/inflammatory mediator
Release mechanism	Passive release and active secretion	Release upon ischemic stress
Receptors	TLR2, TLR4, TLR9, TIM3, and RAGE	TLR4
Regenerative potential	Cardiac and nervous cell regeneration, skin wound healing, bone repair, skeletal muscle repair	Cardiac and nervous cell regeneration
Implication in diseases	Cancer, rheumatoid arthritis stroke, atherosclerosis, sepsis	Cancer, autoimmune diseases, ischemic stroke, viral infections
Additional similarities		
Immunoactive peptides	Hp-106, Hp-31, Hp-91, and Hp-16	ProTα(100–109), proTα (50–89), $T\alpha_1$
Intracellular mobility	Translocation from nucleus to cytoplasm; during apoptosis, translocation into apoptotic cell-derived membranous vesicles	Translocation from nucleus to cytoplasm; during apoptosis, NLS cleavage by caspases and generation of proTα(100–109)

6. PROPOSED SCENARIO FOR THE MECHANISM OF ACTION OF ProTα

Taking all the above into consideration, we propose a scenario depicting the mode of action of proTα. In *normal healthy cells*, proTα is localized in the nucleus (Eschenfeldt & Berger, 1986), albeit a fraction of it is assigned to remain in the cytoplasm (Tsitsiloni et al., 1989). In the nucleus, proTα regulates gene transcription (Karetsou et al., 2002), shapes DNA remodeling, and promotes cell proliferation (Gomez-Marquez & Rodriguez, 1998). In the cytoplasm, proTα protects cells from apoptosis (Jiang et al., 2003) and, due to its negative charge, facilitates the transportation of molecules in the nucleus, protecting the cell from insults, eg, oxidative stress (Niture et al., 2010). In hyperproliferative cells, like *cancer cells*, proTα's gene expression and protein content are increased and this contributes to uncontrolled proliferation and induction of resistance to apoptosis and oxidative stress. Cells receiving death stimuli die violently by necrosis or undergo programmed apoptotic death. *During necrosis*, cell swelling and membrane disruption lead to abrupt uncontrolled release of internal cell components, including proTα, which acts as alarmin, alerts cells of the innate arm of immunity expressing TLR4, and initiates immune responses (Ioannou et al., 2012). *During apoptosis*, the major proportion of proTα is transferred to the cytoplasm, where activated caspases cleave the molecule at its C-terminus, generating among other fragments, the decapeptide proTα(100–109). The negatively charged fragment proTα(1–99) remains in the cytoplasm, where most probably is degraded or, as suggested, is exposed on apoptotic cell surface where it acts as an "eat-me" signal (Evstafieva et al., 2003). The decapeptide proTα(100–109), or a proportion of it, is protected from degradation as it polymerizes adopting a β-sheet conformation (Skopeliti et al., 2009) and, consequently, is excreted from dying cells via an unknown mechanism. In both necrosis and apoptosis, exocytosed proTα and proTα(100–109) are sensed as DAMPs by innate immune cells and ligate TLR4 (Ioannou et al., 2013; Mosoian et al., 2010). Stimulated innate immune cells, such as macrophages, neutrophils, monocytes, and DCs, respond by initiating molecular signaling pathways leading to NF-κB activation, chemokine, IFN, and proinflammatory cytokine secretion. Additionally, stimulated macrophages and neutrophils increase their phagocytosis and produce TNF-α and $\cdot O_2^-$, respectively (Mosoian et al., 2010; Samara, Ioannou, et al., 2013). Stimulation of APCs (DCs and monocytes) increases antigen (eg, shed from dying cancer cells) uptake and presentation through MHC class I molecules directly to CD8+ cytotoxic T cells. Upregulation of MHC class II expression by the thymic peptides on monocytes and DCs increases antigen presentation and their synapse with CD4+ helper T cells, which produce T_H1-type cytokines (eg, IL-2, IFN-γ), providing a favorable environment for enhancing specific

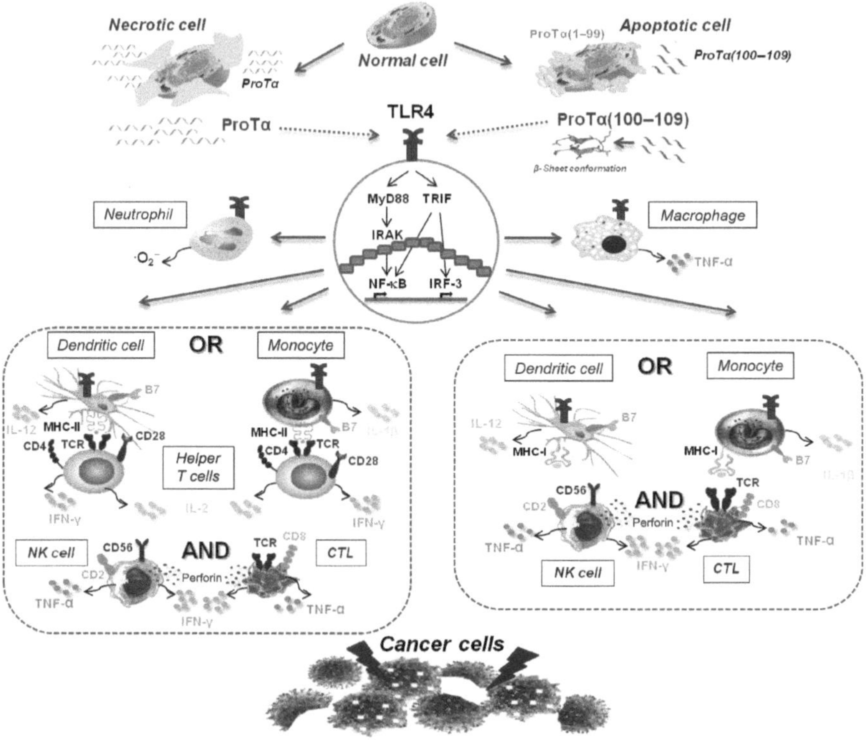

Fig. 2 Proposed scenario on proTα's dual role. In normal cells, proTα regulates gene expression and cell proliferation in the nucleus. Under abnormal conditions, cells die via necrosis or apoptosis. During necrosis, cell membrane is ruptured and intact proTα is released out of the cell. During apoptosis, proTα is relocated to the cytoplasm and is cleaved at its C-terminus by activated caspases, and proTα(100–109) is generated. The decapeptide adopts a β-sheet conformation and is excreted. Extracellularly, both proTα and proTα(100–109) activate innate immune cells expressing TLR4 (macrophages, neutrophils, DCs, and monocytes) and signal through the MyD88 and TRIF pathways. Cytotoxic responses are enhanced through antigen presentation with MHC class II molecules and synapsis with helper T cells which secrete T_H1-type cytokines, and/or with MHC class I molecules, leading to activation of cytotoxic effectors (NK cells and CTLs). In both cases, effector cells upregulate adhesion molecule expression (eg, CD2) and produce lytic molecules (eg, perforin), mediating cell binding and cell destruction, respectively (eg, cancer cell targets, as shown here).

and nonspecific cytotoxicity. CTLs and NK cells produce lytic molecules (perforin), express adhesion molecules (CD2), secrete TNF-α and IFN-γ, and kill cell targets (eg, cancer cells) (Fig. 2).

7. CONCLUSIONS AND FUTURE DIRECTIONS

ProTα, although initially characterized as a thymic hormone, is a ubiquitously expressed polypeptide that exerts pleiotropic immunostimulatory adjuvant effects. Nevertheless, and despite extensive and intensive research on its mode of action, the efficacy of proTα has not yet been tested in humans. In our opinion, the main reason stands in our lack of understanding of its dual role and in elucidating how the same molecule promotes cell proliferation and mediates immunity. This paradox is most surprising and contradictory in the case of cancer. We know that increased proTα gene transcription correlates with carcinogenesis, whereas proTα ligation to TLR4 promotes anticancer-reactive immune responses. In view of recent progress in elucidating the complex machinery of immune responses and the concept that the immune system responds to both pathogens (PAMPs) and danger signals (DAMPs), the adjuvant mode of action of proTα points toward its integration in the family of alarmins. As such, under specific conditions related to danger (resulting from cell damage, destruction, or death), proTα and/or smaller fragments of the molecule [eg, proTα(100–109)] are relocalized extracellularly, ie, at a different site than inside the cell, where they physiologically function. Innate immune cells like DCs that are programmed to sense danger respond to proTα, and the cascade of an immune response is initiated.

Recent evidence additionally suggests that proTα is a TLR agonist. At present, driving closer to optimal orchestration of the immune machinery, several TLR agonists have reached the clinical setting. We believe that proTα has adequately proven its potent immunostimulating capacity in vitro and in animals in vivo, by generating the appropriate cytokine milieu promoting the activation of effector cells. Although it should be studied in more detail, it is also known that the polypeptide when administered in animals at relatively high concentrations does not induce toxicity or severe adverse effects. Therefore, the next step should be to test the adjuvant effectiveness of proTα or smaller immunoactive fragments thereof in clinical studies, aiming to strengthen deficient immune responses.

ACKNOWLEDGMENTS

The authors thank all members of the laboratory for their excellent work, as well as Lilian Williams, Niki Kappa, and Sotirios Fortis for their assistance in preparation of the figures. Research was supported by IKY Fellowships of Excellence for Postgraduate Studies in Greece-Siemens Program to PS, European Union FP7 Capacities Grant REGPOT-CT-2011-284460 (INsPiRE), NATO SfP Project 982838, IKYDA 61/2003, IKYDA 165/2010, Empeirikion Foundation of Athens, and John S. Latsis Public Benefit Foundation. Funding sources had no involvement in manuscript writing.

REFERENCES

Aisenberg, A. C., & Wilkes, B. (1965). Partial immunological restoration of neonatally thymectomized rats with thymus-containing diffusion chambers. *Nature*, *205*, 716–717.

Baxevanis, C. N., Frillingos, S., Seferiadis, K., Reclos, G. J., Arsenis, P., Katsiyiannis, A., et al. (1990). Enhancement of human T lymphocyte function by prothymosin alpha: Increased production of interleukin-2 and expression of interleukin-2 receptors in normal human peripheral blood T lymphocytes. *Immunopharmacology and Immunotoxicology*, *12*, 595–617.

Baxevanis, C. N., Gritzapis, A. D., Dedoussis, G. V., Papadopoulos, N. G., Tsolas, O., & Papamichail, M. (1994). Induction of lymphokine-activated killer activity in mice by prothymosin alpha. *Cancer Immunology, Immunotherapy*, *38*, 281–286.

Baxevanis, C. N., Gritzapis, A. D., Spanakos, G., Tsitsilonis, O. E., & Papamichail, M. (1995). Induction of tumor-specific T lymphocyte responses in vivo by prothymosin alpha. *Cancer Immunology, Immunotherapy*, *40*, 410–418.

Baxevanis, C. N., Reclos, G. J., & Papamichail, M. (1993). Prothymosin alpha restores depressed allogeneic cell-mediated lympholysis and natural-killer-cell activity in patients with cancer. *International Journal of Cancer*, *53*, 264–268.

Baxevanis, C. N., Spanakos, G., Voutsas, I. F., Gritzapis, A. D., Tsitsilonis, O. E., Mamalaki, A., et al. (1999). Increased generation of autologous tumor-reactive lymphocytes by anti-CD3 monoclonal antibody and prothymosin alpha. *Cancer Immunology, Immunotherapy*, *48*, 71–84.

Baxevanis, C. N., Thanos, D., Reclos, G. J., Anastasopoulos, E., Tsokos, G. C., Papamatheakis, J., et al. (1992). Prothymosin alpha enhances human and murine MHC class II surface antigen expression and messenger RNA accumulation. *Journal of Immunology*, *148*, 1979–1984.

Bianchi, M. E. (2007). DAMPs, PAMPs and alarmins: All we need to know about danger. *Journal of Leukocyte Biology*, *81*, 1–5.

Bowick, G. C., Soman, K. V., Wang, H., et al. (2010). Proteomic analysis of Pichindé virus infection identifies differential expression of prothymosin-α. *Journal of Biomedicine and Biotechnology*, *2010*, Article ID: 956823, 9 pages. http://dx.doi.org/10.1155/2010/956823.

Caldarella, J., Goodall, G. J., Felix, A. M., Heimer, E. P., Salvin, S. B., & Horecker, B. L. (1983). Thymosin alpha 11: A peptide related to thymosin alpha 1 isolated from calf thymosin fraction 5. *Proceedings of the National Academy of Sciences of the United States of America*, *80*, 7424–7427.

Camerini, R., & Garaci, E. (2015). Historical review of thymosin α1 in infectious diseases. *Expert Opinion on Biological Therapy*, *1*, S117–S127.

Cannavo, A., Rengo, G., Liccardo, D., Pironti, G., Scimia, M. C., Scudiero, L., et al. (2013). Prothymosin alpha protects cardiomyocytes against ischemia-induced apoptosis via preservation of Akt activation. *Apoptosis*, *18*, 1252–1261.

Carey, K. A., Segal, D., Klein, R., Sanigorski, A., Walder, K., Collier, G. R., et al. (2006). Identification of novel genes expressed during rhabdomyosarcoma differentiation using cDNA microarrays. *Pathology International*, *56*, 246–255.

Chan, J. K., Roth, J., Oppenheim, J. J., Tracey, K. J., Vogl, T., Feldmann, M., et al. (2012). Alarmins: Awaiting a clinical response. *The Journal of Clinical Investigation, 122*, 2711–2719.
Cordero, O. J., Sarandeses, C. S., López, J. L., Cancio, E., Regueiro, B. J., & Nogueira, M. (1991). Prothymosin alpha enhances interleukin 2 receptor expression in normal human T-lymphocytes. *International Journal of Immunopharmacology, 13*, 1059–1065.
Cordero, O. J., Sarandeses, C. S., López, J. L., & Nogueira, M. (1992). Prothymosin alpha enhances human natural killer cell cytotoxicity: Role in mediating signals for NK activity. *Lymphokine and Cytokine Research, 11*, 277–285.
Cordero, O. J., Sarandeses, C., López-Rodríguez, J. L., & Nogueira, M. (1995). The presence and cytotoxicity of CD16 + CD2- subset from PBL and NK cells in long-term IL-2 cultures enhanced by Prothymosin-alpha. *Immunopharmacology, 29*, 215–223.
Cordero, O. J., Sarandeses, C. S., & Nogueira, M. (1996). Binding of 125I-prothymosin alpha to lymphoblasts through the non-thymosin alpha 1 sequence. *Life Sciences, 58*, 1757–1770.
Dauphinee, M. J., Tala, N., Goldstein, A. L., & White, A. (1974). Thymosin corrects the abnormal DNA synthetic response of NZB mouse thymocytes. *Proceedings of the National Academy of Sciences of the United States of America, 71*, 2637–2641.
Dominguez, F., Magdalena, C., Cancio, E., Roson, E., Paredes, J., Loidi, L., et al. (1993). Tissue concentrations of prothymosin alpha: A novel proliferation index of primary breast cancer. *European Journal of Cancer, 29A*, 893–897.
Eckert, K., Garbin, F., Maurer, H. R., Büttner, P., Garbe, C., & Czarnecki, J. (1995). Prothymosin alpha 1 modulates lymphokine-activated killer cell activity and IL-2 production by peripheral blood lymphocytes from melanoma patients in vitro. *International Journal of Immunopharmacology, 17*, 555–561.
Eckert, K., Grünberg, E., Garbin, F., & Maurer, H. R. (1997a). Preclinical studies with prothymosin alpha1 on mononuclear cells from tumor patients. *International Journal of Immunopharmacology, 19*, 493–500.
Eckert, K., Grünberg, E., Immenschuh, P., Garbin, F., Kreuser, E. D., & Maurer, H. R. (1997b). Interleukin-2-activated killer cell activity in colorectal tumor patients: Evaluation of in vitro effects by prothymosin alpha1. *Journal of Cancer Research and Clinical Oncology, 123*, 420–428.
Enkemann, S. A., Wang, R. H., Trumbore, M. W., & Berger, S. L. (2000). Functional discontinuities in prothymosin α caused by caspase cleavage in apoptotic cells. *Journal of Cell Physiology, 182*, 256–268.
Eschenfeldt, W. H., & Berger, S. L. (1986). The human prothymosin alpha gene is polymorphic and induced upon growth stimulation: Evidence using a cloned cDNA. *Proceedings of the National Academy of Sciences of the United States of America, 83*, 9403–9407.
Evstafieva, A. G., Belov, G. A., Rubtsov, Y. P., Kalkum, M., Joseph, B., Chichkova, N. V., et al. (2003). Apoptosis-related fragmentation, translocation, and properties of human prothymosin alpha. *Experimental Cell Research, 284*, 211–223.
Fujita, R., & Ueda, H. (2007). Prothymosin-alpha prevents necrosis and apoptosis following stroke. *Cell Death and Differentiation, 14*, 1839–1842.
Fujita, R., Ueda, M., Fujiwara, K., & Ueda, H. (2009). Prothymosin-alpha plays a defensive role in retinal ischemia through necrosis and apoptosis inhibition. *Cell Death and Differentiation, 16*, 349–358.
Garbin, F., Eckert, K., Büttner, P., Garbe, C., & Maurer, H. R. (1994). Prothymosin alpha augments deficient antitumor activity of monocytes from melanoma patients in vitro. *Anticancer Research, 14*, 2405–2411.
Garbin, F., Eckert, K., Immenschuh, P., Kreuser, E. D., & Maurer, H. R. (1997). Prothymosin alpha 1 effects, in vitro, on the antitumor activity and cytokine production of blood monocytes from colorectal tumor patients. *International Journal of Immunopharmacology, 19*, 323–332.

Gast, K., Damaschun, H., Eckert, K., Schulze-Forster, K., Maurer, H. R., Müller-Frohne, M., et al. (1995). Prothymosin alpha: A biologically active protein with random coil conformation. *Biochemistry*, *34*, 13211–13218.

Goldstein, A. L. (2007). History of the discovery of the thymosins. *Annals of the New York Academy of Sciences*, *1112*, 1–13.

Goldstein, A. L., Low, T. L., Thurman, G. B., Zatz, M. M., Hall, N., Chen, J., et al. (1981). Current status of thymosin and other hormones of the thymus gland. *Recent Progress in Hormone Research*, *37*, 369–415.

Goldstein, A. L., Slater, F. D., & White, A. (1966). Preparation, assay, and partial purification of a thymic lymphocytopoietic factor (thymosin). *Proceedings of the National Academy of Sciences of the United States of America*, *56*, 1010–1017.

Gomez-Marquez, J., & Rodriguez, P. (1998). Prothymosin alpha is a chromatin-remodeling protein in mammalian cells. *Biochemical Journal*, *333*, 1–3.

Good, R. A., Dalmasso, A. P., Martinez, C., Archer, O. K., Pierce, J. C., & Papermaster, B. W. (1962). The role of the thymus in development of immunologic capacity in rabbits and mice. *Journal of Experimental Medicine*, *116*, 773–796.

Ha, S. Y., Song, D. H., Hwang, S. H., Cho, S. Y., & Park, C. K. (2015). Expression of prothymosin alpha predicts early recurrence and poor prognosis of hepatocellular carcinoma. *Hepatobiliary & Pancreatic Diseases International*, *14*, 171–177.

Halder, S. K., Matsunaga, H., Ishii, K. J., & Ueda, H. (2015). Prothymosin-alpha preconditioning activates TLR4-TRIF signaling to induce protection of ischemic retina. *Journal of Neurochemistry*, *135*, 1161–1177.

Halder, S. K., Matsunaga, H., & Ueda, H. (2012). Neuron-specific non-classical release of prothymosin alpha: A novel neuroprotective damage-associated molecular patterns. *Journal of Neurochemistry*, *123*, 262–275.

Halder, S. K., Sugimoto, J., Matsunaga, H., & Ueda, H. (2013). Therapeutic benefits of 9-amino acid peptide derived from prothymosin alpha against ischemic damages. *Peptides*, *43*, 68–75.

Hannappel, E., Davoust, S., & Horecker, B. L. (1982a). Isolation of peptides from calf thymus. *Biochemical and Biophysical Research Communications*, *104*, 266–271.

Hannappel, E., Davoust, S., & Horecker, B. L. (1982b). Thymosins beta 8 and beta 9: Two new peptides isolated from calf thymus homologous to thymosin beta 4. *Proceedings of the National Academy of Sciences of the United States of America*, *79*, 1708–1711.

Hannappel, E., & Huff, T. (2003). The thymosins. Prothymosin alpha, parathymosin, and beta-thymosins: Structure and function. *Vitamins and Hormones*, *66*, 257–296.

Haritos, A. A. (1987). Alpha-thymosins: Relationships in structure, distribution, and function. *Isozymes Current Top Biological & Medicinal Research*, *14*, 123–152.

Haritos, A. A., Goodall, G. J., & Horecker, B. L. (1984). Prothymosin alpha: Isolation and properties of the major immunoreactive form of thymosin alpha 1 in rat thymus. *Proceedings of the National Academy of Sciences of the United States of America*, *81*, 1008–1011.

Haritos, A. A., & Horecker, B. L. (1985). A radioimmunoassay for thymosin alpha 1 that detects the native polypeptide, prothymosin alpha. *Journal of Immunological Methods*, *81*, 199–205.

Haritos, A. A., Tsolas, O., & Horecker, B. L. (1984). Distribution of prothymosin alpha in rat tissues. *Proceedings of the National Academy of Sciences of the United States of America*, *81*, 1391–1393.

Heidecke, H., Eckert, K., Schulze-Forster, K., & Maurer, H. R. (1997). Prothymosin alpha 1 effects in vitro on chemotaxis, cytotoxicity and oxidative response of neutrophils from melanoma, colorectal and breast tumor patients. *International Journal of Immunopharmacology*, *19*, 413–420.

Ho, A. D., Ma, D. D., Price, G., Hunstein, W., & HoVbrand, A. V. (1983). Biochemical and immunological differentiation of human thymocytes induced by thymic hormones. *Immunology*, *50*, 471–476.

Hooper, J. A., McDaniel, M. C., Thurman, G. B., Cohen, G. H., Schulof, R. S., & Goldstein, A. L. (1975). Purification and properties of bovine thymosin. *Annals of the New York Academy of Sciences, 249*, 125–144.

Ioannou, K., Derhovanessian, E., Tsakiri, E., Samara, P., Kalbacher, H., Voelter, W., et al. (2013). Prothymosin α and a prothymosin α-derived peptide enhance T(H)1-type immune responses against defined HER-2/neu epitopes. *BMC Immunology, 14*, 43.

Ioannou, K., Samara, P., Livaniou, E., Derhovanessian, E., & Tsitsilonis, O. E. (2012). Prothymosin alpha: A ubiquitous polypeptide with potential use in cancer diagnosis and therapy. *Cancer Immunology, Immunotherapy, 61*, 599–614.

Jiang, X., Kim, H. E., Shu, H., Zhao, Y., Zhang, H., Kofron, J., et al. (2003). Distinctive roles of PHAP proteins and prothymosin-alpha in a death regulatory pathway. *Science, 299*, 223–226.

Jin, Y., Cao, C., Li, P., Liu, X., Huang, W., Li, C., et al. (2005). Boosting immune response to hepatitis B DNA vaccine by coadministration of Prothymosin alpha-expressing plasmid. *Clinical and Diagnostic Laboratory Immunology, 12*, 1364–1369.

Jou, Y. C., Tung, C. L., Tsai, Y. S., Shen, C. H., Syue-Yi, C., Shiau, A. L., et al. (2009). Prognostic relevance of prothymosin-alpha expression in human upper urinary tract transitional cell carcinoma. *Urology, 74*, 951–957.

Karachaliou, C. E., Liolios, C., Triantis, C., Zikos, C., Samara, P., Tsitsilonis, O. E., et al. (2015). Specific in vitro binding of a new (99m)Tc-radiolabeled derivative of the C-terminal decapeptide of prothymosin alpha on human neutrophils. *International Journal of Pharmaceutics, 486*, 1–12.

Karetsou, Z., Kretsovali, A., Murphy, C., Tsolas, O., & Papamarcaki, T. (2002). Prothymosin alpha interacts with the CREB-binding protein and potentiates transcription. *EMBO Reports, 3*, 361–366.

Kashat, L., So, A. K., Masui, O., Wang, X. S., Cao, J., Meng, X., et al. (2010). Secretome-based identification and characterization of potential biomarkers in thyroid cancer. *Journal of Proteome Research, 9*, 5757–5769.

Khaw, B. A., & Rule, A. H. (1973). Immunotherapy of the Dunning leukemia with thymic extracts. *British Journal of Cancer, 28*, 288–292.

Klein, J. J., Goldstein, A. L., & White, A. (1965). Enhancement of in vivo incorporation of labeled precursors into DNA and total protein of mouse lymph nodes after administration of thymic extracts. *Proceedings of the National Academy of Sciences of the United States of America, 53*, 812–817.

Klimentzou, P., Drougou, A., Fehrenbacher, B., Schalle, M., Voelter, W., Barbatis, C., et al. (2008). Immunocytological and preliminary immunohistochemical studies of prothymosin alpha, a human cancer-associated polypeptide, with a well-characterized polyclonal antibody. *The Journal of Histochemistry and Cytochemistry, 56*, 1023–1031.

Letsas, K. P., Vartholomatos, G., Tsepi, C., Tsatsoulis, A., & Frangou-Lazaridis, M. (2007). Fine-needle aspiration biopsy-RT-PCR expression analysis of prothymosin alpha and parathymosin in thyroid: Novel proliferation markers? *Neoplasma, 54*, 57–62.

Levey, R. H., Trainin, N., & Law, L. W. (1963). Evidence for function of thymic tissue in diffusion chambers implanted in neonatally thymectomized mice. Preliminary report. *Journal of the National Cancer Institute, 31*, 199–217.

Leys, C. M., Nomura, S., LaFleur, B. J., Ferrone, S., Kaminishi, M., Montgomery, E., et al. (2007). Expression and prognostic significance of prothymosin-alpha and ERp57 in human gastric cancer. *Surgery, 141*, 41–50.

Li, K. J., Shiau, A. L., Chiou, Y. Y., Yo, Y. T., & Wu, C. L. (2005). Transgenic overexpression of prothymosin alpha induces development of polycystic kidney disease. *Kidney International, 67*, 1710–1722.

López-Rodríguez, J. L., Cordero, O. J., Sarandeses, C., Viñuela, J., & Nogueira, M. (1994). Interleukin-2 killer cells: In vitro evaluation of combination with prothymosin alpha. *Lymphokine and Cytokine Research, 13*, 175–182.

Low, T. L., Hu, S. K., & Goldstein, A. L. (1981). Complete amino acid sequence of bovine thymosin beta 4: A thymic hormone that induces terminal deoxynucleotidyl transferase activity in thymocyte populations. *Proceedings of the National Academy of Sciences of the United States of America, 78*, 1162–1166.
Magdalena, C., Dominguez, F., Loidi, L., & Puente, J. L. (2000). Tumour prothymosin alpha content, a potential prognostic marker for primary breast cancer. *British Journal of Cancer, 82*, 584–590.
Malinda, K. M., Sidhu, G. S., Banaudha, K. K., Gaddipati, J. P., Maheshwari, R. K., Goldstein, A. L., et al. (1998). Thymosin alpha 1 stimulates endothelial cell migration, angiogenesis, and wound healing. *Journal of Immunology, 160*, 1001–1006.
Manrow, R. E., Sburlati, A. R., Hanover, J. A., & Berger, S. L. (1991). Nuclear targeting of prothymosin alpha. *Journal of Biological Chemistry, 266*, 3916–3924.
Matsunaga, H., & Ueda, H. (2010). Stress-induced non-vesicular release of prothymosin-α initiated by an interaction with S100A13, and its blockade by caspase-3 cleavage. *Cell Death and Differentiation, 17*, 1760–1772.
Miller, J. F. (1961). Immunological function of the thymus. *Lancet, 2*, 748–749.
Miller, J. F. (2002). The discovery of thymus function and of thymus-derived lymphocytes. *Immunology Reviews, 185*, 7–14.
Mori, M., Barnard, G. F., Staniunas, R. J., Jessup, J. M., Steele, G. D., Jr., & Chen, L. B. (1993). Prothymosin-alpha mRNA expression correlates with that of c-myc in human colon cancer. *Oncogene, 8*, 2821–2826.
Mosoian, A., Teixeira, A., Burns, C. S., Sander, L. E., Gusella, G. L., He, C., et al. (2010). Prothymosin-alpha inhibits HIV-1 via Toll-like receptor 4-mediated type I interferon induction. *Proceedings of the National Academy of Sciences of the United States of America, 107*, 10178–10183.
Mosoian, A., Teixeira, A., High, A. A., Christian, R. E., Hunt, D. F., Shabanowitz, J., et al. (2006). Novel function of prothymosin alpha as a potent inhibitor of human immunodeficiency virus type 1 gene expression in primary macrophages. *Journal of Virology, 80*, 9200–9206.
Niture, S. K., Kaspar, J. W., Shen, J., & Jaiswal, A. K. (2010). Nrf2 signaling and cell survival. *Toxicology and Applied Pharmacology, 244*, 37–42.
Omotuyi, O., Matsunaga, H., & Ueda, H. (2015). Evidence for ProTα-TLR4/MD-2 binding: Molecular dynamics and gravimetric assay studies. *Expert Opinion on Biological Therapy, 1*, S223–S229.
Pan, L. X., Haritos, A. A., Wideman, J., Komiyama, T., Chang, M., Stein, S., et al. (1986). Human prothymosin alpha: Amino acid sequence and immunologic properties. *Archives of Biochemistry and Biophysics, 250*, 197–201.
Papanastasiou, M., Baxevanis, C. N., & Papamichail, M. (1992). Promotion of murine antitumor activity by prothymosin alpha treatment: I. Induction of tumoricidal peritoneal cells producing high levels of tumour necrosis factor alpha. *Cancer Immunology, Immunotherapy, 35*, 145–150.
Pavlov, N. A., Cherny, D. I., Heim, G., Jovin, T. M., & Subramaniam, V. (2002). Amyloid fibrils from the mammalian protein prothymosin alpha. *FEBS Letters, 517*, 37–40.
Pawlikowski, M., Radek, M., Kunert-Radek, J., Jaranowska, M., Świętosławski, J., & Winczyk, K. (2014). Overexpression of prothymosin alpha is related to pituitary adenoma recurrence but not to adenoma invasiveness and proliferation. *Endokrynologia Polska, 65*, 382–386.
Petro, T. M., & Watson, R. R. (1982). Resistance to L1210 mouse leukemia cells in moderately protein-malnourished BALB/c mice treated in vivo with thymosin fraction V. *Cancer Research, 42*, 2139–2145.
Piñeiro, A., Begoña Bugia, M., Pilar Arias, M., Cordero, O. J., & Nogueira, M. (2001). Identification of receptors for prothymosin alpha on human lymphocytes. *Biological Chemistry, 382*, 1473–1482.

Piñeiro, A., Cordero, O. J., & Nogueira, M. (2000). Fifteen years of prothymosin alpha: Contradictory past and new horizons. *Peptides*, *21*, 1433–1446.

Qiu, L., Guo, B. Y., Miao, H., Dao, S. Y., Zhang, R., Yuan, P. Q., et al. (2002). Effect of recombinant prothymosin alpha on secretion of IFN-gamma, IFN-alpha and TNF-alpha in vitro. *Yao Xue Xue Bao*, *37*, 326–328.

Romani, L., Moretti, S., Fallarino, F., Bozza, S., Ruggeri, L., Casagrande, A., et al. (2012). Jack of all trades: Thymosin α1 and its pleiotropy. *Annals of the New York Academy of Sciences*, *1269*, 1–6.

Rubtsov, Y. P., Zolotukhin, A. S., Vorobjev, I. A., Chichkova, N. V., Pavlov, N. A., Karger, E. M., et al. (1997). Mutational analysis of human prothymosin alpha reveals a bipartite nuclear localization signal. *FEBS Letters*, *413*, 135–141.

Salgado, F. J., Piñeiro, A., Canda-Sánchez, A., Lojo, J., & Nogueira, M. (2005). Prothymosin alpha-receptor associates with lipid rafts in PHA-stimulated lymphocytes. *Molecular Membrane Biology*, *22*, 163–176.

Samara, P., Ioannou, K., Neagu, M., Arnogiannaki, N., Ardavanis, A., Voelter, W., et al. (2013). The C-terminal decapeptide of prothymosin α is responsible for its stimulatory effect on the functions of human neutrophils in vitro. *International Immunopharmacology*, *15*, 50–57.

Samara, P., Kalbacher, H., Ioannou, K., Radu, D. L., Livaniou, E., Promponas, V. J., et al. (2013). Development of an ELISA for the quantification of the C-terminal decapeptide prothymosin α(100-109) in sera of mice infected with bacteria. *Journal of Immunological Methods*, *395*, 54–62.

Sasaki, H., Nonaka, M., Fujii, Y., Yamakawa, Y., Fukai, I., Kiriyama, M., et al. (2001). Expression of the prothymosin-alpha gene as a prognostic factor in lung cancer. *Surgery Today*, *31*, 936–938.

Sasaki, H., Sato, Y., Kondo, S., Fukai, I., Kiriyama, M., Yamakawa, Y., et al. (2001). Expression of the prothymosin alpha mRNA correlated with that of N-myc in neuroblastoma. *Cancer Letters*, *168*, 191–195.

Sburlati, A. R., Manrow, R. E., & Berger, S. L. (1990). Human prothymosin alpha: Purification of a highly acidic nuclear protein by means of a phenol extraction. *Protein Expression and Purification*, *1*, 184–190.

Sburlati, A. R., Manrow, R. E., & Berger, S. L. (1991). Prothymosin alpha antisense oligomers inhibit myeloma cell division. *Proceedings of the National Academy of Sciences of the United States of America*, *88*, 253–257.

Schiller, M., Heyder, P., Ziegler, S., Niessen, A., Claßen, L., Lauffer, A., et al. (2013). During apoptosis HMGB1 is translocated into apoptotic cell-derived membranous vesicles. *Autoimmunity*, *46*, 342–346.

Skopeliti, M., Iconomidou, V. A., Derhovanessian, E., Pawelec, G., Voelter, W., Kalbacher, H., et al. (2009). Prothymosin alpha immunoactive carboxyl-terminal peptide TKKQKTDEDD stimulates lymphocyte reactions, induces dendritic cell maturation and adopts a beta-sheet conformation in a sequence-specific manner. *Molecular Immunology*, *46*, 784–792.

Skopeliti, M., Kratzer, U., Altenberend, F., Panayotou, G., Kalbacher, H., Stevanovic, S., et al. (2007). Proteomic exploitation on prothymosin alpha-induced mononuclear cell activation. *Proteomics*, 7, 1814–1824.

Skopeliti, M., Voutsas, I. F., Klimentzou, P., Tsiatas, M. L., Beck, A., Bamias, A., et al. (2006). The immunologically active site of prothymosin alpha is located at the carboxy-terminus of the polypeptide. Evaluation of its in vitro effects in cancer patients. *Cancer Immunology, Immunotherapy*, *55*, 1247–1257.

Spangelo, B. L., Farrimond, D. D., Pompilius, M., & Bowman, K. L. (2000). Interleukin-1β and thymic peptide regulation of pituitary and glial cell cytokine expression and cellular proliferation. *Annals of the New York Academy of Sciences*, *917*, 597–607.

Spangelo, B. L., Hall, N. R., & Goldstein, A. L. (1987). Biology and chemistry of thymosin peptides. Modulators of immunity and neuroendocrine circuits. *Annals of the New York Academy of Sciences*, *496*, 196–204.

Spangelo, B. L., Roach, J. D., Hadi, F., Damavandy, A. A., Plieskatt, J., & Badamchian, M. (2007). Thymosin fraction-5 possesses antiproliferative properties in HL-60 human promyelocytic leukemia cells: Characterization of an active peptide. *Annals of the New York Academy of Sciences*, *1112*, 305–316.

Su, Y. C., Ou, H. Y., Wu, H. T., Wu, P., Chen, Y. C., Su, B. H., et al. (2015). Prothymosin-α overexpression contributes to the development of insulin resistance. *The Journal of Clinical Endocrinology and Metabolism*, *100*, 4114–4123.

Su, B. H., Tseng, Y. L., Shieh, G. S., Chen, Y. C., Shiang, Y. C., Wu, P., et al. (2013). Prothymosin α overexpression contributes to the development of pulmonary emphysema. *Nature Communications*. *4*, Article number: 1906. http://dx.doi.org/10.1038/ncomms2906.

Su, B.-H., Tseng, Y.-L., Shieh, G.-S., Chen, Y.-C., Wu, P., Shiau, A.-L., et al. (2016). Over-expression of prothymosin-α antagonizes TGFβ signalling to promote the development of emphysema. *Journal of Pathology*, *238*, 412–422.

Suzuki, S., Takahashi, S., Takahashi, S., Takeshita, K., Hikosaka, A., Wakita, T., et al. (2006). Expression of prothymosin alpha is correlated with development and progression in human prostate cancers. *Prostate*, *66*, 463–469.

Szabo, P., Ehleiter, D., Whittington, E., & Weksler, M. E. (1992). Prothymosin alpha expression occurs during G1 in proliferating B or T lymphocytes. *Biochemical and Biophysical Research Communications*, *185*, 953–959.

Szabo, P., Panneerselvam, C., Clinton, M., Frangou-Lazaridis, M., Weksler, D., Whittington, E., et al. (1993). Prothymosin alpha gene in humans: Organization of its promoter region and localization to chromosome 2. *Human Genetics*, *90*, 629–634.

Talmadge, J. E., Uithoven, K. A., Lenz, B. F., & Chirigos, M. (1984). Immunomodulation and therapeutic characterization of thymosin fraction five. *Cancer Immunology, Immunotherapy*, *18*, 185–194.

Teixeira, A., Yen, B., Gusella, G. L., Thomas, A. G., Mullen, M. P., Aberg, J., et al. (2015). Prothymosin α variants isolated from CD8+ T cells and cervicovaginal fluid suppress HIV-1 replication through type I interferon induction. *Journal of Infectious Diseases*, *211*, 1467–1475.

Telusma, G., Datta, S., Mihajlov, I., Ma, W., Li, J., Yang, H., et al. (2006). Dendritic cell activating peptides induce distinct cytokine profiles. *International Immunology*, *18*, 1563–1573.

Tripathi, S. C., Matta, A., Kaur, J., Grigull, J., Chauhan, S. S., Thakar, A., et al. (2011). Overexpression of prothymosin alpha predicts poor disease outcome in head and neck cancer. *PLoS One*, *6*, e19213.

Tsai, Y. S., Jou, Y. C., Tung, C. L., Lin, C. T., Shen, C. H., Chen, S. Y., et al. (2014). Loss of nuclear prothymosin-α expression is associated with disease progression in human superficial bladder cancer. *Virchows Archiv*, *464*, 717–724.

Tsitsiloni, O. E., Stiakakis, J., Koutselinis, A., Gogas, J., Markopoulos, C., Yialouris, P., et al. (1993). Expression of alpha-thymosins in human tissues in normal and abnormal growth. *Proceedings of the National Academy of Sciences of the United States of America*, *90*, 9504–9507.

Tsitsiloni, O. E., Yialouris, P. P., Sekeri-Pataryas, K., & Haritos, A. A. (1989). Prothymosin alpha is not a nuclear polypeptide. *Experientia*, *45*, 332–334.

Tsitsilonis, O. E., Bekris, E., Voutsas, I. F., Baxevanis, C. N., Markopoulos, C., Papadopoulos, S. A., et al. (1998). The prognostic value of alpha-thymosins in breast cancer. *Anticancer Research*, *18*, 1501–1508.

Ueda, H. (2009). Prothymosin alpha and cell death mode switch, a novel target for the prevention of cerebral ischemia-induced damage. *Pharmacology & Therapeutics*, *123*, 323–333.

Ueda, H., Fujita, R., Yoshida, A., Matsunaga, H., & Ueda, M. (2007). Identification of prothymosin-alpha1, the necrosis-apoptosis switch molecule in cortical neuronal cultures. *Journal of Cell Biology*, *176*, 853–862.

Voutsas, I. F., Baxevanis, C. N., Gritzapis, A. D., Missitzis, I., Stathopoulos, G. P., Archodakis, G., et al. (2000). Synergy between interleukin-2 and prothymosin alpha for the increased generation of cytotoxic T lymphocytes against autologous human carcinomas. *Cancer Immunology, Immunotherapy*, *49*, 449–458.

Voutsas, I. F., Pistamaltzian, N., Tsiatas, M. L., Skopeliti, M., Katsila, T., Mavrothalassiti, I., et al. (2013). Ovarian malignant ascites-derived lymphocytes stimulated with prothymosin α or its immunoactive decapeptide lyse autologous tumour cells in vitro and retard tumour growth in SCID mice. *European Journal of Cancer*, *49*, 1706–1714.

Wara, D. W., & Ammann, A. J. (1975). Activation of T-cell rosettes in immunodeficient patients by thymosin. *Annals of the New York Academy of Sciences*, *249*, 308–315.

Welsh, R. M., Bahl, K., Marshall, H. D., & Urban, S. L. (2012). Type 1 interferons and antiviral CD8 T-cell responses. *PLoS Pathogen*, *8*, e1002352.

Wu, C. G., Habib, N. A., Mitry, R. R., Reitsma, P. H., van Deventer, S. J., & Chamulueau, R. A. (1997). Overexpression of hepatic prothymosin alpha, a novel marker for hepatocellular carcinoma. *British Journal of Cancer*, *76*, 1199–1204.

Wu, C. L., Shiau, A. L., & Lin, C. S. (1997). Prothymosin alpha promotes cell proliferation in NIH3T3 cells. *Life Sciences*, *61*, 2091–2101.

Zhang, M., Cu, F., Lu, S., Lu, H., Jiang, T., Chen, J., et al. (2014). Increased expression of prothymosin-α, independently or combined with TP53, correlates with poor prognosis in colorectal cancer. *International Journal of Clinical and Experimental Pathology*, *7*, 4867–4876.

Zhang, B. C., & Sun, L. (2015). Tongue sole (Cynoglossus semilaevis) prothymosin alpha: Cytokine-like activities associated with the intact protein and the C-terminal region that lead to antiviral immunity via Myd88-dependent and -independent pathways respectively. *Developmental & Comparative Immunology*, *53*, 96–104.

CHAPTER NINE

Cardioprotection by Thymosin Beta 4

G.T. Pipes[1], J. Yang
Cardiovascular Drug Discovery, Discovery Biology Research & Development, Bristol-Myers Squibb, Pennington, NJ, United States
[1]Corresponding author: e-mail address: gordon.pipes@bms.com

Contents

Abstract

Treatment with thymosin beta 4 (Tβ4) reduces infarct volume and preserves cardiac function in preclinical models of cardiac ischemic injury. These effects stem in part from decreased infarct size, but additional benefits are likely due to specific antifibrotic and proangiogenic activities. Injected or transgenic Tβ4 increase blood vessel growth in large and small animal models, consistent with Tβ4 converting hibernating myocardium to an actively contractile state following ischemia. Tβ4 and its degradation products have antifibrotic effects in in vitro assays and in animal models of fibrosis not related to cardiac injury. This large number of pleiotropic effects results from Tβ4's many interactions with cellular signaling pathways, particularly indirect regulation of cellular motility and movement via the SRF–MRTF–G-actin transcriptional pathway. Variation in effects and effect sizes in animal models may potentially be due to variable distribution of Tβ4. Preclinical studies of PK/PD relationships and a reliable pharmacodynamic biomarker would facilitate clinical development of Tβ4.

Vitamins and Hormones, Volume 102
ISSN 0083-6729
http://dx.doi.org/10.1016/bs.vh.2016.04.004

1. INTRODUCTION

Loss of contractile myocardium during ischemia or ischemia–reperfusion injury, incurred during myocardial infarction (MI), coronary artery bypass graft surgery, or cardiac transplantation, is a significant cause of morbidity and mortality worldwide (Sluijter et al., 2014). Unless practical cardiac regeneration can be brought into widespread clinical use, novel strategies will be needed to protect the myocardium during ischemic injury, and also to restore function to injured myocardium.

In 2004, Bock-Marquette and colleagues demonstrated dramatic cardioprotection resulting from injection of thymosin beta 4 (Tβ4). Since then, there have been a large number of studies of cardioprotection by Tβ4 in a variety of settings. This review summarizes these findings, focusing on preclinical studies defining mechanism of action, and also discusses experimental subtleties of some of the animal models used in these studies.

2. BASIC PROPERTIES OF Tβ4: EXPRESSION, STRUCTURE, AND INTERACTION WITH ACTIN

Tβ4 was initially purified in 1966, as part of a mixture of putative hormones from calf thymus (Goldstein, Slater, & White, 1966). Cloning of the human gene encoding Tβ4, TMSB4X (Clauss et al., 1991), and subsequent molecular analyses determined that it is expressed in almost every cell type in mammals. The Tβ4 peptide is short (43 residues) and lacks intrinsic structure in water when not bound to any other proteins, but assumes a conformation with two-alpha helices in some solutions (Zarbock et al., 1990). The protein is abundantly expressed: up to 1% of protein synthesis is of Tβ4, and it is present at ~16 μg/mL in whole blood (Hannappel, 2007).

Tβ4 plays a central role in regulation of actin dynamics (Safer, Elzinga, & Nachmias, 1991; Sanders, Goldstein, & Wang, 1992). Tβ4 binds to actin monomers with moderate avidity; however, given that the concentration of Tβ4 in the cytoplasm is high, the Tβ4:actin complex is thus likely to function as reservoir of actin monomers (Husson et al., 2010). Evidence from X-ray crystallographic studies indicates that the binding interaction between Tβ4 and actin monomers can happen in two modes: one has Tβ4 in a two-alpha helix conformation and is likely to block both the barbed and pointed ends of G-actin. The other, in which Tβ4 has one alpha helix, may permit exchange of actin between Tβ4 and other actin-regulating proteins such as profilin (Xue, Leyrat, Grimes, & Robinson, 2014).

Tβ4 is extensively posttranslationally modified, and these modifications impact its biochemical activities. Tβ4 is always acetylated at its N-terminus in vivo, and often has additional acetyls on multiple internal lysine residues (Hannappel, 2010). Cleavage by endogenous proteases then gives rise to the acetylated, 4-residue peptide Ac-SDKP (Grillon et al., 1990), which is clearly antifibrotic in vivo in preclinical animal models (Peng et al., 2014; Zuo et al., 2013). Tβ4 can be oxidized on methionine residue 6. Tβ4-Met-6-Sulfoxide has been isolated in vivo from teleost (Evans et al., 2013) and is generated by mammalian neutrophils following treatment with glucocorticoids (Young et al., 1999). Exogenous Tβ4-Met-6-Sulfoxide appears to be antiinflammatory in vivo, and this activity could play a role in the cardioprotective effects of Tβ4 (Evans et al., 2013).

3. SIGNALING PATHWAYS DOWNSTREAM OF Tβ4

Tβ4 has a very large interactomic footprint, exerting multiple effects on cytoplasmic signaling, and binding to multiple partners. Here we give a brief overview of actin-dependent, actin-independent, and extracellular receptor-mediated signaling downstream of Tβ4, giving special emphasis to pathways relevant to the cardioprotective effects of Tβ4.

3.1 Actin-Dependent

Tβ4 levels have been shown to impact transcription indirectly, via myocardin-related transcription factor A (MRTF-A, also known as MAL). MRTF-A activity is regulated by binding to actin monomers: MRTF-A is sequestered in the cytoplasm by binding to G-actin, preventing it from entering the nucleus to stimulate transcription (Posern, Miralles, Guettler, & Treisman, 2004; Vartiainen, Guettler, Larijani, & Treisman, 2007). MRTF-A competes with Tβ4 for the same binding site on G-actin, and addition of Tβ4 effectively competes off MRTF-A from actin, resulting in nuclear translocation of MRTF-A and activation of cytoskeletal genes regulated by MRTF-A (Morita & Hayashi, 2013). This aspect of Tβ4 signaling, via G-actin and MRTF-A, has been shown to be relevant in vivo, for coronary angiogenesis in the pig (Hinkel et al., 2014). These activities are evidently not required for cardioprotection per se, as a recent study showed that a C-terminal truncation of Tβ4 consisting of the residues AGES, which is not known to bind actin, is sufficient to reduce infarct size in mouse and pig animal models (Hinkel et al., 2015).

MRTF-A interacts with and regulates the activity of multiple transcription factors, including SRF, FOXO4, and GATA-4 (Mikhailov & Torrado,

2012). These transcription factors in turn play central roles in cardiac development, growth, muscle differentiation, and vascular structure (Estrella & Naya, 2014; Franco & Li, 2009; Maiese, Chong, Hou, & Shang, 2009; Xin et al., 2006). Thus, the Tβ4-G-actin aspect of Tβ4 signaling is likely to indirectly impact myriad cellular and physiological processes.

3.2 Actin-Independent

Beyond actin, multiple binding interactions between Tβ4 and other proteins have been identified, including the Ku80 subunit of DNA helicase (Bednarek et al., 2008), stabilin-2 (Lee, So, Park, & Kim, 2008), PINCH-1, integrin-linked kinase (ILK) (Bock-Marquette, Saxena, White, Dimaio, & Srivastava, 2004), and hMLH1 (Brieger, Plotz, Zeuzem, & Trojan, 2007). ILK is itself a central signaling node, indirectly regulating a wide variety of cellular processes including motility, adhesion, migration, and proliferation (Legate, Montanez, Kudlacek, & Fassler, 2006). The Tβ4–ILK complex furthermore activates AKT (Cheng et al., 2008; Fan, Gong, Ghosh, Graham, & Fox, 2009; Sopko et al., 2011), which in turn has been shown to regulate cell survival, proliferation, angiogenesis, and epithelial-to-mesenchymal transitions (Cheng et al., 2008; Legate et al., 2006)—all of which play significant roles in the mechanisms by which Tβ4 protects the myocardium during injury (Fig. 1). In contrast, it is not currently clear whether the interactions between Tβ4 and Ku80, stabilin-2 and hMLH1 are engaged in vivo in the injured myocardium following administration of Tβ4.

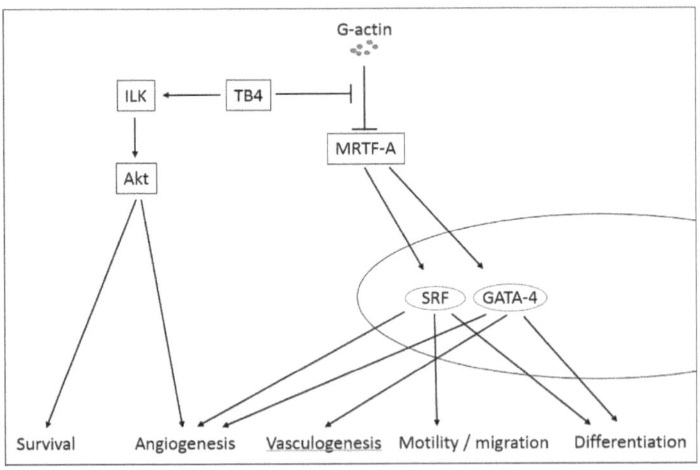

Fig. 1 Simplified diagram depicting linkage between the MRTF-actin pathway and the ILK-Akt pathway via Tβ4, and subsequent effects at the cellular and tissue levels. (See the color plate.)

3.3 Extracellular Receptor-Mediated

Tβ4 has been shown to be protective in cardiovascular injury models using transgenic overexpression and also using injected protein. Transgenic overexpression would result in increased Tβ4 levels intracellularly, while injection of Tβ4 protein would result in increased Tβ4 levels in the extracellular matrix around cells. Tβ4 is not likely to spontaneously diffuse through cell membranes, based on its size and the number of polar residues in its primary amino acid sequence. Therefore, there may be an extracellular receptor for Tβ4, which either internalizes Tβ4 and makes it available to the cytoplasmic G-actin pool or transduces a signal to the cytoplasm in response to the increased Tβ4 levels. Given that intracellular levels of Tβ4 are already very high, exogenously added Tβ4 likely acts at least partly via receptor-mediated signaling rather than just via increasing the intracellular pool of Tβ4.

One report indicates that Tβ4 signals indirectly through the cardioprotective P2X4 purinergic receptor. Tβ4 binds to the alpha and beta subunits of ATP synthase, and in so doing stimulates migration of HUVEC endothelial cells. Migration of HUVECs is inhibited by siRNA directed against P2X4 and by addition of apyrase, consistent with Tβ4 binding causing a local increase in ATP levels which then activates P2X4 (Freeman, Bowman, & Zetter, 2011). Thus, extracellular Tβ4 may activate a neoangiogenic program. Furthermore, activation of P2X4 may protect hypoxic cardiac myocytes from cell death (Yang, Sonin, Jones, Barry, & Liang, 2004).

There is limited evidence for internalization of Tβ4 in specific cultured cell types. Tβ4 applied to corneal cells in culture is internalized in an actin-dependent manner (Ho et al., 2007), and exogenous Tβ4 is internalized and localized to the actin cytoskeleton in EA.hy926 cells within 45 min (Cierniewski, Sobierajska, Selmi, Kryczka, & Bednarek, 2012). Additionally, Tβ4 applied to HUVECs is either bound to the cells or internalized in a saturable manner; this binding is specific insofar as it can be competed away with unlabeled Tβ4 (Grant et al., 1999).

4. CELL AND TISSUE EFFECTS RELEVANT TO CARDIOPROTECTION

During the acute phase of response to injury, Tβ4 promotes survival of myocardial tissues and enhances epicardial activation. Following this period, beginning roughly 1–3 days after injury, Tβ4 likely reduces inflammation. At this point, Tβ4 treatment has preserved myocardial tissues through or into an ischemic period; if the tissue remains ischemic but is alive, it will be hibernating myocardium. Depending on where in the heart this tissue is located, the location of scar tissue and the stresses it is subjected to by neighboring contractile myocardium, the hibernating tissue may still contribute passively to hemodynamic performance. During the late phase, after inflammatory cells have removed debris from dead tissues and an appropriate scar has formed, prolonged Tβ4 treatment is likely to convert hibernating myocardium into contractile myocardium by restoring blood supply via neoangiogenesis and potentially vasculogenesis.

A major mechanism of cardioprotection by Tβ4 is a reduction in cell death during ischemia or reperfusion injury. This effect is tissue wide: survival of cardiomyocytes, fibroblasts, epicardial cells, and endothelial progenitor cells (EPCs) are all improved following Tβ4 treatment or overexpression. Multiple studies have demonstrated a reduction in cardiomyocyte TUNEL staining following permanent ischemia in the mouse (Bock-Marquette et al., 2004; Peng et al., 2014), and in cardiomyocytes and endothelial cells in ischemia–reperfusion in the pig (Hinkel et al., 2008) resulting from Tβ4 treatment before or just after injury. Suppression of other mechanisms of cell death besides apoptosis has not been investigated, but may also be involved given the large amount of myocardial tissue protected.

In addition to preventing myocardial tissues from dying during ischemia, Tβ4 also promotes blood vessel growth. In the chick chorioallantoic membrane assay, Tβ4 induced angiogenesis at all concentrations tested, from 0.05 to 20 μg per membrane (Koutrafouri et al., 2001). The potency of Tβ4 cannot be determined from these studies, as the effect size was roughly the same at all concentrations tested. In normal mice (ie, no cardiac injury), repeated IP dosing of Tβ4 increased expression of endothelial and SMC marker proteins throughout a 28-day time course (Smart et al., 2010). Multiple studies in the MI mouse model report increased capillary densities in the region neighboring the infarct, with effect sizes ranging from 60% to 250% (Bock-Marquette et al., 2009; Peng et al., 2014; Smart et al., 2010) (Table 1). Transgenic Tβ4 (which is predicted to increase Tβ4 levels in the cytoplasm of the target cells directly, rather than relying on signaling through a transmembrane receptor or internalization of Tβ4) results in increased blood vessel densities, perfusion, and function in the mouse hind limb and porcine myocardium (Hinkel et al., 2014; Postrach et al., 2014).

Evidently, multiple cellular mechanisms underlie the angiogenic and vasculogenic effects of Tβ4. The epicardium, a niche of vascular progenitor cells in the adult heart, mounts a profound response to Tβ4 in injured hearts. Following Tβ4 treatment, infarcted hearts display enhanced epicardial thickening, beyond the amount of epicardial thickening normally observed following injury. In vitro, Tβ4 treatment of epicardial explants induces migration of multipotent "epicardium-derived cells" which express epicardial markers and are capable of differentiating into smooth muscle cells, fibroblasts, and endothelial cells (Smart et al., 2010, 2007a, 2007b). Tβ4's role as an actin-binding protein may play in to these effects via MRTFs (Trembley, Velasquez, de Mesy Bentley, & Small, 2015). Lineage tracing studies indicate that epicardial-derived cells proliferate, giving rise to alpha smooth muscle actin-expressing vascular cells (Bock-Marquette et al., 2009; Smart et al., 2010, 2007a, 2007b).

Another cellular mechanism contributing to activation of new vascular growth by Tβ4 is secretion by EPCs. EPCs home in on injured regions and promote neovascularization, likely via local secretion of growth and trophic factors (Zhang, Malik, & Rehman, 2014). In a porcine cardiac ischemia/reperfusion model (Hinkel et al., 2008) and in a rabbit hind limb ischemia model (Hinkel, Bock-Marquette, Hatzopoulos, & Kupatt, 2010) infusion of EPCs isolated from embryos significantly increased blood vessel growth near the site of injury. However, RNAi vs. Tβ4 reduced this effect in both models. The authors argue that this effect is related to paracrine, non-cell autonomous roles of Tβ4 (ie, secretion of Tβ4 by EPCs rather than expression of Tβ4 within the cytoplasm of EPCs) because direct application of exogenous Tβ4 protein was able to mimic many of the beneficial effects of EPC infusion (Hinkel et al., 2008).

Tβ4-activated cells originating from the epicardium were reported in one study to differentiate into cardiomyocytes (Smart et al., 2011), but these findings are in conflict with similar mouse studies (Zhou et al., 2012). Additionally, measurement of the rate at which de novo cardiomyogenesis occurs (see, for example, van Berlo et al., 2014) in response to Tβ4 treatment have not been reported, so it cannot be determined whether the amount of cardiomyogenesis induced would be sufficient to have any effect on heart function. Additionally, the authors of the initial 2004 Tβ4 cardioprotection study provided evidence that the amount of de novo cardiomyogenesis and neo cardiomyogenesis occurring in response to Tβ4 treatment is negligible: coexpression of stem cell markers and cardiac markers was not increased in Tβ4-treated MI mouse hearts, and the number of immature cardiomyocytes did not increase in response to Tβ4 treatment (Bock-Marquette et al., 2004).

Tβ4-treated MI mice have reduced myocardial fibrosis (Peng et al., 2014; Sopko et al., 2011). However, the reduction in infarct size in these animals preserves cardiac function and reduces wall stress, which indirectly reduces profibrotic stimuli and confounds direct measurement of Tβ4's antifibrotic effects in the heart in ischemic models in vivo. Similarly, in models of diffuse myocardial injury, induced by galectin-3 administration (Liu et al., 2009) or diabetes (Castoldi et al., 2010), improvement in fibrosis resulting from Ac-SDKP administration is concomitant with preservation of cardiac function. In injury models independent of cardiac function, Tβ4 or Ac-SDKP reduced interstitial fibrosis and fibrotic signaling in the kidney (Zuo et al., 2013), Ac-SDKP reduced fibrosis in rats treated with subpressor doses of AngII (Rasoul et al., 2004), and reduced fibrosis in bleomycin-induced lung injury (Conte et al., 2012). It is not clear whether Ac-SDKP accounts for all of the antifibrotic effect for Tβ4. Additionally, data from in vitro studies indicate that Tβ4 or Ac-SDKP directly reduces profibrotic signaling and myofibroblast conversion (Kumar & Gupta, 2011; Peng, Carretero, Peterson, & Rhaleb, 2010). Taken together, these results are consistent with Tβ4's preservation of cardiac function following injury being due in part to specific antifibrotic activities.

Following acute myocardial injuries which cause cell death, inflammatory cells infiltrate the damaged tissue, remove debris from dead and damaged cells, activate the wound healing process, and promote remodeling of dead tissue into a fibrotic scar. Benefit from Tβ4 in cardiac injury models may be due in part to antiinflammatory activities. Systemic inflammation is also a characteristic of chronic heart failure and may play a role in the development and progression of the disease. As such, targeting inflammatory responses acutely or chronically may provide new approaches to treat acute MI and prevent progression of heart failure. In the mouse, dosing Tβ4 intraperitoneally following LPS challenge dose dependently improved survival and reduced inflammatory cytokine induction (Badamchian et al., 2003). Tβ4-Met-6-Sulfoxide reduced inflammatory infiltrate in mice with experimentally induced MIs, but, as noted in Evans et al., this effect may be secondary to the reduction in infarct volume caused by Tβ4. In contrast, supporting the idea that one of the primary effects of Tβ4 is antiinflammatory, Tβ4-Met-6-Sulfoxide treatment of zebrafish with similarly sized infarcts resulted in reduced numbers of leukocytes in the heart (Evans et al., 2013). Similarly, in mice with MIs treated with continuous i.p. Tβ4 via minipump, infarct sizes were similar, but infiltration of inflammatory cells was reduced (Peng et al., 2014). In the pig, virally transduced Tβ4 prolonged survival of transplanted hearts, partly via reduction in inflammation (Postrach et al., 2014), and retroinfusion of Tβ4 into infarcted hearts reduced infiltration of inflammatory cells (Hinkel et al., 2008).

5. CARDIOPROTECTION IN PRECLINICAL ANIMAL MODELS

Data from preclinical studies describing cardioprotection by Tβ4 are tabulated in Table 1. In the first description of cardioprotection by Tβ4, by Bock-Marquette et al. (2004), the authors used a mouse permanent MI model, in which the left anterior descending (LAD) coronary artery is permanently ligated. Tβ4 injection at the time of injury resulted in reduced scar

Table 1 Tβ4 Dosing Regimens and Impacts on Capillary Density, Cardiac Function, and Myocardial Protection

References	Model	Species	Dose	Endpoint and Result
Bock-Marquette et al. (2004)	MI	Mouse	6 mpk IP bolus every 3rd day for 28 days; or 6 mpk intramyocardially; or both	54% reduction in infarct size; %EF improved by 20% points
Bock-Marquette et al. (2009)	MI	Mouse	6 mpk IP bolus; analyses on day 1 or 3	Epicardial thickening; 60% increase in capillary density
Sopko et al. (2011)	MI	Mouse	0.02 mpk intramyocardially and 8 mpk IP bolus every 3rd day for 7 days	38% reduction in infarct size; %EF improved by 40% points
Smart et al. (2010)	MI	Mouse	6 mpk IP bolus every 2nd day for 7 days	250% increase in capillary density
	No injury	Mouse	6 mpk IP bolus every 3rd day for 28 days	Epicardial thickening; increased endothelial and VSMC marker expression
Peng et al. (2014)	MI	Mouse	1.6 mpk/day IP minipump for 7 days prior to MI, then for 7 days	13% reduction in infarct size; 3% points increase in %EF
			1.6 mpk/day IP minipump for 5 weeks post-MI	82% increase in capillary density; 12% reduction in infarct size (ns); 17% points increase in %EF

Continued

Table 1 Tβ4 Dosing Regimens and Impacts on Capillary Density, Cardiac Function, and Myocardial Protection—cont'd

References	Model	Species	Dose	Endpoint and Result
Hinkel et al. (2015)	MI	Mouse	150 μg IP bolus every 3rd day for 21 days; or 200 ng intramyocardially	81% reduction in infarct size; %EF improved by 11% points
	I/R	Pig	20 mg intracoronary retroinfustion	38% reduction in infarct size; improved SES under load
Bao et al. (2013)	MI	Rat	5.4 mpk IP bolus for 3 days, then every 3rd day for 28 days	43% reduction in infarct size; 5% increase in %EF (ns); no increase in capillary density
	I/R	Rat	5.4 mpk IP bolus during ischemia	24% reduction in infarct size
Hinkel et al. (2008)	I/R	Pig	15 mg intracoronary retroinfusion	36% reduction in infarct size; improved SES under load and rest

% Change infarct sizes calculated as the absolute value of (value in uninjured animals − value in injured animals treated with Tβ4)/(value in uninjured animals − value in injured animals treated with vehicle) × 100%. Change for capillary densities calculated as the absolute value of (Tβ4-treated animals − vehicle-treated animals)/vehicle-treated animals × 100. Change for %EF is not % change; rather, it is the simple number of % points difference between injury + vehicle and injury + Tβ4. Values used were estimated by visual inspection of published figures when exact numbers were not provided in the text.

volume and preserved cardiac function. Many studies using the same mouse MI model replicated the finding of cardioprotection resulting from Tβ4 treatment, but with significant variation in effect sizes. %EF in the mouse permanent MI model ranges from a nonstatistically significant 5% increase to a 49% increase, and the percent reduction in infarct volume or scar volume ranges from 81% to 13%. Part of this variation may be due to the technical difficulty of measuring %EF by echocardiography, especially in small animals with a high heart rate such as the mouse. Additional variation in this model is probably due to variation of coronary artery anatomy (compare Icardo & Colvee, 2001 and Salto-Tellez et al., 2004) causing injuries of differing severity. Implanting the ligation at the same anatomical location on different mouse hearts can fail to capture a major tributary of the LAD, reducing infarct sizes in a subset of mice.

In order to more precisely measure the effect sizes of cardioprotection by Tβ4, Bao et al. (2013) tested Tβ4 in a rat permanent MI model. The rat has more consistent coronary artery anatomy and greater blood volume than the mouse, facilitating consistent surgical injury and allowing for analysis of plasma biomarkers of cardiac function. Additionally, rather than using echocardiography to measure %EF, the authors used MRI, which is generally more accurate (Gardner, Bingham, Allen, Blatter, & Anderson, 2009). Despite showing a 43% reduction in infarct size, MI rats treated with Tβ4 showed only a nonstatistically significant 5% points increase in %EF. Some hemodynamic measurements taken at end-of-study were improved in the Tβ4 treatment group. Interestingly, no increase in capillary density was observed following repeated IP bolus Tβ4 at doses similar to those shown to be efficacious in the mouse (Bao et al., 2013). These findings are consistent with Tβ4 protecting a large volume of myocardium into permanent ischemia, but not activating growth of blood vessels, resulting in the majority of the preserved myocardial tissue being noncontractile.

The bases for these differences may be anatomical differences between species (rat hearts have a larger volume of myocardial tissue served by any given coronary artery than in the mouse; hence, there are fewer places for medium-diameter arterioles to collateralize from, and thus fewer places for capillaries to grow out from), differences in the duration of the angiogenic effect, or differences in drug exposure levels. Additionally, for unknown reasons, the level of Akt activation the rat model was lower than that described in the mouse: using electrochemiluminesence to quantify phosho- and total-Akt levels in I/R rats, the degree of Akt activation was barely significant in the postischemic region and was actually decreased in the nonischemic remote region. While difficult to compare these specific results directly to data from the mouse given the different technologies used, it is clear that the level of Akt activation in the mouse was significantly higher. Based on these data, it is tempting to speculate that Akt activation may correlate with degree of angiogenesis, and thus serve as a pharmacodynamic marker of Tβ4 activity in the heart.

Data from large animals are currently limited to acute studies in the pig. 15 or 20 mg Tβ4 dosed via retroinfusion during I/R injury reduced infarct sizes by 36–38%, protected endothelial cells, and preserved subendocardial segment shortening under load (Hinkel et al., 2015, 2008). The differences between the large and small animal studies are notable: in the pig studies, Tβ4 was dosed via intracoronary retroinfusion, rather than IP or IV, as has been done in rodents. Additionally, in contrast to the published pig studies, rodent %EF data were collected from anesthetized animals under minimal cardiac load. These data are consistent with the data generated using the mouse and provide a rationale for clinical testing of Tβ4.

6. CLINICAL STUDIES

Despite mounting preclinical data suggesting an important role of Tβ4 in cardiac repair, the clinical relevance of Tβ4 in human cardiac disease is largely unknown. In patients with coronary artery disease, lower serum Tβ4 levels were reported in patients with poor collateralization (1024.50 ± 516.92 ng/mL) compared to those with collateralization (1373.00 ± 1082.88 ng/mL) (Lv, Cheng, Xu, Wang, & Xu, 2011). Similar observation of significantly lower Tβ4 levels in poorly collateralized vs well-collateralized subjects was reported in patients presenting with severely stenotic coronary artery disease (Bicer et al., 2011). Recently, plasma levels of Tβ4 were measured in ischemic heart failure patients enrolled on the REGENERATE-IHD cell therapy trial (Choudry, Yeo, Mozid, Martin, & Mathur, 2015). In this study, the plasma levels of Tβ4 were 570.0 ng/mL (329.6–1148 ng/mL) at the baseline and increased at 24 h after bone-marrow-derived stem cell treatment. Interestingly, patients with $\geq$1-fold increase of Tβ4 levels after cell infusion reported to have experienced improvement of NYHA symptom class at 6 months, although no change in NT-proBNP levels was observed.

An injectable form of synthetic Tβ4 (RGN-352) was developed by RegeneRx and tested in healthy volunteers in a randomized, double-blinded, and placebo-controlled Phase I study (Ruff, Crockford, Girardi, & Zhang, 2010). In this study, Tβ4 was first given at 42, 140, 420, or 1260 mg intravenously as a single bolus injection ($n=10$ per group). Following safety review, the same dose regimen of Tβ4 was then given daily for 14 days ($n=10$ per group). Overall, Tβ4 infusion over the tested dose range was well tolerated with no evidence of dose limiting toxicity. The pharmacokinetic profile from the single dose study suggested dose proportional plasma exposures of Tβ4. The estimated mean terminal half-life of Tβ4 was dose dependent (0.95, 1.2, 1.9, and 2.1 h at 42, 140, 420, and 1260 mg doses, respectively). This study demonstrated feasibility and tolerability of repeat dosing of Tβ4 in healthy human subjects. Unfortunately, a

Phase II study of RGN-352 in acute MI was placed on clinical hold due to manufacturing issues in 2011, and there has been no further information provided regarding clinical study for cardiac indications.

7. CONCLUSIONS

A relatively large body of preclinical evidence suggests that Tβ4 represents a potential new treatment for injuries related to myocardial ischemia. Tβ4 exhibits pleiotropic effects on a variety of cell types involved in the heart's responses to injury. These include antiinflammatory effects on white blood cells, angiogenic/vasculogenic effects on epicardial, endothelial or smooth muscle cells, antifibrotic effects on the epicardium and cardiac fibroblasts, and antiapoptotic effects on cardiomyocytes, endothelial cells and fibroblasts. These effects of Tβ4 are orchestrated by a high number of interactions with a variety of signaling pathways and with the cytoskeleton. This large constellation of activities would be hard to believe were it not for the facts that Tβ4 is expressed at high levels in almost all cells and has been present evolutionarily since the emergence of multicellularity (Herrmann, Hatta, & Hoffmeister-Ullerich, 2005).

Therapeutic approaches to raise Tβ4 levels therefore hold promise for a number of cardiovascular indications. For example, administration of Tβ4 in acute MI, heart transplantation (Postrach et al., 2014), or other types of ischemic injury would be predicted to preserve myocardial tissue through the ischemic period, until adequate blood supply can be restored through surgical or medical means. Prolonged Tβ4 treatment after the ischemic period may furthermore promote restoration of blood supply via increased collateralization and angiogenesis, and prevent fibrotic remodeling. If Tβ4 treatment were effective at increasing blood flow and preventing fibrosis in human myocardial tissue, as has been shown in preclinical models, it may also improve the outcome of heart failure with preserved ejection fraction, diabetic microangiopathy, or other indications in which capillary rarefaction contributes to pathology.

However, significant translational gaps remain between preclinical data and clinical development. First, angiogenic and antiapoptotic activities need to be verified in human cells or tissue, preferably coupled to changes in intracellular signaling pathways known to promote angiogenesis and survival. Second, the optimal dose or therapeutic modality remains to be determined. Related to this issue, the absorption and distribution of

Tβ4 following a variety of dosing regimens is unknown, even in preclinical models. Last, development of translational biomarkers providing pharmacodynamic evidence of Tβ4 activity in the target tissue or cell type would facilitate dose selection.

REFERENCES

Badamchian, M., Fagarasan, M. O., Danner, R. L., Suffredini, A. F., Damavandy, H., & Goldstein, A. L. (2003). Thymosin beta(4) reduces lethality and down-regulates inflammatory mediators in endotoxin-induced septic shock. *International Immunopharmacology, 3*(8), 1225–1233.

Bao, W., Ballard, V. L., Needle, S., Hoang, B., Lenhard, S. C., Tunstead, J. R., et al. (2013). Cardioprotection by systemic dosing of thymosin beta four following ischemic myocardial injury. *Frontiers in Pharmacology, 4*, 149.

Bednarek, R., Boncela, J., Smolarczyk, K., Cierniewska-Cieslak, A., Wyroba, E., & Cierniewski, C. S. (2008). Ku80 as a novel receptor for thymosin beta4 that mediates its intracellular activity different from G-actin sequestering. *The Journal of Biological Chemistry, 283*(3), 1534–1544.

Bicer, A., Karakurt, O., Akdemir, R., Erden, G., Yildiz, A., Ozcan, O., et al. (2011). Thymosin beta 4 is associated with collateral development in coronary artery disease. *Scandinavian Journal of Clinical and Laboratory Investigation, 71*(8), 625–630.

Bock-Marquette, I., Saxena, A., White, M. D., Dimaio, J. M., & Srivastava, D. (2004). Thymosin beta4 activates integrin-linked kinase and promotes cardiac cell migration, survival and cardiac repair. *Nature, 432*(7016), 466–472.

Bock-Marquette, I., Shrivastava, S., Pipes, G. C., Thatcher, J. E., Blystone, A., Shelton, J. M., et al. (2009). Thymosin beta4 mediated PKC activation is essential to initiate the embryonic coronary developmental program and epicardial progenitor cell activation in adult mice in vivo. *Journal of Molecular and Cellular Cardiology, 46*(5), 728–738.

Brieger, A., Plotz, G., Zeuzem, S., & Trojan, J. (2007). Thymosin beta 4 expression and nuclear transport are regulated by hMLH1. *Biochemical and Biophysical Research Communications, 364*(4), 731–736.

Castoldi, G., di Gioia, C. R., Bombardi, C., Perego, C., Perego, L., Mancini, M., et al. (2010). Prevention of myocardial fibrosis by N-acetyl-seryl-aspartyl-lysyl-proline in diabetic rats. *Clinical Science (London, England), 118*(3), 211–220.

Cheng, G. Z., Park, S., Shu, S., He, L., Kong, W., Zhang, W., et al. (2008). Advances of AKT pathway in human oncogenesis and as a target for anti-cancer drug discovery. *Current Cancer Drug Targets, 8*(1), 2–6.

Choudry, F. A., Yeo, C., Mozid, A., Martin, J. F., & Mathur, A. (2015). Increases in plasma Tbeta4 after intracardiac cell therapy in chronic ischemic heart failure is associated with symptomatic improvement. *Regenerative Medicine, 10*(4), 403–410.

Cierniewski, C. S., Sobierajska, K., Selmi, A., Kryczka, J., & Bednarek, R. (2012). Thymosin beta4 is rapidly internalized by cells and does not induce intracellular Ca^{2+} elevation. *Annals of the New York Academy of Sciences, 1269*, 44–52.

Clauss, I. M., Wathelet, M. G., Szpirer, J., Islam, M. Q., Levan, G., Szpirer, C., et al. (1991). Human thymosin-beta 4/6-26 gene is part of a multigene family composed of seven members located on seven different chromosomes. *Genomics, 9*(1), 174–180.

Conte, E., Genovese, T., Gili, E., Esposito, E., Iemmolo, M., Fruciano, M., et al. (2012). Protective effects of thymosin beta4 in a mouse model of lung fibrosis. *Annals of the New York Academy of Sciences, 1269*, 69–73.

Estrella, N. L., & Naya, F. J. (2014). Transcriptional networks regulating the costamere, sarcomere, and other cytoskeletal structures in striated muscle. *Cellular and Molecular Life Sciences, 71*(9), 1641–1656.

Evans, M. A., Smart, N., Dube, K. N., Bollini, S., Clark, J. E., Evans, H. G., et al. (2013). Thymosin beta4-sulfoxide attenuates inflammatory cell infiltration and promotes cardiac wound healing. *Nature Communications*, *4*, 2081.

Fan, Y., Gong, Y., Ghosh, P. K., Graham, L. M., & Fox, P. L. (2009). Spatial coordination of actin polymerization and ILK-Akt2 activity during endothelial cell migration. *Developmental Cell*, *16*(5), 661–674.

Franco, C. A., & Li, Z. (2009). SRF in angiogenesis: Branching the vascular system. *Cell Adhesion & Migration*, *3*(3), 264–267.

Freeman, K. W., Bowman, B. R., & Zetter, B. R. (2011). Regenerative protein thymosin beta-4 is a novel regulator of purinergic signaling. *The FASEB Journal*, *25*(3), 907–915.

Gardner, B. I., Bingham, S. E., Allen, M. R., Blatter, D. D., & Anderson, J. L. (2009). Cardiac magnetic resonance versus transthoracic echocardiography for the assessment of cardiac volumes and regional function after myocardial infarction: An intrasubject comparison using simultaneous intrasubject recordings. *Cardiovascular Ultrasound*, 7, 38.

Goldstein, A. L., Slater, F. D., & White, A. (1966). Preparation, assay, and partial purification of a thymic lymphocytopoietic factor (thymosin). *Proceedings of the National Academy of Sciences of the United States of America*, *56*(3), 1010–1017.

Grant, D. S., Rose, W., Yaen, C., Goldstein, A., Martinez, J., & Kleinman, H. (1999). Thymosin beta4 enhances endothelial cell differentiation and angiogenesis. *Angiogenesis*, *3*(2), 125–135.

Grillon, C., Rieger, K., Bakala, J., Schott, D., Morgat, J. L., Hannappel, E., et al. (1990). Involvement of thymosin beta 4 and endoproteinase Asp-N in the biosynthesis of the tetrapeptide AcSerAspLysPro a regulator of the hematopoietic system. *FEBS Letters*, *274*(1–2), 30–34.

Hannappel, E. (2007). beta-Thymosins. *Annals of the New York Academy of Sciences*, *1112*, 21–37.

Hannappel, E. (2010). Thymosin beta4 and its posttranslational modifications. *Annals of the New York Academy of Sciences*, *1194*, 27–35.

Herrmann, D., Hatta, M., & Hoffmeister-Ullerich, S. A. (2005). Thypedin, the multi copy precursor for the hydra peptide pedin, is a beta-thymosin repeat-like domain containing protein. *Mechanisms of Development*, *122*(11), 1183–1193.

Hinkel, R., Ball, H. L., DiMaio, J. M., Shrivastava, S., Thatcher, J. E., Singh, A. N., et al. (2015). C-terminal variable AGES domain of Thymosin beta4: The molecule's primary contribution in support of post-ischemic cardiac function and repair. *Journal of Molecular and Cellular Cardiology*, *87*, 113–125.

Hinkel, R., Bock-Marquette, I., Hatzopoulos, A. K., & Kupatt, C. (2010). Thymosin beta4: A key factor for protective effects of eEPCs in acute and chronic ischemia. *Annals of the New York Academy of Sciences*, *1194*, 105–111.

Hinkel, R., El-Aouni, C., Olson, T., Horstkotte, J., Mayer, S., Muller, S., et al. (2008). Thymosin beta4 is an essential paracrine factor of embryonic endothelial progenitor cell-mediated cardioprotection. *Circulation*, *117*(17), 2232–2240.

Hinkel, R., Trenkwalder, T., Petersen, B., Husada, W., Gesenhues, F., Lee, S., et al. (2014). MRTF-A controls vessel growth and maturation by increasing the expression of CCN1 and CCN2. *Nature Communications*, *5*, 3970.

Ho, J. H., Chuang, C. H., Ho, C. Y., Shih, Y. R., Lee, O. K., & Su, Y. (2007). Internalization is essential for the antiapoptotic effects of exogenous thymosin beta-4 on human corneal epithelial cells. *Investigative Ophthalmology & Visual Science*, *48*(1), 27–33.

Husson, C., Cantrelle, F. X., Roblin, P., Didry, D., Le, K. H., Perez, J., et al. (2010). Multifunctionality of the beta-thymosin/WH2 module: G-actin sequestration, actin filament growth, nucleation, and severing. *Annals of the New York Academy of Sciences*, *1194*, 44–52.

Icardo, J. M., & Colvee, E. (2001). Origin and course of the coronary arteries in normal mice and in iv/iv mice. *Journal of Anatomy*, *199*(Pt. 4), 473–482.

Koutrafouri, V., Leondiadis, L., Avgoustakis, K., Livaniou, E., Czarnecki, J., Ithakissios, D. S., et al. (2001). Effect of thymosin peptides on the chick chorioallantoic membrane angiogenesis model. *Biochimica et Biophysica Acta*, *1568*(1), 60–66.

Kumar, S., & Gupta, S. (2011). Thymosin beta 4 prevents oxidative stress by targeting antioxidant and anti-apoptotic genes in cardiac fibroblasts. *PLoS One*, *6*(10), e26912.

Lee, S. J., So, I. S., Park, S. Y., & Kim, I. S. (2008). Thymosin beta4 is involved in stabilin-2-mediated apoptotic cell engulfment. *FEBS Letters*, *582*(15), 2161–2166.

Legate, K. R., Montanez, E., Kudlacek, O., & Fassler, R. (2006). ILK, PINCH and parvin: The tIPP of integrin signalling. *Nature Reviews. Molecular Cell Biology*, 7(1), 20–31.

Liu, Y. H., D'Ambrosio, M., Liao, T. D., Peng, H., Rhaleb, N. E., Sharma, U., et al. (2009). N-acetyl-seryl-aspartyl-lysyl-proline prevents cardiac remodeling and dysfunction induced by galectin-3, a mammalian adhesion/growth-regulatory lectin. *American Journal of Physiology. Heart and Circulatory Physiology*, *296*(2), H404–H412.

Lv, S., Cheng, G., Xu, Y., Wang, Y., & Xu, G. (2011). Relationship between serum thymosin beta4 levels and coronary collateral development. *Coronary Artery Disease*, *22*(6), 401–404.

Maiese, K., Chong, Z. Z., Hou, J., & Shang, Y. C. (2009). The "O" class: Crafting clinical care with FoxO transcription factors. *Advances in Experimental Medicine and Biology*, *665*, 242–260.

Mikhailov, A. T., & Torrado, M. (2012). In search of novel targets for heart disease: Myocardin and myocardin-related transcriptional cofactors. *Biochemistry Research International*, *2012*, 973723.

Morita, T., & Hayashi, K. (2013). G-actin sequestering protein thymosin-β4 regulates the activity of myocardin-related transcription factor. *Biochemical & Biophysical Research Communications*, *437*(3), 331–335.

Peng, H., Carretero, O. A., Peterson, E. L., & Rhaleb, N. E. (2010). Ac-SDKP inhibits transforming growth factor-beta1-induced differentiation of human cardiac fibroblasts into myofibroblasts. *American Journal of Physiology. Heart and Circulatory Physiology*, *298*(5), H1357–H1364.

Peng, H., Xu, J., Yang, X. P., Dai, X., Peterson, E. L., Carretero, O. A., et al. (2014). Thymosin-beta4 prevents cardiac rupture and improves cardiac function in mice with myocardial infarction. *American Journal of Physiology Heart and Circulatory Physiology*, *307*(5), H741–H751.

Posern, G., Miralles, F., Guettler, S., & Treisman, R. (2004). Mutant actins that stabilise F-actin use distinct mechanisms to activate the SRF coactivator MAL. *The EMBO Journal*, *23*(20), 3973–3983.

Postrach, J., Schmidt, M., Thormann, M., Thein, E., Burdorf, L., Reichart, B., et al. (2014). Adeno-associated viral vector 2.9 thymosin ss4 application attenuates rejection after heart transplantation: Results of a preclinical study in the pig. *Transplantation*, *98*(8), 835–843.

Rasoul, S., Carretero, O. A., Peng, H., Cavasin, M. A., Zhuo, J., Sanchez-Mendoza, A., et al. (2004). Antifibrotic effect of Ac-SDKP and angiotensin-converting enzyme inhibition in hypertension. *Journal of Hypertension*, *22*(3), 593–603.

Ruff, D., Crockford, D., Girardi, G., & Zhang, Y. (2010). A randomized, placebo-controlled, single and multiple dose study of intravenous thymosin beta4 in healthy volunteers. *Annals of the New York Academy of Sciences*, *1194*, 223–229.

Safer, D., Elzinga, M., & Nachmias, V. T. (1991). Thymosin beta 4 and Fx, an actin-sequestering peptide, are indistinguishable. *The Journal of Biological Chemistry*, *266*(7), 4029–4032.

Salto-Tellez, M., Yung Lim, S., El-Oakley, R. M., Tang, T. P., ALmsherqi, Z. A., & Lim, S. K. (2004). Myocardial infarction in the C57BL/6 J mouse: A quantifiable and highly reproducible experimental model. *Cardiovascular Pathology*, *13*(2), 91–97.

Sanders, M. C., Goldstein, A. L., & Wang, Y. L. (1992). Thymosin beta 4 (Fx peptide) is a potent regulator of actin polymerization in living cells. *Proceedings of the National Academy of Sciences of the United States of America*, *89*(10), 4678–4682.

Sluijter, J. P., Condorelli, G., Davidson, S. M., Engel, F. B., Ferdinandy, P., Hausenloy, D. J., et al. (2014). Novel therapeutic strategies for cardioprotection. *Pharmacology & Therapeutics*, *144*(1), 60–70.

Smart, N., Bollini, S., Dube, K. N., Vieira, J. M., Zhou, B., Davidson, S., et al. (2011). De novo cardiomyocytes from within the activated adult heart after injury. *Nature*, *474*(7353), 640–644.

Smart, N., Risebro, C. A., Clark, J. E., Ehler, E., Miquerol, L., Rossdeutsch, A., et al. (2010). Thymosin beta4 facilitates epicardial neovascularization of the injured adult heart. *Annals of the New York Academy of Sciences*, *1194*, 97–104.

Smart, N., Risebro, C. A., Melville, A. A., Moses, K., Schwartz, R. J., Chien, K. R., et al. (2007a). Thymosin beta4 induces adult epicardial progenitor mobilization and neovascularization. *Nature*, *445*(7124), 177–182.

Smart, N., Risebro, C. A., Melville, A. A., Moses, K., Schwartz, R. J., Chien, K. R., et al. (2007b). Thymosin beta-4 is essential for coronary vessel development and promotes neovascularization via adult epicardium. *Annals of the New York Academy of Sciences*, *1112*, 171–188.

Sopko, N., Qin, Y., Finan, A., Dadabayev, A., Chigurupati, S., Qin, J., et al. (2011). Significance of thymosin beta4 and implication of PINCH-1-ILK-alpha-parvin (PIP) complex in human dilated cardiomyopathy. *PloS One*, *6*(5), e20184.

Trembley, M. A., Velasquez, L. S., de Mesy Bentley, K. L., & Small, E. M. (2015). Myocardin-related transcription factors control the motility of epicardium-derived cells and the maturation of coronary vessels. *Development*, *142*(1), 21–30.

van Berlo, J. H., Kanisicak, O., Maillet, M., Vagnozzi, R. J., Karch, J., Lin, S. C., et al. (2014). c-kit + cells minimally contribute cardiomyocytes to the heart. *Nature*, *509*(7500), 337–341.

Vartiainen, M. K., Guettler, S., Larijani, B., & Treisman, R. (2007). Nuclear actin regulates dynamic subcellular localization and activity of the SRF cofactor MAL. *Science*, *316*(5832), 1749–1752.

Xin, M., Davis, C. A., Molkentin, J. D., Lien, C. L., Duncan, S. A., Richardson, J. A., et al. (2006). A threshold of GATA4 and GATA6 expression is required for cardiovascular development. *Proceedings of the National Academy of Sciences of the United States of America*, *103*(30), 11189–11194.

Xue, B., Leyrat, C., Grimes, J. M., & Robinson, R. C. (2014). Structural basis of thymosin-beta4/profilin exchange leading to actin filament polymerization. *Proceedings of the National Academy of Sciences of the United States of America*, *111*(43), E4596–E4605.

Yang, A., Sonin, D., Jones, L., Barry, W. H., & Liang, B. T. (2004). A beneficial role of cardiac P2X4 receptors in heart failure: Rescue of the calsequestrin overexpression model of cardiomyopathy. *American Journal of Physiology. Heart and Circulatory Physiology*, *287*(3), H1096–H1103.

Young, J. D., Lawrence, A. J., MacLean, A. G., Leung, B. P., McInnes, I. B., Canas, B., et al. (1999). Thymosin beta 4 sulfoxide is an anti-inflammatory agent generated by monocytes in the presence of glucocorticoids. *Nature Medicine*, *5*(12), 1424–1427.

Zarbock, J., Oschkinat, H., Hannappel, E., Kalbacher, H., Voelter, W., & Holak, T. A. (1990). Solution conformation of thymosin beta 4: A nuclear magnetic resonance and simulated annealing study. *Biochemistry*, *29*(34), 7814–7821.

Zhang, M., Malik, A. B., & Rehman, J. (2014). Endothelial progenitor cells and vascular repair. *Current Opinion in Hematology*, *21*(3), 224–228.

Zhou, B., Honor, L. B., Ma, Q., Oh, J. H., Lin, R. Z., Melero-Martin, J. M., et al. (2012). Thymosin beta 4 treatment after myocardial infarction does not reprogram epicardial cells into cardiomyocytes. *Journal of Molecular and Cellular Cardiology*, *52*(1), 43–47.

Zuo, Y., Chun, B., Potthoff, S. A., Kazi, N., Brolin, T. J., Orhan, D., et al. (2013). Thymosin beta4 and its degradation product, Ac-SDKP, are novel reparative factors in renal fibrosis. *Kidney International*, *84*(6), 1166–1175.

CHAPTER TEN

Thymosin β4: Roles in Development, Repair, and Engineering of the Cardiovascular System

E.D. Marks, A. Kumar*
Nanomedicine Research Laboratory, University of Delaware, Newark, DE, United States
*Corresponding author: e-mail address: arunk@udel.edu

Contents

Abstract

The burden of cardiovascular disease is a growing worldwide issue that demands attention. While many clinical trials are ongoing to test therapies for treating the heart after myocardial infarction (MI) and heart failure, there are few options doctors able to currently give patients to repair the heart. This eventually leads to decreased ventricular contractility and increased systemic disease, including vascular disorders that could result in stroke. Small peptides such as thymosin β4 (Tβ4) are upregulated in the cardiovascular niche during fetal development and after injuries such as MI, providing increased neovasculogenesis and paracrine signals for endogenous stem cell recruitment to aid in wound repair. New research is looking into the effects of in vivo administration of Tβ4 through injections and coatings on implants, as well as its effect on cell differentiation. Results so far demonstrate Tβ4 administration leads to robust increases in angiogenesis and wound healing in the heart after MI and the brain after stroke, and

Vitamins and Hormones, Volume 102
ISSN 0083-6729
http://dx.doi.org/10.1016/bs.vh.2016.04.010

can differentiate adult stem cells toward the cardiac lineage for implantation to the heart to increase contractility and survival. Future work, some of which is currently in clinical trials, will demonstrate the in vivo effect of these therapies on human patients, with the goal of helping the millions of people worldwide affected by cardiovascular disease.

1. INTRODUCTION

The circulatory system, first described in English physician William Harvey's seminal work "On the Motion of the Heart and Blood in Animals," published in 1628 (Harvey, 1628), pumps approximately 5 L of blood per minute around the body (termed cardiac output). As the heart beats faster through such physiological drivers as exercise or stress, cardiac output can significantly increase and over time, if the physiological drivers of increased heart rate are coupled with progressive worsening of artery blockages, the heart muscle cells (cardiomyocytes) begin to hypertrophy to compensate for the increased pressure against which the heart pumps. When the flow of oxygenated blood to the heart cannot keep up with physiological need of the cardiomyocytes, ischemia occurs and the tissue begins to die. As the dead contractile muscle is replaced with fibrotic scar tissue, the heart can no longer provide a sufficient flow of blood to meet the body's demands and heart failure develops.

The natural process of cardiac hypertrophy can be genetic or pathological. For example, autosomal dominant inheritance of the gene causing familial hypertrophic cardiomyopathy can occur in a healthy athlete (Cirino & Ho, 2014), and the heart muscle can appear clinically similar to someone who has had a massive heart attack. In young athletes, this condition can lead to sudden cardiac death, where the increased strain on the heart from exercise causes the individual to die on the playing field, despite defibrillation.

Why, then, could an obese individual survive a massive heart attack necessitating a quadruple bypass, where a healthy young athlete dies suddenly on the field? What are the intrinsic biological healing mechanisms involved that could save either person? And what is being done in laboratories around the world to heal not only cardiac tissue after a heart attack but also stroke and genetic malformations of the heart? In this chapter, we review cardiac development and healing, with specific focus on nanomedicine and other translational therapies that employ tissue engineering techniques. The cardiac protein thymosin beta4 will highlighted in particular.

2. DEVELOPMENT OF THE CARDIAC SYSTEM

2.1 General Timeline

The first rudimentary timeline of fetal heart development was documented by Hippocrates in 300–400 BC, but it would take almost 2000 years for the first dissections and wax sculptures to be documented for intense scientific inquiry (Dhanantwari et al., 2009; Kramer, 1942; Silverman, Grove, & Upshaw, 2006). Since these discoveries, beginning in 1839 with Dr. Jan Evangelista Purkinje's elucidation of his namesake cardiac conduction fibers, technological advances have allowed researchers to learn much more about fetal cardiac development. Though many techniques still rely on donor embryos and mouse gestation for histological and microscopic data, importantly, the use of high-resolution imaging such as MRI and ultrasound allows for noninvasive analysis of fetal heart development.

Cardiac fetal development begins as early as Day 19 (Carnegie stage [CS] 7), when the formation of blood necessitates the development of a vessel system (Rones, McLaughlin, Raffin, & Mercola, 2000), and contraction of the heart tube begins. By Days 25–27 (CS 9), the rotation of the cardiac anlage results in its lying ventral to the pericardial cavity; this cavity, which will eventually envelop the cardiac system, secretes trophic factors that aid in fetal cardiac development (Darland & D'Amore, 2001; Rones et al., 2000). Growth and expansion of smooth muscle cells and the endothelium of the coronary arteries are supported by proepicardial serosal cells, which also play a tempering role in myocardial development (Wilm, Ipenberg, Hastie, Burch, & Bader, 2005; Xavier-Neto, Shapiro, Houghton, & Rosenthal, 2000). At the end of the first month of development (CS 12), the fetal heart contains an epicardium and an inner myocardial mantle surrounding cardiac jelly; the cardiac lumen contains endocardial cells.

Completion of cardiac development occurs in a matter of weeks, from ~Day 32 to 56 (CS 13–23) (Dhanantwari et al., 2009). On ~Day 33 (CS 14), four ventricles have appeared, with septation of the atria from the ventricles occurring in the 7th and 8th weeks (CS 14–18), a full 2 weeks before closure of the interventricular foramen and complete formation of the ventricular septum (Dhanantwari et al., 2009; Yamada et al., 2010). During this process, the heart is also getting bigger as the cells begin to divide (cardiac hyperplasia) while slowly dropping into the thoracic cavity as the fetus' brain develops and takes up more intrauterine space. By Day 56 (CS 23) the heart is at a critical size and centrally located within the fetus. Once the baby is out of the womb, the heart quickly transitions from a hyperplastic to a hypertrophic phenotype, characterized by elimination of cell division but an increase in cell size (Li, Wang, Capasso, & Gerdes, 1996). With some exception, this phenotype persists into adulthood (Bergmann et al., 2009; Frey, Katus, Olson, & Hill, 2004).

Genetic and morphological malformations can lead to defects in heart formation during development, some of which can plague patients throughout their lives (Wraya & Sensky, 2001). Imaging of the fetus during midgestation (18–24 weeks) using echocardiography is common (Simpson, 2009), mainly for early diagnosis of serious physiological anomalies such as ventricular septal defects (Gómez et al., 2014). The fetal heart has an enormous capacity to regenerate lost heart tissue to compensate for these congenital heart defects (Gómez et al., 2014; Sturzu et al., 2015), but often perinatal surgeons need to correct the malformations before the baby is born in order to improve the chances of survival after birth (Yuan, 2014). Postnatal treatment options are much more plentiful, encompassing surgical, biomaterial, and tissue engineering strategies. For example, to repair ventricular septal defects, researchers at Rice University have developed cardiac patches that mechanically and electrically couple to native tissue matrices (Pok, Benavides, Hallal, & Jacot, 2014). Patches like this provide a supporting growth surface for natural repair of the defect. When coupled with stem cells, these tissue engineering techniques can seed the defected area with new tissue, decreasing the burden of the newborn heart to repair significant damages on its own (Avolio, Caputo, & Madeddu, 2015; Petsche, Camci-Unal, Khademhosseini, & Jacot, 2013). Most importantly, these therapies can be applied with minimally invasive surgeries, and do not require follow-up procedures since the biomaterial slowly degrades as the tissue takes over.

2.2 Role of Thymosin β4 in Fetal Development

β-Thymosins are small, acidic peptides stemming from prohormones (Hannappel, 2007; Huff, Müller, Otto, Netzker, & Hannappel, 2001). Chief among the β-thymosins is thymosin β4 (Tβ4), a 43 amino acid G-actin sequestering peptide (Goldschmidt-Clermont et al., 1992) ubiquitously spread throughout the body (Huff et al., 2001; Nemolato et al., 2009). Tβ4 is found mainly in the cytoplasm of cells where it associates with G-actin in a 1:1 ratio; at higher concentrations (>100 μM), this association can also extend to F-actin (Carlier et al., 1996). Slight changes in the amino acid structure, particularly on the N-terminal end, can significantly alter the binding efficiency of Tβ4 to actin bundles. In the adult, Tβ4 has been implicated in an immense array of biological processes: stimulating migration of human umbilical vein endothelial cells (Malinda, Goldstein, & Kleinman, 1997); matrix metalloproteinase (MMP) induction (Grant et al., 1995); enhancing angiogenesis (Bock-Marquette, Saxena, White, Dimaio, & Srivastava, 2004); hair follicle stem cell activation (Philp, Goldstein, & Kleinman, 2004); acting as a chemoattractant (Huff et al., 2001); and repair of cornea (Sosne et al., 2001), brain (Smart et al., 2007), and heart (Rossdeutsch, Smart, & Riley, 2008).

Tβ4 is crucial in fetal development. Despite some evidence to the contrary (Banerjee et al., 2012), the overwhelming majority of reports point to significant fetal cardiac malformations and death in Tβ4 knockout murine models. By embryonic day 9.5 (E9.5), Tβ4 gene and protein expression can be observed in highly proliferative areas of the developing cardiac system, including the cardiac outflow tract, the endocardial cushion, and the compact layer, a fast growing area of the fetal myocardium (Bock-Marquette et al., 2004). Using an interfering RNA strategy to knock down Tβ4 expression in the developing mouse, Smart et al. (2007) showed pericardial edema by E10.5 and early signs of necrosis and leftward cardiac displacement by E12.5. Issues persisted through E16.5 owing to decreases in vasculogenesis, a crucial process in embryonic development that is mediated by VEGF and Tβ4. The ability of embryonic Tβ4 to bolster migration of cardiomyocytes and endothelial progenitor cells can also be observed after cardiac damage in the adult heart (described in Section 3) and is currently being pursued as a therapeutic strategy in cardiac tissue engineering (Section 4).

Interestingly, Tβ4 has also been explicitly linked to fetal development of many other organ systems (Choi, Noh, Kim, Sun, & Kim, 2007; Dathe & Brand-Saberi, 2004; Nemolato et al., 2009; Wang et al., 2012) and adult tumorigenesis. Tβ4 along with Tβ10 and other β-thymosins are released from developing salivary glands of the neonate and act in a paracrine manner to induce differentiation of the oral cavity. In the adult, oral β-thymosins are only readily observed during development of urothelial carcinomas (Faa et al., 2012; Wang et al., 2012) (normal postgestational production never exceeds 0.1 μmol/L saliva (Nemolato et al., 2009) compared to over 2 μmol/L in pregestational development and carcinomas). These values also align with Tβ4-induced modulation of the ILK/AKT/β-catenin signaling pathway, leading to invasion and metastasis of colorectal cancers (Cierniewski et al., 2010; Piao, Hong, Jung, Choi, & Park, 2014). Tβ4's ability to revert certain cells back to the epithelial–mesenchymal transition stage, thus increasing tumor invasion and migration, is presumed to be the cause for the similarities in these two carcinomas (Huang et al., 2007; Wang et al., 2012).

3. PROGRESSION AND REPAIR AFTER DISEASE

3.1 Pathogenesis of Heart Attack

Cardiovascular disease is responsible for one out of every three deaths worldwide, making it the leading cause of mortality. Continuing decreases in smoking prevalence have reduced the number of attributable cardiac deaths though the number is relatively stagnant due to the increases in obesity and poor nutrition (Mozaffarian et al., 2015). Arterial blockages, more common in smokers and the obese, impede blood flow and any downstream tissues will begin to die if not quickly reperfused. When this blockage occurs in an artery supplying blood to the heart muscle tissue, such as the left anterior descending artery, the resulting pathological condition is known as a myocardial infarction (MI), or heart attack.

MI follows a fairly predictable healing path: acute ischemia, followed by necrotic, fibrotic, and remodeling phases (Holmes, Borg, & Covell, 2005). During acute ischemia, the size and mechanical function of the infarct is dependent on rate of reperfusion. Within 40 min, coupling of cardiomyocytes to the extracellular matrix (ECM) at Z-bands are destroyed, with associated disappearance of cardiomyocyte striations and ECM glycoproteins in the following hour (Caulfield, Tao, & Nachtigal, 1985; Sato, Ashraf, Millard, Fujiwara, & Schwartz, 1983). Within 2 h, significant increases in soluble and insoluble matrix metalloproteinases (MMPs) lead to continued destruction of the ECM (Peterson et al., 2001). The necrotic phase begins within hours after acute ischemia and can be identified by wavy cardiomyocyte actin fibers (indicative of edema) and continued destruction of collagen fibers by MMP-1, -2, and -9 (Arai, 2015). Other structural proteins, including fibronectin (Ulrich et al., 1997), laminin, and collagen type IV (Morishita, Kusachi, Yamasaki, Kondo, & Tsuji, 1996), begin to take the place of native collagen; native procollagen mRNA expression is detected slightly later (Cleutjens, Verluyten, Smiths, & Daemen, 1995). From 1 to 3 weeks after infarct, significant increases in collagen matrix deposition and fibroblast infiltration to the infarct zone signal transition to the fibrotic phase. Layers of type III collagen provide a basis for anisotropic alignment and cross-linking of massive type I collagen fibers; most infarcts are stiffest during this period in direct correlation to the amount of deposited, cross-linked collagen (Voorhees et al., 2015). The final remodeling phase is described as occurring for ~6 weeks after the fibrotic phase, though there is rarely a definitive "healed" state of a biologically active infarct (Holmes et al., 2005). Collagen content continues to increase, while, paradoxically, infarct stiffness begins to decrease. This decrease in infarct size and stiffness may be due to cardiac stem cells becoming new cardiomyocytes coupled with resident cardiomyocytes reentering the cell cycle, both instances spurred by factors such as Tβ4 (Bergmann & Jovinge, 2014; Chen, Shimoda, Chen, & Grayburn, 2013).

Resulting ventricular remodeling leading to lack of contractile tissue is mainly due to the extremely slow proliferation rate of cardiomyocytes. As shown by Bergmann et al. (2009), cardiomyocytes can derive from division of preexisting cardiomyocytes, meaning cardiomyocyte death as a result of MI decreases the progenitor pool from which proliferation can occur. Complete reperfusion, regardless of the method (stenting or coronary artery bypass grafting), is associated with decreased long-term mortality and significantly decreases the likelihood of another MI and the need for additional revascularization compared to incomplete reperfusion (Garcia et al., 2013). Immediate and complete reperfusion following MI leads to decreased MMP release and increased survival of native cardiomyocytes, decreasing infarct size, and resulting cardiac dysfunction (Michelis, Boehm, & Kovacic, 2014). Increased vascularization also provides a delivery system for systemic release of cytokines and signaling peptides to the body for recruitment of immune cells and resident stem cells to heal the infarct (Rennert, Sorkin, Garg, & Gurtner, 2012) (Fig. 1).

3.2 Pathogenesis of Stroke and Subsequent Repair

Stroke is defined simply as a lack of oxygen being delivered to a particular brain region. Ischemic stroke, by far the most common, occurs when blood vessels to portions of the brain become clogged by plaques or hemorrhaging preventing the delivery of oxygenated blood to downstream tissues. Affecting over 795,000 people each year, ischemic strokes cost the U.S. healthcare system $34 billion annually and are one of the leading causes of serious long-term disability (Mozaffarian et al., 2015).

Similar to heart attacks, time between disease onset and treatment is crucial. Two scoring systems, the Rankin Scale (Rankin, 1957) and the National Institutes of Health Stroke Scale (Lyden et al., 2001), help determine the extent of damage when a patient is brought to the clinic and include assessments for determining patient status at discharge. Based on

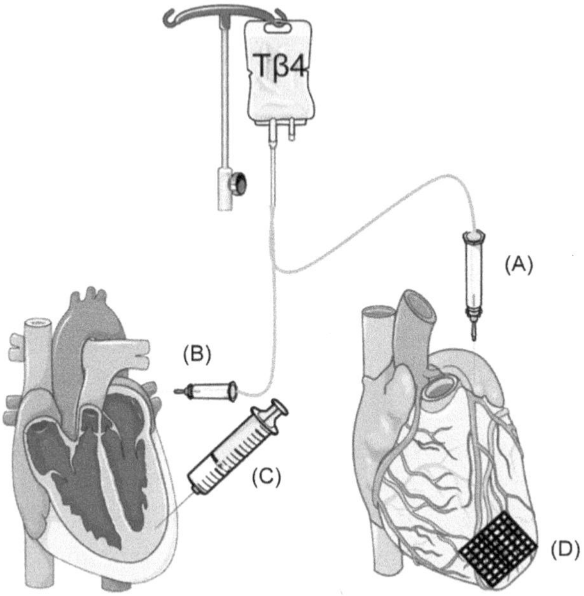

Fig. 1 Schematic of administration routes for Tβ4. (A) Systemic administration via IV delivers peptide to the damaged left ventricle through the left anterior descending (LAD) artery. (B) Catheterization to the left ventricle allows for more specific delivery. (C) Direct intramuscular injections. (D) Drug-eluting patches can be applied to the epicardium or directly within the myocardium. Direct injections and patches require less overall peptide for therapeutic efficiency, but can be more invasive than catheterization or systemic administration through IV. (See the color plate.)

these scales and others (Kasner, 2006), patients are put into one of two categories: progressive cerebral infarction (P) or nonprogressive cerebral infarction (NP) (Sumer, Ozdemir, & Erturk, 2003). NP group members are classified as neurologically normal by 24–48 h, whereas P group's members, comprising 10–31% of clinical patients, significantly worsen and have an overall lower chance of survival and higher rate of numbness and paralysis (DeGraba, Hallenbeck, Pettigrew, Dutka, & Kelly, 1999; Kwan & Hand, 2006; Sumer et al., 2003).

Surprisingly, reperfusion of the anoxic brain tissue is known to cause severe, often deadly complications (Karapanayiotides et al., 2005). Unlike myocardial ischemia which benefits greatly from vascular reperfusion (Michelis et al., 2014), the extensive flow of blood back to the ischemic brain tissue overwhelms the system with inflammatory cytokines, damaging recovering cells and causing the release of reactive oxygen species. Tissue plasminogen activator (tPA), the only FDA-approved stroke treatment that cleaves the Arg561-Val562 peptide bond of plasminogen, promoting its conversion to plasmin to begin the anticlotting process, can exacerbate the issue by causing hyperfibrinolysis, or excessive bleeding at the site of enzyme activity. Standard blood clotting procedures also involve Tβ4, which is incorporated into the fibrin(ogen) αC-domains by Factor XIIIa (Makogonenko, Goldstein, Bishop, & Medved, 2004). In normal clotting, the influx of Tβ4 leads to increases in angiogenesis, wound healing, and inflammatory responses. The resulting increase in endothelial permeability coupled with cytokine release into the new vessels recruits immune cells to defend and repair the damaged area. Current research is looking into using Tβ4 as a cell-level stimulator of axonal and vascular growth (more in Section 4), though so far increasing vasculogenesis has resulted in cytokine-mediated death of the tissue by uncontrolled immune cell activity.

4. TISSUE ENGINEERING USING Tβ4

4.1 Direct Injections

Injections of Tβ4 have been given as far back as the 1960s, when Asanuma, Goldstein, and White (1970) injected Tβ4 into neonatally thymectomized mice and found that it increased the rate of weight gain and survival, as well as the numbers of lymphoid cells in the blood, spleen, and lymph nodes compared to controls. A follow-up study (Low, Hu, & Goldstein, 1981) determined the thymic mechanism by showing injected Tβ4 regulates the early maturation of thymus-dependent lymphocytes by controlling stem cells in the lymph nodes.

Injections of Tβ4 into the cardiac niche have been studied extensively. Previous reports have shown endogenous upregulation of Tβ4 in the heart after injury (Bock-Marquette et al., 2004; Gehrig, Koopman, Naim, Tjoakarfa, & Lynch, 2010), leading others to postulate that direct injections would increase the therapeutic potential of this peptide. In both injured and control hearts, a report showed intraperitoneal injections of 1.5 μg/μL Tβ4 promoted formation of a neovascular microenvironment by recruiting and organizing epicardium-derived cells (Smart, Dube, & Riley, 2010; Smart, Risebro, et al., 2010). This neovascularization effect is due to: (1) the upstream regulation of Tβ4 by the transcription factor Hand1 (Smart, Dube, et al., 2010), the absence of which is embryonic lethal; and (2) downstream activation of the Akt/PI3K pathways (Trenkwalder et al., 2015). Interestingly, the Akt/PI3K pathway has been implicated in the paracrine-mediated effects of cardiac improvement seen after injection of mesenchymal stem cells (MSCs) (Gnecchi et al., 2006; Noiseux et al., 2006), hinting at a therapeutic potential of a combined MSC/Tβ4 intramyocardial injection.

Tβ4 is found normally in both growing and adult brains (Gomez-Marquez & Anadon, 2002) and, like the cardiac niche, levels of gene transcripts increase after injuries such as cerebral ischemia (Vartiainen, Pyykonen, Hokfelt, & Koistinaho, 1996). Owing to this fact, and the lack of approved treatments and the risk of intracranial hemorrhaging with the only approved therapy (tPA), Tβ4 injections have also been studied as a treatment for stroke (Morris, Chopp, Zhang, & Zhang, 2010). A recent dose–response study (Morris et al., 2014) using intraperitoneal injections of Tβ4 at 0, 2, 12, and 18 mg/kg in a rat model of embolic stroke found mid-level dosages (average of 3.75 mg/kg) significantly increased cognitive function over 3 months compared to controls and higher level doses. The neuroprotective effect could be due to intracranial regulation of the p38 MAP kinases following injury (Santra et al., 2012), while observed neuroregenerative effects could result from increased miRNA-146a expression, a known driver of oligodendrocyte differentiation and expansion (Santra et al., 2014). There are also numerous studies on nonstroke-related neuroprotective and neurorestorative effects of Tβ4 injections (Morris et al., 2012; Xiong, Mahmood, et al., 2012; Xiong, Zhang, et al., 2012), but these will not be covered here.

Considering the far reaching and beneficial effects of Tβ4 on the in vivo system, including increasing angiogenesis and immune cell migration to sites of injury, injections of the peptide have been studied in illnesses beyond the cardiac and neural niche. For example, Vos et al. showed injections of Tβ4 restored some immune function after toxic doses of 2,3,7,8-tetrachlorodibenzo-*p*-dioxin (TCDD), a known carcinogen and teratogen. This protection in turn prevented the mice from being fully colonized by pathogenic Salmonella or Listeria species (Vos, Kreeftenberg, Engel, Minderhoud, & Van Noorle Jansen, 1978). After injecting mice with lipopolysaccharide as an endotoxin, Badamchian et al. (2003) showed systemic circulation of Tβ4 decreased septic shock and increased survival; this effect was also observed clinically in septic humans where Tβ4 prevented the often deadly cytokine storm. Overall, most studies testing injected administrations of Tβ4 look for antiinflammatory efficacy, with a vast majority showing positive effects on immune cell upregulation and recruitment to sites of injury (Goldstein, Guha, Zatz, Hardy, & White, 1972; Scheinberg, Goldstein, & Cathcart, 1976; Wara, Goldstein, Doyle, & Ammann, 1975) (Fig. 2).

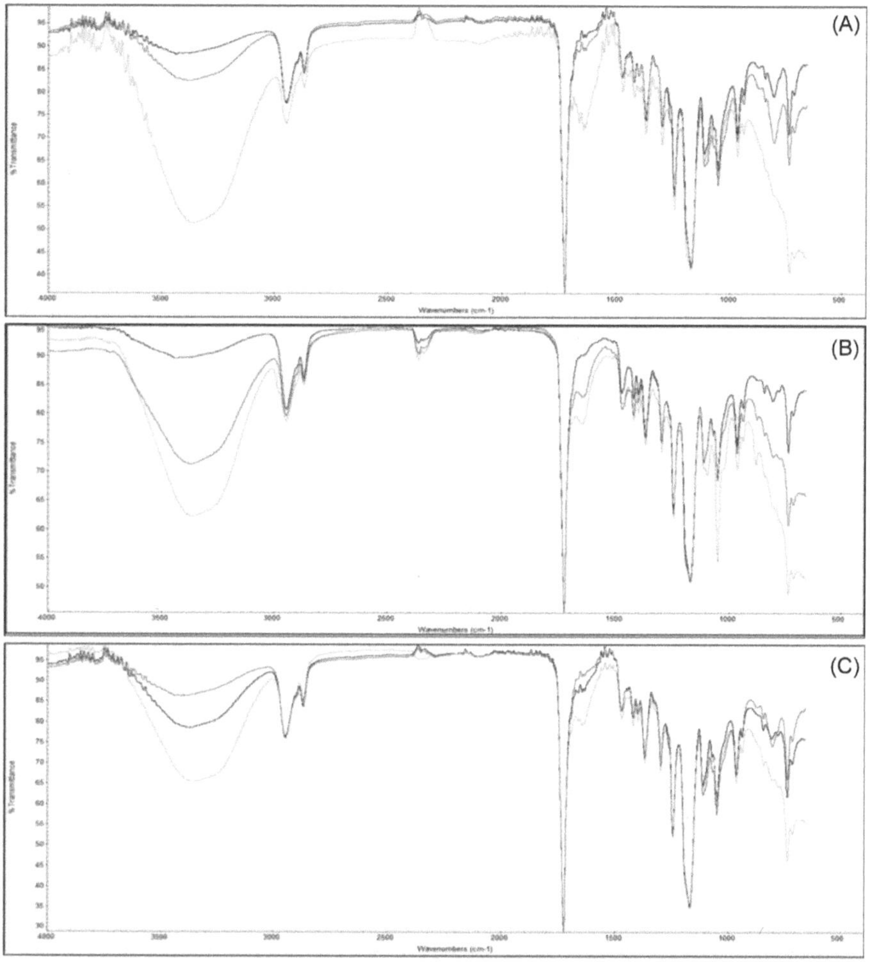

Fig. 2 Drug-eluting nanoscaffolds release peptide and small molecules in a concentration-dependent manner. (A) Large quantities are detected immediately after coating with high concentration coatings, but quickly degrade. (B) Mid-range concentrations are encapsulated with high efficiency and release peptide more slowly. (C) Low concentrations do not stay within the scaffold to any appreciable degree. (See the color plate.)

4.2 Tβ4 as a Scaffold Coating

Issues with injections of Tβ4, including leaking from muscle, slow or non-existent migration to sites of damage, and unknown therapeutic concentration, have led us and others to use scaffolding systems to house and control Tβ4 release (Chiu, Reis, & Radisic, 2012; Kumar, Patel, Duvalsaint,

Desai, & Marks, 2014; Ti et al., 2015). Depending on the application, peptide release could occur from days to months. For example, healing after a MI takes 7 days on average in a human before remodeling occurs; it would therefore be unnecessary to use extended release protocols beyond 2 weeks as most vascularization and cell recruitment has occurred within the first 5–7 days (Arai, 2015). In contrast, translational skin tissue engineering would require extensive angiogenesis to continually feed the upper layers to prevent shrinkage and promote fusion with the native tissue (Ti et al., 2015); in this case, remodeling does not occur for up to 3 weeks after injury and further healing timelines are dependent on skin type and location (Thomas & Frost, 1993), necessitating an adaptive scaffold with long-lasting release parameters.

The Radisic lab has done extensive work on controlled release of Tβ4 from hydrogels (Chiu, Montgomery, Liang, Liu, & Radisic, 2012; Chiu & Radisic, 2011; Chiu, Reis, Momen, & Radisic, 2012). Focusing on angiogenesis for cardiac tissue engineering, they have shown progressive seepage of Tβ4 from hydrogels that subsequently aid in increasing vasculature density (6–8 branches/mm^2) and length (400–550 μm) compared to control (1–4 branches/mm^2; 75–300 μm), with no change in branch morphology. Micropatterned grooves within a culture surface measuring 50 μm wide significantly increased density and length of vessels compared to 25 and 100 μm grooves, a finding consistent with previous reports on scaffold architecture effects on cell growth (Bettinger, Langer, & Borenstein, 2009). For direct healing after MI, these hydrogels releasing Tβ4 for less than 1 week showed significantly more blood vessels >50 μm in diameter compared with controls ($p < 0.0001$).

Hydrogels and thicker gel composites are well known to increase ventricular wall strength and decrease collagen-driven remodeling (Bertassoni et al., 2014; Li & Guan, 2011), but their density often prevents cell and vasculature infiltration and growth. Further, slow gelation times allow leakage of embedded cells and compounds such as Tβ4 out of the matrix, compromising therapeutic efficiency and decreasing the longevity of sustained release. To combat these and other issues, including in vivo delivery, our group has developed a polycaprolactone (PCL) nanofiber scaffolding system housing Tβ4. Our system is capable of releasing Tβ4 (and other small molecules; see Section 4.3) over a time period commensurate with peptide concentration. Tβ4 is incorporated to the nanoscaffold in one of two ways: (1) Tβ4 is embedded in the nanoscaffold design

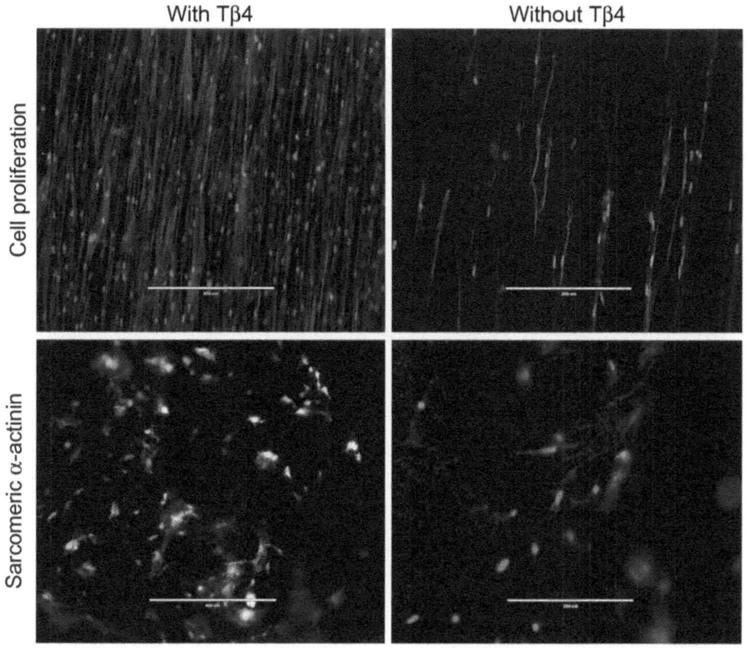

Fig. 3 Tβ4 coatings promote proliferation and differentiation of stem cells toward the cardiac lineage. On coated nanoscaffolds, stem cells proliferate to produce a phenotype similar to myocardial tissue. Immunostaining for sarcomeric α-actinin, a biomarker for cardiomyocyte differentiation, is significantly increased in cells grown on coated nanoscaffolds. (See the color plate.)

by electrospinning with PCL; or (2) the nanoscaffold is coated with and subsequently covalently bonded to Tβ4 postelectrospinning. The efficiency of coating approaches 70%, and usable concentrations degrade at a rate of $12 \pm 2\%$ per week in active culture, continuing to release measurable quantities of Tβ4 for up to 10 weeks. Future tests will determine the rate of release in in vivo cardiac systems (Fig. 3).

4.3 Tβ4 and Cell Differentiation

The effect of Tβ4 on the epithelial–mesenchymal transition (EMT) cannot be understated (see Section 2.2). While the overwhelming majority of studies have focused on Tβ4's effects on EMT in relation to the proliferation and metastasis of cancers (particularly colorectal) (Faa et al., 2012; Fu et al., 2015; Hong, Lee, Hong, & Hong, 2016; Huang et al., 2007; Nemolato et al.,

2012; Wang et al., 2012), EMT effects have also been investigated in progression of lung fibrosis (Rodriguez, Nathan, Chinna, Leema, & Grant, 2015), and fetal development of the tooth oral cavity (Ookuma et al., 2013), blood vessels, and heart (Dathe & Brand-Saberi, 2004). These cell transitional events are promoted almost entirely by E-cadherin and β-catenin signaling. Intercellular β-catenin acts as an actin regulator and transcription factor, turning on Wnt target genes (c-myc, c-jun, cycD, etc.) using TCF/Lef1 proteins, which promotes mesoderm and stem cell differentiation, cell proliferation, migration, and adhesion, and the transition from epithelium to mesenchyme. This process can be mediated by intracellular E-cadherins that sequester β-catenin and force the organization of the actin cytoskeleton. Upregulation of Tβ4 during fetal and tumor development activates integrin-linked kinase (ILK), a process that increases production of transcription factors Snail and ZEB1, decreasing both the production of E-cadherins and the expression of epithelial-related genes (McPhee, McDonald, Oloumi, & Dedhar, 2008). This results in increased Wnt and β-catenin signaling. Together with the known facts that Tβ4 increases cytoskeletal rearrangement by sequestering actin and promotes cytosolic accumulation of β-catenin through inhibition of GSK-3β, these results show the importance of leveraging the in vivo expression of Tβ4 to enhance specific pathway signaling and gene transcription.

The ability of Tβ4 to promote proliferation, differentiation, adhesion, and survival of cells has been used to push stem cells toward the cardiac lineage, both in vivo and in vitro. As previously mentioned, Tβ4 can be embedded within or used to coat scaffolds, on which cells are subsequently deposited. As these cells grow along the scaffolding, extracellular phosphatidylinositol 3-kinase (PI3K) signaling is induced through interaction with Tβ4, promoting the ILK-driven signaling cascade described earlier. Downstream activation of endothelial nitric oxide synthase 3 (eNOS) further increases cellular proliferation; in vivo, eNOS is crucial for regulation of vascular tone and promotion of angiogenesis, helping to explain how Tβ4 increases vessel formation and stability (Cooke & Losordo, 2002; Forstermann & Munzel, 2006). Though there is still some speculation as to the surface receptor for Tβ4, it is widely believed to be the purinergic receptor P2X4 (Freeman, Bowman, & Zetter, 2011), a receptor implicated in, among other biological processes, inflammasome activation after injury and ATP-stimulated electrical currents in ventricular myocytes (de Rivero Vaccari et al., 2012; Shen, Pappano, & Liang, 2006).

In vitro and in vivo cell differentiation studies have been performed using Tβ4 as a primary or secondary differentiation factor. The Srivastava lab has worked extensively on characterizing Tβ4 action for cell fate determination and differentiation, and organogenesis. A recent paper by the lab shows that cardiac fibroblasts can be differentiated to functional cardiomyocytes in vivo using the transcription factors Gata4, Mef2c, and Tbx5, a process greatly enhanced by follow-up injection with Tβ4. Injections of Tβ4 into the injured heart recruit resident cardiac stem cells to the site of injury where they differentiate into functional cardiomyocytes to aid in ventricular repair (Smart et al., 2007; Tateishi, Takehara, Matsubara, & Oh, 2008), though whether Tβ4 directly induces differentiation or whether induced changes in the cardiac microenvironment are the cause of differentiation is not yet clear (Zhou et al., 2012). Qiu et al. used a transwell migration assay to show Tβ4 is able to induce migration and differentiation of endothelial progenitor cells in vitro through the aforementioned ILK/eNOS pathway (Qiu, Song, Zheng, Zhao, & Fu, 2009). Human umbilical vein endothelial cells, an in vitro model of angiogenesis, doubled vessel area and formation when exposed to 100 ng of Tβ4, with subsequent increases occurring in a concentration-dependent manner (Grant et al., 1999).

We (Kumar et al., 2014) and others (Jeon et al., 2013; Ye et al., 2013) have shown the effect of Tβ4 on MSC proliferation and differentiation. Stimulation of the NF-kB pathway through phosphorylation of ERK1/2 promotes IL-8-mediated migration of MSCs toward the wound bed, which further promotes MSC cytokine release. The in vivo combination of Tβ4 and MSCs, which have been shown to evade the immune system by down-regulating tissue recruitment of B cells and lymphocytes (Ankrum, Ong, & Karp, 2014), can decrease systemic inflammation and promote cell survival. Indeed, MSCs coinjected with Tβ4 to a rat model of MI significantly decrease cell damage and death, and increase cardiac metrics such as ejection fraction and fractional shortening (Ye et al., 2013). Our group has primed MSCs in vitro to differentiate toward the cardiac lineage, showing that both ex vivo murine cardiac stem cells and human MSCs grown on nanoscaffolds benefit from increasing concentrations of Tβ4. The actin sequestration of Tβ4 allows the use of stiffer scaffolds that would otherwise prevent expression of cardiac proteins such as troponin I, desmin, and α-actinin. Combining Tβ4 with small molecule Wnt effectors significantly increases MSC proliferation and differentiation toward the cardiac lineage, with cardiac gene and protein expression increased in as little as 24 h compared to Tβ4 alone (Fig. 4).

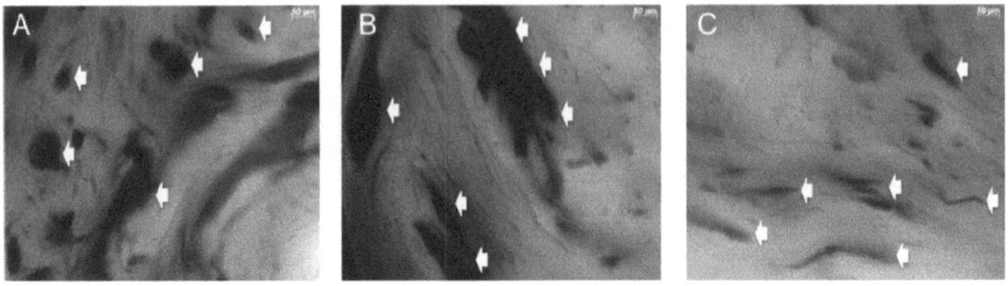

Fig. 4 Nanoscaffolds successfully engraft to myocardium after injection. (A and B) Large and small pieces of scaffold system engraft from epi- to endocardium. (C) Smaller scaffold pieces assist in diffuse spread of cells and factors and are more likely to align to the native myocardial tissue structure. All images taken 48 h after MI induction and 1 h after scaffold injection. (See the color plate.)

5. CONCLUSION AND FUTURE DIRECTIONS

The elucidation of Tβ4 as the major biologically active thymic fraction has resulted in an immense increase in therapeutic strategies in cardiology, neurology, and oncology. In both prenatal development and wound healing, vasculogenesis is quickly becoming a major factor in determining proper outcomes. In development, stunted neovasculogenesis mediated by Tβ4 results in deformed hearts, if not complete embryonic death. Upregulation of Tβ4 after wound inducement, particularly in the cardiac niche, results in increased vessel formation to provide the wound with immune cells and stem cells that can heal the damaged tissue. By harnessing the biological power of peptides like Tβ4, researchers are learning how to heal wounds in vivo in a noninvasive manner, decreasing the need for extensive surgeries and painful engraftments. Using materials science and nanomedicine approaches is pushing the boundary ever forward; encapsulation and sustained release of Tβ4 and small molecules coupled with stem cells has allowed healing of the myocardium, a feat that could result in saving hundreds of thousands of lives a year if put into clinical practice. Indeed, RegeneRx has two clinical trials ongoing to determine safety (NCT00743769) and efficacy (NCT01311518; clinicaltrials.gov) of injectable Tβ4 for MI. We and others are currently moving through preclinical small and large animal studies demonstrating the efficacy of various delivery models of Tβ4. These treatment strategies will be tested against a range of pathologies, with the goal of pushing therapies into clinical trials in the near future for the ever-growing number of suffering patients with no other therapies to turn to.

REFERENCES

Ankrum, J. A., Ong, J. F., & Karp, J. M. (2014). Mesenchymal stem cells: Immune evasive, not immune privileged. *Nature Biotechnology*, *32*, 252–260.

Arai, A. E. (2015). Healing after myocardial infarction: A loosely defined process. *JACC. Cardiovascular Imaging*, *8*, 680–683.

Asanuma, Y., Goldstein, A. L., & White, A. (1970). Reduction in the incidence of wasting disease in neonatally thymectomized CBA-W mice by the injection of thymosin. *Endocrinology*, *86*, 600–610.

Avolio, E., Caputo, M., & Madeddu, P. (2015). Stem cell therapy and tissue engineering for correction of congenital heart disease. *Frontiers in Cell and Developmental Biology*, *30*, 3–39.

Badamchian, M., Fagarasan, M. O., Danner, R. L., Suffredini, A. F., Damavandy, H., & Goldstein, A. L. (2003). Thymosin beta(4) reduces lethality and down-regulates inflammatory mediators in endotoxin-induced septic shock. *International Immunopharmacology*, *3*, 1225–1233.

Banerjee, I., Zhang, J., Moore-Morris, T., Lange, S., Shen, T., Dalton, N. D., et al. (2012). Thymosin beta 4 is dispensable for murine cardiac development and function. *Circulation Research*, *110*, 456–464.

Bergmann, O., & Jovinge, S. (2014). Cardiac regeneration in vivo: Mending the heart from within? *Stem Cell Research*, *13*, 523–531.

Bergmann, O., Bhardwaj, R. D., Bernard, S., Zdunek, S., Barnabé-Heider, F., Walsh, S., et al. (2009). Evidence for cardiomyocyte renewal in humans. *Science*, *324*, 98–102.

Bertassoni, L. E., Cecconi, M., Manoharan, V., Nikkhah, M., Hjortnaes, J., Cristino, A. L., et al. (2014). Hydrogel bioprinted microchannel networks for vascularization of tissue engineering constructs. *Lab on a Chip*, *14*, 2202–2211.

Bettinger, C. J., Langer, R., & Borenstein, J. T. (2009). Engineering substrate topography at the micro- and nanoscale to control cell function. *Angewandte Chemie*, *48*, 5406–5415.

Bock-Marquette, I., Saxena, A., White, M. D., Dimaio, J. M., & Srivastava, D. (2004). Thymosin beta4 activates integrin-linked kinase and promotes cardiac cell migration, survival and cardiac repair. *Nature*, *432*, 466–472.

Carlier, M. F., Didry, D., Erk, I., Lepault, J., Vantroys, M. L., Vandekerckhove, J., et al. (1996). Tβ4 is not a simple G-actin sequestering protein and interacts with F-actin at high concentration. *Journal of Biological Chemistry*, *271*, 9231–9239.

Caulfield, J. B., Tao, S. B., & Nachtigal, M. (1985). Ventricular collagen matrix and alterations. *Advances in Myocardiology*, *5*, 257–269.

Chen, S., Shimoda, M., Chen, J., & Grayburn, P. A. (2013). Stimulation of adult resident cardiac progenitor cells by durable myocardial expression of thymosin beta 4 with ultrasound-targeted microbubble delivery. *Gene Therapy*, *20*, 225–233.

Chiu, L. L., & Radisic, M. (2011). Controlled release of thymosin beta4 using collagen-chitosan composite hydrogels promotes epicardial cell migration and angiogenesis. *Journal of Controlled Release*, *155*, 376–385.

Chiu, L. L., Montgomery, M., Liang, Y., Liu, H., & Radisic, M. (2012). Perfusable branching microvessel bed for vascularization of engineered tissues. *Proceedings of the National Academy of Sciences of the United States of America*, *109*, E3414–E3423.

Chiu, L. L., Reis, L. A., Momen, A., & Radisic, M. (2012). Controlled release of thymosin beta4 from injected collagen-chitosan hydrogels promotes angiogenesis and prevents tissue loss after myocardial infarction. *Regenerative Medicine*, 7, 523–533.

Chiu, L. L., Reis, L. A., & Radisic, M. (2012). Controlled delivery of thymosin beta4 for tissue engineering and cardiac regenerative medicine. *Annals of the New York Academy of Sciences*, *1269*, 16–25.

Choi, S. Y., Noh, M. R., Kim, D. K., Sun, W., & Kim, H. (2007). Neuroprotective function of thymosin-beta and its derivative peptides on the programmed cell death of chick and rat neurons. *Biochemical and Biophysical Research Communications*, *363*, 587–593.

Cierniewski, C. S., Papiewska-Pajak, I., Malinowski, M., Sacewicz-Hofman, I., Wiktorska, M., Kryczka, J., et al. (2010). Thymosin beta4 regulates migration of colon cancer cells by a pathway involving interaction with Ku80. *Annals of the New York Academy of Sciences, 1191*, 60–71.

Cirino, A. L., & Ho, C. (2014). *GeneReviews: Hypertrophic cardiomyopathy overview*. Seattle, WA: University of Washington.

Cleutjens, J. P., Verluyten, M. J., Smiths, J. F., & Daemen, M. J. (1995). Collagen remodeling after myocardial infarction in the rat heart. *The American Journal of Pathology, 147*, 325–338.

Cooke, J. P., & Losordo, D. W. (2002). Nitric oxide and angiogenesis. *Circulation, 105*, 2133–2135.

Darland, D. C., & D'Amore, P. A. (2001). Cell-cell interactions in vascular development. *Current Topics in Developmental Biology, 52*, 107–149.

Dathe, V., & Brand-Saberi, B. (2004). Expression of thymosin beta4 during chick development. *Anatomy and Embryology, 208*, 27–32.

de Rivero Vaccari, J. P., Bastien, D., Yurcisin, G., Pineau, I., Dietrich, W. D., De Koninck, Y., et al. (2012). P2X4 receptors influence inflammasome activation after spinal cord injury. *The Journal of Neuroscience, 32*, 3058–3066.

DeGraba, T. J., Hallenbeck, J. M., Pettigrew, K. D., Dutka, A. J., & Kelly, B. J. (1999). Progression in acute stroke: Value of the initial NIH stroke scale score on patient stratification in future trials. *Stroke; A Journal of Cerebral Circulation, 30*, 1208–1212.

Dhanantwari, P., Lee, E., Krishnan, A., Samtani, R., Yamada, S., Anderson, S., et al. (2009). Human cardiac development in the first trimester: A high-resolution magnetic resonance imaging and episcopic fluorescence image capture atlas. *Circulation, 120*, 343–351.

Faa, G., Nemolato, S., Cabras, T., Fanni, D., Gerosa, C., Fanari, M., et al. (2012). Thymosin β4 expression reveals intriguing similarities between fetal and cancer cells. *Annals of the New York Academy of Sciences, 1269*, 53–60.

Forstermann, U., & Munzel, T. (2006). Endothelial nitric oxide synthase in vascular disease: From marvel to menace. *Circulation, 113*, 1708–1714.

Freeman, K. W., Bowman, B. R., & Zetter, B. R. (2011). Regenerative protein thymosin beta-4 is a novel regulator of purinergic signaling. *FASEB Journal, 25*, 907–915.

Frey, N., Katus, H. A., Olson, E. N., & Hill, J. A. (2004). Hypertrophy of the heart: A new therapeutic target? *Circulation, 109*, 1580–1589.

Fu, X., Cui, P., Chen, F., Xu, J., Gong, L., Jiang, L., et al. (2015). Thymosin beta4 promotes hepatoblastoma metastasis via the induction of epithelial-mesenchymal transition. *Molecular Medicine Reports, 12*, 127–132.

Garcia, S., Sandoval, Y., Roukoz, H., Adabag, S., Canoniero, M., Yannopoulos, D., et al. (2013). Outcomes after complete versus incomplete revascularization of patients with multivessel coronary artery disease: A meta-analysis of 89,883 patients enrolled in randomized clinical trials and observational studies. *Journal of the American College of Cardiology, 62*, 1421–1431.

Gehrig, S. M., Koopman, R., Naim, T., Tjoakarfa, C., & Lynch, G. S. (2010). Making fast-twitch dystrophic muscles bigger protects them from contraction injury and attenuates the dystrophic pathology. *The American Journal of Pathology, 176*, 29–33.

Gnecchi, M., He, H., Noiseux, N., Liang, O. D., Zhang, L., Morello, F., et al. (2006). Evidence supporting paracrine hypothesis for akt-modified mesenchymal stem cell-mediated cardiac protection and functional improvement. *FASEB Journal, 20*(6), 661–669.

Goldschmidt-Clermont, P. J., Furman, M. I., Wachsstock, D., Safer, D., Nachmias, V. T., & Pollard, T. D. (1992). The control of actin nucleotide exchange by thymosin beta 4 and profilin. A potential regulatory mechanism for actin polymerization in cells. *Molecular Biology of the Cell, 3*, 1015–1024.

Goldstein, A. L., Guha, A., Zatz, M. M., Hardy, M. A., & White, A. (1972). Purification and biological activity of thymosin, a hormone of the thymus gland. *Proceedings of the National Academy of Sciences of the United States of America*, *69*, 1800–1803.

Gómez, O., Martínez, J. M., Olivella, A., Bennasar, M., Crispi, F., Masoller, N., et al. (2014). Isolated ventricular septal defects in the era of advanced fetal echocardiography: Risk of chromosomal anomalies and spontaneous closure rate from diagnosis to age of 1 year. *Ultrasound in Obstetrics & Gynecology*, *43*, 65–71.

Gomez-Marquez, J., & Anadon, R. (2002). The beta-thymosins, small actin-binding peptides widely expressed in the developing and adult cerebellum. *Cerebellum*, *1*, 95–102.

Grant, D. S., Kinsella, J. L., Kibbey, M. C., LaFlamme, S., Burbelo, P. D., Goldstein, A. L., et al. (1995). Matrigel induces thymosin beta 4 gene in differentiating endothelial cells. *Journal of Cell Science*, *108*, 3685–3694.

Grant, D. S., Rose, W., Yaen, C., Goldstein, A., Martinez, J., & Kleinman, H. (1999). Thymosin beta4 enhances endothelial cell differentiation and angiogenesis. *Angiogenesis*, *3*, 125–135.

Hannappel, E. (2007). Beta-thymosins. *Annals of the New York Academy of Sciences*, *1112*, 21–37.

Harvey, W. (1628). *On the moton of the heart and blood in animals*: *Vol. 38*. New York: Collier & Son (Part 3).

Holmes, J. W., Borg, T. K., & Covell, J. W. (2005). Structure and mechanics of healing myocardial infarcts. *Annual Review of Biomedical Engineering*, 7, 223–253.

Hong, H. O., Lee, J. I., Hong, S., & Hong, S. (2016). Thymosin β 4 induces proliferation, invasion, and epithelial to mesenchymal transition of oral squamous cell carcinoma. *Amino Acids*, *48*, 117–127.

Huang, H. C., Hu, C. H., Tang, M. C., Wang, W. S., Chen, P. M., & Su, Y. (2007). Thymosin beta4 triggers an epithelial-mesenchymal transition in colorectal carcinoma by upregulating integrin-linked kinase. *Oncogene*, *26*, 2781–2790.

Huff, T., Müller, C. S., Otto, A. M., Netzker, R., & Hannappel, E. (2001). β-Thymosins, small acidic peptides with multiple functions. *The International Journal of Biochemistry & Cell Biology*, *33*, 205–220.

Jeon, B. J., Yang, Y., Kyung Shim, S., Yang, H. M., Cho, D., & Ik Bang, S. (2013). Thymosin beta-4 promotes mesenchymal stem cell proliferation via an interleukin-8-dependent mechanism. *Experimental Cell Research*, *319*, 2526–2534.

Karapanayiotides, T., Meuli, R., Devuyst, G., Piechowski-Jozwiak, B., Dewarrat, A., Ruchat, P., et al. (2005). Postcarotid endarterectomy hyperperfusion or reperfusion syndrome. *Stroke; A Journal of Cerebral Circulation*, *36*, 21–26.

Kasner, S. E. (2006). Clinical interpretation and use of stroke scales. *The Lancet Neurology*, *5*, 603–612.

Kramer, T. (1942). The partitioning of the truncus and conus and the formation of the membranous portion of the interventricular septum in the human heart. *The American Journal of Anatomy*, *71*, 343–370.

Kumar, A., Patel, A., Duvalsaint, L., Desai, M., & Marks, E. D. (2014). Thymosin beta4 coated nanofiber scaffolds for the repair of damaged cardiac tissue. *Journal of Nanobiotechnology*, *12*, 10.

Kwan, J., & Hand, P. (2006). Early neurological deterioration in acute stroke: Clinical characteristics and impact on outcome. *QJM: Monthly Journal of the Association of Physicians*, *99*, 625–633.

Li, Z., & Guan, J. (2011). Hydrogels for cardiac tissue engineering. *Polymers*, *3*, 740–762.

Li, F., Wang, X., Capasso, J. M., & Gerdes, A. M. (1996). Rapid transition of cardiac myocytes from hyperplasia to hypertrophy during postnatal development. *Journal of Molecular and Cellular Cardiology*, *28*, 1737–1746.

Low, T. L., Hu, S. K., & Goldstein, A. L. (1981). Complete amino acid sequence of bovine thymosin beta 4: A thymic hormone that induces terminal deoxynucleotidyl transferase activity in thymocyte populations. *Proceedings of the National Academy of Sciences of the United States of America*, *78*, 1162–1166.

Lyden, P. D., Lu, M., Levine, S. R., Brott, T. G., Broderick, J., & NINDS rtPA Stroke Study Group. (2001). A modified national institutes of health stroke scale for use in stroke clinical trials: Preliminary reliability and validity. *Stroke; A Journal of Cerebral Circulation*, *32*, 1310–1317.

Makogonenko, E., Goldstein, A. L., Bishop, P. D., & Medved, L. (2004). Factor XIIIa incorporates thymosin beta4 preferentially into the fibrin(ogen) alphaC-domains. *Biochemistry*, *43*, 10748–10756.

Malinda, K. M., Goldstein, A. L., & Kleinman, H. K. (1997). Thymosin beta 4 stimulates directional migration of human umbilical vein endothelial cells. *FASEB Journal*, *11*, 474–481.

McPhee, T. R., McDonald, P. C., Oloumi, A., & Dedhar, S. (2008). Integrin-linked kinase regulates E-cadherin expression through PARP-1. *Developmental Dynamics*, *237*, 2737–2747.

Michelis, K. C., Boehm, M., & Kovacic, J. C. (2014). New vessel formation in the context of cardiomyocyte regeneration—The role and importance of an adequate perfusing vasculature. *Stem Cell Research*, *13*, 666–682.

Morishita, N., Kusachi, S., Yamasaki, S., Kondo, J., & Tsuji, T. (1996). Sequential changes in laminin and type IV collagen in the infarct zone—Immunohistochemical study in rat myocardial infarction. *Japanese Circulation Journal*, *60*, 108–114.

Morris, D. C., Chopp, M., Zhang, L., & Zhang, Z. G. (2010). Thymosin beta4: A candidate for treatment of stroke? *Annals of the New York Academy of Sciences*, *1194*, 112–117.

Morris, D. C., Zhang, Z. G., Zhang, J., Xiong, Y., Zhang, L., & Chopp, M. (2012). Treatment of neurological injury with thymosin beta4. *Annals of the New York Academy of Sciences*, *1269*, 110–116.

Morris, D. C., Cui, Y., Cheung, W. L., Lu, M., Zhang, L., Zhang, Z. G., et al. (2014). A dose–response study of thymosin beta4 for the treatment of acute stroke. *Journal of the Neurological Sciences*, *345*, 61–67.

Mozaffarian, D., Benjamin, E. J., Go, A. S., Arnett, D. K., Blaha, M. J., Cushman, M., et al. (2015). Heart disease and stroke statistics—2015 update: A report from the American heart association. *Circulation*, *131*, e29–e322.

Nemolato, S., Messana, I., Cabras, T., Manconi, B., Inzitari, R., Fanali, C., et al. (2009). Thymosin β4 and β10 levels in pre-term newborn oral cavity and foetal salivary glands evidence a switch of secretion during foetal development. *PloS One*, *4*e5109.

Nemolato, S., Restivo, A., Cabras, T., Coni, P., Zorcolo, L., Orru, G., et al. (2012). Thymosin beta 4 in colorectal cancer is localized predominantly at the invasion front in tumor cells undergoing epithelial mesenchymal transition. *Cancer Biology & Therapy*, *13*, 191–197.

Noiseux, N., Gnecchi, M., Lopez-Ilasaca, M., Zhang, L., Solomon, S. D., Deb, A., et al. (2006). Mesenchymal stem cells overexpressing akt dramatically repair infarcted myocardium and improve cardiac function despite infrequent cellular fusion or differentiation. *Molecular Therapy*, *14*, 840–850.

Ookuma, Y. F., Kiyoshima, T., Kobayashi, I., Nagata, K., Wada, H., Fujiwara, H., et al. (2013). Multiple functional involvement of thymosin beta-4 in tooth germ development. *Histochemistry and Cell Biology*, *139*, 355–370.

Peterson, J. T., Hallak, H., Johnson, L., Li, H., O'Brien, P. M., Sliskovic, D. R., et al. (2001). Matrix metalloproteinase inhibition attenuates left ventricular remodeling and dysfunction in a rat model of progressive heart failure. *Circulation*, *103*, 2303–2309.

Petsche, C. J., Camci-Unal, G., Khademhosseini, A., & Jacot, J. G. (2013). Amniotic fluid-derived stem cells for cardiovascular tissue engineering applications. *Tissue Engineering Part B*, *19*, 368–379.

Philp, D., Goldstein, A. L., & Kleinman, H. K. (2004). Thymosin beta4 promotes angiogenesis, wound healing, and hair follicle development. *Mechanisms of Ageing and Development, 125*, 113–115.

Piao, Z., Hong, C., Jung, M., Choi, C., & Park, Y. (2014). Thymosin β4 induces invasion and migration of human colorectal cancer cells through the ILK/AKT/β-catenin signaling pathway. *Biochemical and Biophysical Research Communications, 452*, 858–864.

Pok, S., Benavides, O. M., Hallal, P., & Jacot, J. G. (2014). Use of myocardial matrix in a chitosan-based full-thickness heart patch. *Tissue Engineering. Part A, 20*, 1877–1887.

Qiu, F. Y., Song, X. X., Zheng, H., Zhao, Y. B., & Fu, G. S. (2009). Thymosin beta4 induces endothelial progenitor cell migration via PI3K/akt/eNOS signal transduction pathway. *Journal of Cardiovascular Pharmacology, 53*, 209–214.

Rankin, J. (1957). Cerebral vascular accidents in patients over the age of 60. II. Prognosis. *Scottish Medical Journal, 2*, 200–215.

Rennert, R. C., Sorkin, M., Garg, R. K., & Gurtner, G. C. (2012). Stem cell recruitment after injury: Lessons for regenerative medicine. *Regenerative Medicine, 7*, 833–850.

Rodriguez, L., Nathan, S., Chinna, M., Leema, M., & Grant, G. (2015). A novel role for thymosin beta4 in lung fibrosis via TGF-β mediated epithelial mesenchymal transition. C64 having a blast with fibroblasts. *American Journal of Respiratory and Critical Care Medicine, 191*, A4920. http://www.atsjournals.org/doi/abs/10.1164/ajrccm-conference.2015.191.1_MeetingAbstracts.A4920.

Rones, M. S., McLaughlin, K. A., Raffin, M., & Mercola, M. (2000). Serrate and Notch specify cell fates in the heart field by suppressing cardiomyogenesis. *Development, 127*, 3865–3876.

Rossdeutsch, A., Smart, N., & Riley, P. R. (2008). Thymosin beta4 and ac-SDKP: Tools to mend a broken heart. *Journal of Molecular Medicine, 86*, 29–35.

Santra, M., Chopp, M., Zhang, Z. G., Lu, M., Santra, S., Nalani, A., et al. (2012). Thymosin beta 4 mediates oligodendrocyte differentiation by upregulating p38 MAPK. *Glia, 60*, 1826–1838.

Santra, M., Zhang, Z. G., Yang, J., Santra, S., Santra, S., Chopp, M., et al. (2014). Thymosin beta4 up-regulation of microRNA-146a promotes oligodendrocyte differentiation and suppression of the toll-like proinflammatory pathway. *The Journal of Biological Chemistry, 289*, 19508–19518.

Sato, S., Ashraf, M., Millard, R. W., Fujiwara, H., & Schwartz, A. (1983). Connective tissue changes in early ischemia of porcine myocardium: An ultrastructural study. *Journal of Molecular and Cellular Cardiology, 15*, 261–275.

Scheinberg, M. A., Goldstein, A. L., & Cathcart, E. S. (1976). Thymosin restores T cell function and reduces the incidence of amyloid disease in casein-treated mice. *Journal of Immunology, 116*, 156–158.

Shen, J. B., Pappano, A. J., & Liang, B. T. (2006). Extracellular ATP-stimulated current in wild-type and P2X4 receptor transgenic mouse ventricular myocytes: Implications for a cardiac physiologic role of P2X4 receptors. *FASEB Journal, 20*, 277–284.

Silverman, M. E., Grove, D., & Upshaw, C. B., Jr. (2006). Why does the heart beat? The discovery of the electrical system of the heart. *Circulation, 113*, 2775–2781.

Simpson, J. M. (2009). Fetal cardiac interventions: Worth it? *Heart, 95*, 1653–1655.

Smart, N., Risebro, C. A., Melville, A. A., Moses, K., Schwartz, R. J., Chien, K. R., et al. (2007). Thymosin beta4 induces adult epicardial progenitor mobilization and neovascularization. *Nature, 445*, 177–182.

Smart, N., Dube, K. N., & Riley, P. R. (2010). Identification of thymosin beta4 as an effector of Hand1-mediated vascular development. *Nature Communications, 1*, 46.

Smart, N., Risebro, C. A., Clark, J. E., Ehler, E., Miquerol, L., Rossdeutsch, A., et al. (2010). Thymosin beta4 facilitates epicardial neovascularization of the injured adult heart. *Annals of the New York Academy of Sciences, 1194*, 97–104.

Sosne, G., Chan, C. C., Thai, K., Kennedy, M., Szliter, E. A., Hazlett, L. D., et al. (2001). Thymosin beta 4 promotes corneal wound healing and modulates inflammatory mediators in vivo. *Experimental Eye Research*, *72*, 605–608.

Sturzu, A. C., Rajarajan, K., Passer, D., Plonowska, K., Riley, A., Tan, T. C., et al. (2015). Fetal mammalian heart generates a robust compensatory response to cell loss. *Circulation*, *132*, 109–121.

Sumer, M., Ozdemir, I., & Erturk, O. (2003). Progression in acute ischemic stroke: Frequency, risk factors and prognosis. *Journal of Clinical Neuroscience*, *10*, 177–180.

Tateishi, K., Takehara, N., Matsubara, H., & Oh, H. (2008). Stemming heart failure with cardiac- or reprogrammed-stem cells. *Journal of Cellular and Molecular Medicine*, *12*, 2217–2232.

Thomas, J. R., & Frost, T. W. (1993). Immediate versus delayed repair of skin defects following resection of carcinoma. *Otolaryngologic Clinics of North America*, *26*, 203–213.

Ti, D., Hao, H., Xia, L., Tong, C., Liu, J., Dong, L., et al. (2015). Controlled release of thymosin beta 4 using a collagen-chitosan sponge scaffold augments cutaneous wound healing and increases angiogenesis in diabetic rats with hindlimb ischemia. *Tissue Engineering. Part A*, *21*, 541–549.

Trenkwalder, T., Deindl, E., Bongiovanni, D., Lee, S., Schunkert, H., Kupatt, C., et al. (2015). Thymosin-beta4-mediated therapeutic neovascularization: Role of the PI3K/AKT pathway. *Expert Opinion on Biological Therapy*, *15*, 175–185.

Ulrich, M. M., Janssen, A. M., Daemen, M. J., Rappaport, L., Samuel, J. L., Contard, F., et al. (1997). Increased expression of fibronectin isoforms after myocardial infarction in rats. *Journal of Molecular and Cellular Cardiology*, *29*, 2533–2543.

Vartiainen, N., Pyykonen, I., Hokfelt, T., & Koistinaho, J. (1996). Induction of thymosin beta(4) mRNA following focal brain ischemia. *Neuroreport*, *7*, 1613–1616.

Voorhees, A. P., DeLeon-Pennell, K. Y., Ma, Y., Halade, G. V., Yabluchanskiy, A., Iyer, R. P., et al. (2015). Building a better infarct: Modulation of collagen cross-linking to increase infarct stiffness and reduce left ventricular dilation post-myocardial infarction. *Journal of Molecular and Cellular Cardiology*, *85*, 229–239.

Vos, J. G., Kreeftenberg, J. G., Engel, H. W., Minderhoud, A., & Van Noorle Jansen, L. M. (1978). Studies on 2,3,7,8-tetrachlorodibenzo-p-dioxin induced immune suppression and decreased resistance to infection: Endotoxin hypersensitivity, serum zinc concentrations and effect of thymosin treatment. *Toxicology*, *9*, 75–86.

Wang, Z., Zeng, F., Zhu, Z., Jiang, G., Lv, L., Wan, F., et al. (2012). Evaluation of thymosin β4 in the regulation of epithelial-mesenchymal transformation in urothelial carcinoma. *Urologic Oncology*, *30*, 167–176.

Wara, D. W., Goldstein, A. L., Doyle, N. E., & Ammann, A. J. (1975). Thymosin activity in patients with cellular immunodeficiency. *The New England Journal of Medicine*, *292*, 70–74.

Wilm, B., Ipenberg, A., Hastie, N. D., Burch, J. B., & Bader, D. M. (2005). The serosal mesothelium is a major source of smooth muscle cells of the gut vasculature. *Development*, *132*, 5317–5328.

Wraya, J., & Sensky, T. (2001). Congenital heart disease and cardiac surgery in childhood: Effects on cognitive function and academic ability. *Heart*, *85*, 687–691.

Xavier-Neto, J., Shapiro, M. D., Houghton, L., & Rosenthal, N. (2000). Sequential programs of retinoic acid synthesis in the myocardial and epicardial layers of the developing avian heart. *Developmental Biology*, *219*, 129–141.

Xiong, Y., Mahmood, A., Meng, Y., Zhang, Y., Zhang, Z. G., Morris, D. C., et al. (2012). Neuroprotective and neurorestorative effects of thymosin beta4 treatment following experimental traumatic brain injury. *Annals of the New York Academy of Sciences*, *1270*, 51–58.

Xiong, Y., Zhang, Y., Mahmood, A., Meng, Y., Zhang, Z. G., Morris, D. C., et al. (2012). Neuroprotective and neurorestorative effects of thymosin beta4 treatment initiated 6 hours after traumatic brain injury in rats. *Journal of Neurosurgery*, *116*, 1081–1092.

Yamada, S., Samtani, R. R., Lee, E. S., Lockett, E., Uwabe, C., Shiota, K., et al. (2010). Developmental atlas of the early first trimester human embryo. *Developmental Dynamics*, *239*, 1585–1595.

Ye, L., Zhang, P., Duval, S., Su, L., Xiong, Q., & Zhang, J. (2013). Thymosin beta4 increases the potency of transplanted mesenchymal stem cells for myocardial repair. *Circulation*, *128*, S32–S41.

Yuan, S. M. (2014). Fetal cardiac interventions: An update of therapeutic options. *Revista Brasileira de Cirurgia Cardiovascular*, *29*, 388–395.

Zhou, B., Honor, L. B., Ma, Q., Oh, J. H., Lin, R. Z., Melero-Martin, J. M., et al. (2012). Thymosin beta 4 treatment after myocardial infarction does not reprogram epicardial cells into cardiomyocytes. *Journal of Molecular and Cellular Cardiology*, *52*, 43–47.

CHAPTER ELEVEN

Thymosin β4 Promotes Dermal Healing

H.K. Kleinman*[,1], G. Sosne[†]
*George Washington University, Washington, DC, United States
[†]Kresge Eye Institute, Wayne State University School of Medicine, Detroit, MI, United States
[1]Corresponding author: e-mail address: hyndakk@aol.com

Contents

Abstract

No agent has been identified that significantly accelerates the repair of chronic dermal wounds in humans. Thymosin beta 4 (Tβ4) is a small, abundant, naturally occurring regenerative protein that is found in body fluids and inside cells. It was found to have angiogenic and antiinflammatory activity and to be high in platelets that aggregate at the wound site. Thus we used Tβ4 initially in dermal healing. It has since been shown to have many activities important in tissue protection, repair, and regeneration. Tβ4 increases the rate of dermal healing in various preclinical animal models, including

Vitamins and Hormones, Volume 102
ISSN 0083-6729
http://dx.doi.org/10.1016/bs.vh.2016.04.005

diabetic and aged animals, and is active for burns as well. Tβ4 also accelerated the rate of repair in phase 2 trials with patients having pressure ulcers, stasis ulcers, and epidermolysis bullosa wounds. It is safe and well tolerated and will likely have additional uses in the skin and in injured organs for tissue repair and regeneration.

1. INTRODUCTION

Thymosin beta 4 (Tβ4) is a highly conserved, naturally occurring, water-soluble regenerative peptide that is found in all tissues and in all cell types, except red blood cells (Goldstein, Hannappel, Sosne, & Kleinman, 2012; Goldstein & Kleinman, 2015). It is also found in the blood and in other body fluids, including tears, saliva, cerebrospinal fluid, and wound fluids (Badamchian et al., 2007; Huang, Wang, Barnes, & Elmets, 2006; Mohring, Kellmann, Jurgens, & Schrader, 2005). Both platelets and leukocytes release Tβ4 into the wound fluid such that the final concentration is 13 μg/mL (Fromm, Gunne, Bergman, et al., 1996; Hannappel & van Kampen, 1987).

Although Tβ4 contains only 43 amino acids, it appears to have a wide range of regenerative activities and specific sites on the molecule have been shown to mediate these effects (Goldstein & Kleinman, 2015; Sosne, Qiu, Goldstein, & Wheater, 2010). Both chemically synthesized and recombinant forms have shown efficacy for dermal healing in preclinical models and in human patients (Ehrlich & Hazard, 2012; Kim & Kwon, 2014, 2015; Malinda et al., 1999; Philp, Badamchian, et al., 2003; Philp & Kleinman, 2010; Philp et al., 2006; Ti et al., 2015; Treadwell et al., 2012). A dimeric form has been found to accelerate the rate of dermal healing in an animal model more rapidly than that of the parent molecule (Xu et al., 2013). Tβ4 has also shown repair and regenerative activity in a number of other injury models, such as traumatic brain injury, spinal cord injury, stroke, a model of multiple sclerosis, ischemic limbs, and cardiac damage due to ischemia (Bock-Marquette, Saxena, White, Dimaio, & Srivastava, 2004; Cheng, Kuang, Zhang, Ju, & Wang, 2014; Dube, Bollini, Smart, & Riley, 2012; Morris, Chopp, Zhang, Lu, & Zhang, 2010; Morris et al., 2014; Philp & Kleinman, 2010; Postrach et al., 2014; Smart et al., 2007; Sopko et al., 2011; Ti et al., 2015, Wang et al., 2012; Wei, Kim, Li, Wu, & Gupta, 2014; Xiong, Mahmood, Meng, et al., 2011; Zhang, Zhang, Morris, et al., 2009; Zuo et al., 2013). The processes and pathways for Tβ4-mediated repair are similar in these various tissues and support the observed promotion of dermal healing.

The first study to show that Tβ4-promoted tissue repair was a dermal study performed in rats (Malinda et al., 1999). It had previously been found to promote angiogenesis and was reported to be high in platelets (Grant et al., 1995; Hannappel & van Kampen, 1987; Malinda, Goldstein, & Kleinman, 1997; Philp, Huff, Gho, Hannappel, & Kleinman, 2003). Since platelets are the first cells to enter a wound, it was clear that Tβ4 should be tested in dermal wounds in an animal model (Malinda et al., 1997, 1999; Philp, Badamchian, et al., 2003). In the first dermal study using 8 mm full-thickness punch wounds in rats, Tβ4 at 5 μg/50 μL of phosphate-buffered saline was found to accelerate wound closure, increase angiogenesis, and accelerate collagen deposition (Malinda et al., 1999). Tβ4 was only applied at the time of injury and at 48 h since after that the crust had formed. Visible macroscopic improvement was seen in the treated group by day 4. The study also found that Tβ4 promoted keratinocyte migration in vitro with activity in the picogram range. The findings were confirmed in various additional animal models (Table 1) and led to the clinical trials for hard to heal wound in patients as detailed in Table 2.

Table 1 Animal Studies: Increased Dermal Healing

Model	Dose	Outcome	Mechanism	References
Rat: skin flaps	5 mg/kg IP 2×/day 7 days	Flap survival	Increased VEGF Decreased SOD Increased malondialdehyde	Lin, Lin, Lin, Huang, and Cao (2015)
Mouse: burn	5.30 mg/kg Intradermal 2×/week 2 weeks	Increased closure Healing Angiogenesis HSP70	Increased actin remodeling	Kim and Kwon (2015)
Mouse: burn Diabetic	5 mg/kg Intradermal 2×/week 2 weeks	Increased closure Granulation Angiogenesis	Decreased RAGE	Kim and Kwon (2015)
Rat: hind limb Ischemia diabetes	Tβ4 in collagen/ chitosan sponge 21 days	Increased angiogenesis Healing	Decreased TNF IL (inflammation) Increased VEGF PDGF (angiogenesis)	Ti et al. (2015)

Continued

Table 1 Animal Studies: Increased Dermal Healing—cont'd

Model	Dose	Outcome	Mechanism	References
Rat: punch wound	Recombinant dimer at 0.5, 0.25, 0.125 mg/mL 2×/day 21 days	Increased healing Angiogenesis	Increased cell migration	Xu et al. (2013)
Mouse: punch wound	Recombinant 5 μg/50 μL PBS days 0, 2	Increased healing	Cell migration	Li et al. (2007)
Rat: punch wound	5 μg/50 μL topical days 0, 2	Increased healing Angiogenesis	Cell migration	Philp, Badamchian, et al. (2003) and Philp, Huff, et al. (2003)
Rat: punch wound	60 μg IP days 0, 2	Improved healing	Cell migration	Malinda et al. (1999)

Table 2 Human Studies: Increased Trend in Rate of Dermal Healing

Wound Type	Patients (#)	Study Duration (Days)	Optimal Dose (%)[a]	Outcome
Pressure ulcers (stages III and IV)	72	84	0.02	⇑ Healing rate trend[b,c]
Stasis ulcers	72	84	0.03	⇑ Healing rate trend[b,d]
Epidermolysis bullosa	28	56	0.03	⇑ Healing rate trend[e,f,g]

[a]Three doses and placebo were tested in all three phase 2 trials. Dose and frequency: 0. 0.01%, 0.02%, or 0.03% and 0.10% were given every day for the study duration.
[b]Data are from Treadwell et al. (2012).
[c]Mean healing rate 22 days vs 57 days for placebo, not statistically significant.
[d]Mean healing rate 39 days vs 71 days for placebo, not statistically significant.
[e]Data analyzed by diving wound size at postbaseline visit by mean wound size at baseline.
[f]Unpublished.
[g]Not statistically significant.

2. HUMAN DERMAL HEALING

Human dermal healing is a highly complex but efficient process in most situations especially in young people requiring little attention beyond cleaning and protection from infections (Gould, Abadir, Brem, et al., 2015). In contrast, the elderly, diabetics, and immobilized patients have a reduced rate of healing and often suffer from chronic wounds that are costly to treat and can compromise their quality of life (Gould et al., 2015). The wounds are chronic due to prolonged inflammation, increased proteases and elastases, lower expression of transforming growth factor beta (TGF-β), and diminished matrix production. In particular, these patients suffer from venous legs ulcers, pressure ulcers, diabetic foot ulcers, and limb ischemia. No effective treatment exists at this time, and this population of patients is increasing as longevity increases, so there is an unmet need to find new therapeutics that accelerate the healing of these chronic wounds.

In the absence of an effective drug, simple cleansing, debriding, compression bandaging where appropriate, and good wound care are the major treatments generally used for chronic wounds. There has been recently excitement about the use of stem cells and/or fibroblasts alone or in combination with extracellular matrix components for the treatment of chronic wounds. However, these treatments are costly and are often not covered by insurance. In addition, they show limited efficacy. Clearly, stem cells are an important part of the regenerative process in dermal healing. Tβ4 has been shown to recruit stem cells during hair growth and neural repair in vivo and promote stem cell migration in vitro (Morris et al., 2010; Philp et al., 2004; Xiong et al., 2011). In the central nervous system (CNS) and in the heart, Tβ4 promotes tissue repair and survival via stem cell recruitment (Bock-Marquette et al., 2009; Bollini, Riley, & Smart, 2015; Chopp & Zhang, 2015). When stem cells overexpressing Tβ4 are injected into animals with cardiac injury, repair is accelerated over that seen with stem cells alone. Furthermore, a single local injection of Tβ4 in a pig model of reperfusion injury shows the same cardioprotection as a stem cell injection (Kupatt, Bosk-Marquette, & Bockstegers, 2008). Finally, stem cells genetically programmed to have reduced (silenced) Tβ4 show a reduced ability to repair heart tissue over that of control stem cells (Hinkel, El-Aouni, Olson, Horstkotte, et al., 2008). Thus Tβ4 is a trophic factor for stem cells and likely functions in dermal repair to recruit stem cells. It should also be noted that the mechanism by which stem cells promote tissue regeneration is through their secreted factors and the actual stem cells themselves do not replace damaged tissue to a significant extent (Gallina, Turunetto, & Giachino, 2015). These findings provide the rationale for concluding that Tβ4 may be as effective as or better than stem cell transplants for both tissue repair and regeneration, and it may also be less costly.

3. STAGES IN DERMAL HEALING

There are three stages in the complex cascade of healing dermal wounds, including inflammation, proliferation, and remodeling (Fig. 1; Reiger, Zhao, Martin, Abe, & Lisse, 2015; Singer & Clark, 1999; Woodley, Wysong, DeClerck, Chen, & Li, 2015). These stages overlap and are interdependent. The duration of each of these stages and of the healing process depends on many factors, including the type, width, depth, and location of the wound, age of the person, general health of the person, treatment, etc. In general, dermal healing is rapid and efficient for young and healthy people. However, there are many patients whose wounds do not heal for a long period of time, and their quality of life is significantly impaired (Gould et al., 2015). Such patients include diabetics, the elderly, those who are bedridden, patients with epidermolysis bullosa, etc. Healing of burns is also problematic due to the pain, large areas to be healed, and scar formation.

3.1 Inflammation

Immediately after dermal injury, the wounded area is filled with blood consisting of cells and proteins from the circulation, proteins released from the injured tissue, and vasoactive amines from the locally damaged blood vessels and from the dermal keratinocytes. A clot forms as platelets aggregate. Platelets secrete chemotactic growth factors, such as platelet-derived growth

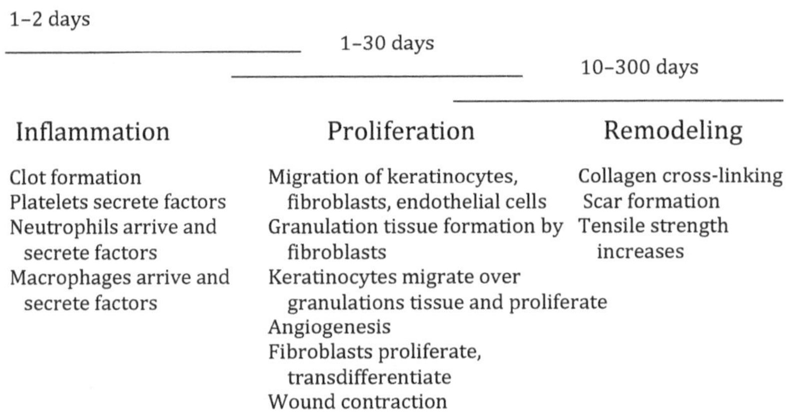

Fig. 1 Steps in dermal healing.

factor (PDGF) and TGF-β1 that increase vascular permeability and attract initially neutrophils and then monocytes to the wound site. By 24–48 h after dermal injury, the neutrophils have undergone apoptosis, while the macrophages begin removing debris and secrete chemotactic and repair factors as well.

3.2 Proliferation

Endothelial cells are recruited by released chemotactic factors into the clot, where they proliferate and differentiate to form new blood vessels. Fibroblasts also migrate in response to factors secreted by the macrophages into the wounded area, proliferate, and deposit a provisional extracellular matrix termed the "granulation tissue." Keratinocytes migrate along the injured dermis above the provisional matrix, begin to cover the wound, and proliferate in a location distant from the lead migrating cells. By 3–10 days after injury, the wound is filled with granulation tissue, and the fibroblasts begin to differentiate into myofibroblasts. The wound begins to contract and collagen is laid down by the myofibroblasts.

3.3 Remodeling

At the remodeling phase, collagen synthesis is reduced approximately 2 weeks postinjury. Collagen cross-linking and reorganization begin and can occur for months/years after injury to restore tissue structure and function. In scar tissue, collagen alignment occurs in a single direction. The tensile strength of the tissue increases over time.

4. Tβ4 ACTIVITY

4.1 Activity in Wounds

Tβ4 is found in almost all cells (Goldstein et al., 2012). Tβ4 has activity in the three major aspects of tissue regeneration that are important in dermal repair and in regeneration: protection, repair, and regeneration (Fig. 2). In the early inflammation stage of dermal healing, it is highly significant for dermal repair that Tβ4 is enriched in platelets because these are the first cells to arrive at a site of injury where these cells release various factors, including Tβ4, that initiate the repair cascade (Hannappel & van Kampen, 1987; Kaur & Mutus, 2012). Tβ4 is chemotactic for the directional migration of both endothelial cells and keratinocytes, which migrate rapidly into the wound site, allowing the healing process to begin (Grant et al., 1995; Malinda et al., 1997; Roy et al., 2001).

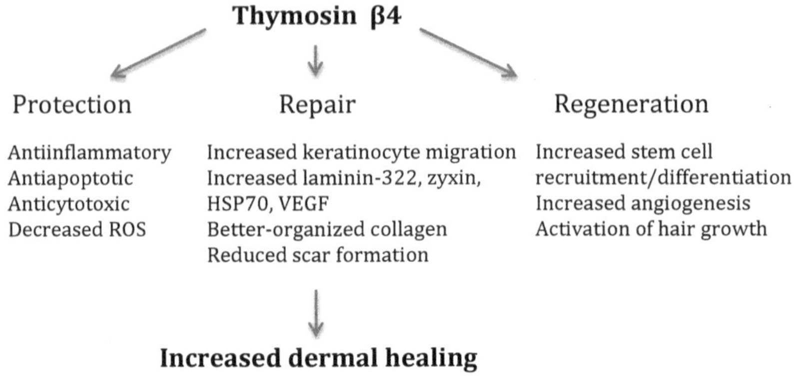

Fig. 2 Activity in tissue protection, repair, and regeneration. (See the color plate.)

Tβ4 is increased significantly in wound fluids following surgery and during ischemic injury to the heart. Thus, increases in Tβ4 are part of the natural early steps in the normal dermal healing process. Tβ4 protects the injured tissue by various mechanisms (Fig. 2). Tβ4 is also antiapoptotic and antiinflammatory and thus protects the injured tissue from further injury due to the debris released from the damaged cells (Badamchian et al., 2003; Ho et al., 2007; Popoli et al., 2007; Sosne, Siddiqi, & Kurpakus-Wheater, 2004; Sosne et al., 2002; Young et al., 1999). Direct studies on toxicity using glutamate, benzalkonium, and ethanol have demonstrated the anticytotoxic effects of Tβ4 (Popoli et al., 2007; Sosne et al., 2010). Hypoxia in the wound also increases the expression of Tβ4. It has been shown to reduce reactive oxygen species (ROS) in some cells, and this could also occur in and be beneficial to protection and dermal healing (Ho et al., 2008; Kumar & Gupta, 2011; Wei, Kumar, Kim, & Gupta, 2012).

During the proliferation phase of dermal healing, Tβ4 promotes the migration of various cells into the wound, including fibroblasts, keratinocytes, and endothelial cells. The fibroblasts proliferate and deposit the granulation tissue. Tβ4 increases keratinocyte migration across the granulation tissue that is important in reestablishing a barrier to prevent fluid loss and in infection. It also stimulates angiogenesis that helps protect the wound by removing debris and by increasing nutrients for repair, respectively (Grant et al., 1995, Malinda et al., 1997, 1999). Stem cells are likely recruited to the dermal wound and differentiate as shown in injury models of other tissues (Hinkel et al., 2008; Kupatt et al., 2008; Philp et al., 2004). The rate of collagen synthesis and the more organized formation of the collagen fibers are increased by Tβ4 (Ehrlich & Hazard, 2012). Other proteins important in wound repair are also increased by Tβ4, including laminin-332, HGF, VEGF, inflammatory mediators, etc. (Jo et al., 2010; Sosne, Chrispopherson, Barrett, & Fridman, 2005; Sosne et al., 2004; Sosne, Qiu, Christopherson, & Wheater, 2007). Tβ4 has been shown to reduce the formation of myofibroblasts, and this likely explains its ability to reduce fibrosis/scar formation (Ehrlich & Hazard, 2010, 2012). Tβ4 appears to act by disrupting microtubules and altering TGF-β-Smad signaling.

During the remodeling phase, Tβ4 acts to better organize the collagen fibers and likely reduces scar formation. Various studies in nondermal tissues have shown that Tβ4 and its component N-terminal peptide SDKP not only reduce scar formation/fibrosis but also reverse it (Castoldi et al., 2013; Cavasin, 2006; Cavasin, Liao, Yang, Yang, & Carretero, 2007; Gonzalez et al., 2014; Wang, Liu, Jia, Mu, & Xie, 2010; Zuo et al., 2013). Finally, it has been shown to promote hair growth in various animal models, including rats, mice, aged mice, and mice with hair loss induced by chemotherapeutic agents (Goldstein et al., 2012; Goldstein & Kleinman, 2015; Philp et al., 2004). Mice overexpressing Tβ4 also have an increased amount of hair and mice underexpressing Tβ4 have reduced hair growth (Cha et al., 2010; Gao et al., 2015).

In summary, these studies show that Tβ4 is active in all phases of dermal wound healing. It is naturally present in wounds and in wound fluids. Related studies have reported the beneficial effects of naturally elevated endogenous Tβ4 levels. For example, patients with severe coronary artery disease were found to have better collateral vessel development if the serum Tβ4 levels were elevated (Bicer, Karakurt, Akdemir, Erden, et al., 2011). Also patients with scleroderma were found to have less tissue fibrosis if their levels of Tβ4 were elevated (De Santis, Inzitari, Bosello, Peluso, et al., 2011). Thus, Tβ4 is an endogenous repair/regeneration factor that when added exogenously increases the rate of healing and most importantly decreases scarring.

4.2 Tβ4 Activity in Dermal Burns

Dermal burn wounds are highly susceptible to infection, are very painful, can result in prolonged hospitalization, can heal slowly, and can often cause considerable scarring. Finding an effective agent for dermal burns would be a major medical advance and greatly benefit the patients. Tβ4 is an excellent candidate as a therapeutic for dermal burns due to its many activities (Fig. 2) in dermal repair. Two studies have directly examined the effect of Tβ4 on dermal burns in preclinical animal models. Both found an improved rate of wound closure and increased angiogenesis (Kim & Kwon, 2014, 2015). These studies also pointed to important mechanisms involved in the Tβ4-mediated healing process for burns. In the diabetic preclinical db/db mouse model, Tβ4 was found to reduce the receptor of advanced glycation end products (RAGE) (Kim & Kwon, 2014). In a normal mouse study, the same authors found that when comparing Tβ4 given at 5 and 30 mg/kg (given intradermally twice a week), the higher dose was more effective by day 7. They observed complete healing by day 10 with Tβ4, while the control group was still not healed by day 14 (Kim & Kwon, 2015). These results demonstrate that Tβ4 accelerates the healing of dermal burns in animal models. They found an increase in cell viability and a decrease in the level of intracellular ROS similar to other groups (Kumar & Gupta, 2011; Wei et al., 2012). These findings further demonstrate the cytoprotective activity of Tβ4. They also found increased expression of heat-shock protein 70 (HSP70), while other HSPs were not upregulated. HSPs generally function as chaperones that are increased in the presence of heat and facilitate the refolding of proteins denatured by the heat. Since they also found that the Tβ4-induced HSP70 affected the expression of both p-AKT and VEGF, they concluded that Tβ4-induced HSP70 acts to reorganize the cytoskeleton and affects the signaling pathways of p-AKT. Increased VEGF and p-AKT have also been observed in response to Tβ4 in other types of pathologies and injuries (Cha, Jeong, & Kleinman, 2003; Gao et al., 2015; Lin et al., 2015; Trenkwalder et al., 2015). These studies show that Tβ4 increases HSP70 that regulates the actin cytoskeleton and thus plays an important role in burn healing. Tβ4 has been shown also to directly bind to actin (Goldstein et al., 2012).

4.3 Active Sites in Tβ4 and Cell Surface Receptor

Given the various biological activities of Tβ4, it is not surprising that multiple active sites have been identified (Goldstein et al., 2012). In general, these sites have been identified using synthetic peptides and in some cases natural fragments (Fig. 3; Philp, Huff, et al., 2003; Sosne et al., 2010).

4.3.1 Peptide 1–4

The small amino terminal peptide 1–4 (Ac-SDKP, aspartate–serine–lysine–proline) contains the antifibrotic activity as determined in vivo and antiinflammatory activity as determined in vitro (Cavasin, 2006;

Amino acids 1–4 (Ac-SDKP)
Antifibrotic
Antiinflammatory
Stimulates epicardium-derived stem cells
Inhibits bone marrow differentiation
Decreases TNF-alpha release by macrophages
Suppress Smad activation (blocks TGF-beta signaling)

Amino acids 1–15 (SDKPMAEIEKFDKS)
Antiapoptotic
Protects cells from chemical injury

Amino acids 17–23 (LKKTETQ)
Actin binding
Promotes cell migration
Promotes hair growth
Promotes angiogenesis
Promotes dermal healing
Promotes mast cells exocytosis

Amino acids 40–43(AGES)
Increases myocyte survival via Akt signaling
Increases vessel growth
Decreases inflammation

Activities not yet defined by specific sequences in thymosin β_4
Laminin-332 and VEGF production
Better organized collagen fibrils
Decreased ROS (reactive oxygen species)
Stem cell recruitment

Fig. 3 Active sites on thymosin β4.

Sosne et al., 2010). Various studies have also shown that this site can reverse fibrosis in the heart and kidney (Castoldi et al., 2013; Cavasin, 2006; Cavasin et al., 2007; Gonzalez et al., 2014; Wang et al., 2010; Zuo et al., 2013). SDKP is a naturally occurring peptide fragment of Tβ4 and is found in the blood. It is generated by the activity of prolyl oligopeptidase (POP). SDKP is further degraded by angiotensin-converting enzyme (ACE). Interestingly, inhibitors of POP decrease the levels of circulating SDKP and increase fibrosis. In contrast, ACE inhibitors increase the levels of circulating SDKP up to five-fold, and it is also elevated in the heart, kidney, and urine. ACE inhibitors decrease fibrosis, and it is possible that the tissue protection observed with ACE inhibitors is due to increased SDKP levels. SDKP has also been found to decrease fibroblast proliferation, inhibit bone marrow cell differentiation, decrease TNF-alpha release by macrophages, and suppress Smad activation (blocks TGF-β signaling). Some studies also report angiogenic activity (Wang, Carretero, et al., 2004). In dermal healing, this site is likely important in reduction of both inflammation and scar formation.

4.3.2 Peptide 1–15

The active site for antiapoptosis and cytoprotection is contained in the amino terminal 15 amino acids. When fibroblasts were treated with various chemicals, such as benzalkonium, ethanol, and chlorhexidine, Tβ4 was found to protect the cells from dying (Sosne et al., 2010). This is important as such activity may protect wounded tissue from naturally toxic materials present in the wound due to the large amount of debris released from damaged cells and even infection.

4.3.3 Peptide 17–23

The central actin binding sequence (LKKTET) is present in this peptide. This peptide has been shown to promote dermal repair when tested alone (Huang et al., 2006; Philp, Badamchian, et al., 2003; Sosne et al., 2010; Wyczoikowska, Walczak-Drzewiecka, Wagner, & Dastych, 2007). It has many important activities for dermal repair, including cell migration, angiogenesis, and actin binding (Goldstein et al., 2012). It also promotes hair growth and mast cell exocytosis (Philp et al., 2004). This peptide is not naturally occurring. The signaling via this peptide appears to involve purinergic receptors that have also been described for the intact molecule as well (Freeman, Bowman, & Zetter, 2011; Huang et al., 2006). This is the only receptor system that has been defined for Tβ4, but others likely exist given the multiple active sites and different activities.

4.3.4 Peptide 40–43 (AGES)

The carboxy terminal peptide of Tβ4 is unique among the different thymosin and has recently been found to have activity with hypoxic heart tissue (Hinkel, Ballm, DiMiado, Srivastava, et al., 2015). It increases myocyte survival via Akt signaling, vessel growth, and decreases inflammation. It has not yet been found to have activity in dermal wounds, but this recent finding in the cardiac system may stimulate such a study.

4.3.5 Other Activities in Unknown Sequences

Additional activities of Tβ4 remain to be defined in terms of specific peptide sequences. These include the production of laminin-332 and VEGF, the decrease in ROS, better organization of collagen fibrils, and stem cell recruitment. While peptide 17–23 can promote dermal healing, its ability to promote ocular, cardiac, and CNS healing has not yet been tested. Clearly all of the active peptides should be tested in different injured tissues to determine the exact active site for each tissue or if multiple specific peptides are needed.

4.3.6 Cell Surface Receptor ATP Synthase

Purinergic signaling has been shown to be involved in Tβ4 activity (Freeman et al., 2011; Huang et al., 2006). Not only does Tβ4 bind to ATP synthase, but antibodies and antagonists to ATP synthase block the Tβ4-induced increase in cell surface ATP levels. Finally, using a silencing approach, the ATP-responsive P2X4 receptor is required for Tβ4-induced cell migration. These studies show that Tβ4 regulates purinergic signaling through ATP synthase. Energy production, mainly as ATP, is required for proper wound healing so this is not an unexpected finding. Furthermore, dysfunctional purine metabolism has been shown to impair diabetic dermal healing (Weinstein, Lalezarzadeh, Soares, Saddeh, & Ceradini, 2015) where there is local ischemia due to vascular occlusion and increases in inflammatory cytokines and ROS. Increases in ROS are linked to overproduction of xanthine oxidase (Parihar, Parihar, Miner, & Bhat, 2008). Reduction of this enzyme by specific inhibitors or by silencing RNA decreases oxidative stress and accelerates diabetic wound healing possibly by decreasing the overproduction of ROS (Weinstein et al., 2015). Such findings suggest a role for Tβ4 in the natural healing process through purinergic signaling.

4.4 Tβ4 Does Not Promote Cell Growth

Although many of the factors released by the platelets are important in cell growth, Tβ4 is not a growth factor; ie, it does not promote cell growth (Goldstein & Kleinman, 2015). Tβ4 is smaller than standard growth factors, which are generally similar in size (4964 Da vs 14,000–16,000 Da, respectively). Also, unlike many growth factors, it does not bind to heparin that is ubiquitously present in tissues; therefore, Tβ4 can freely diffuse deeply into tissues to promote angiogenesis, cell migration, reepithelialization, and downregulate inflammation, among other effects. Furthermore, Tβ4 is present inside all cells and is not secreted, while growth factors are secreted, stored in the extracellular matrices outside of the cells, and only produced by certain cells. Growth factors act in a cell-type-specific manner in their interactions and activity and mainly only promote cell growth and migration, while Tβ4 acts on many cell types and has many different biological effects (Goldstein & Kleinman, 2015). Finally, there are many growth factors and variants. For example, there are more than 28 fibroblast growth factors (FGFs) and 5 different FGF receptors (Gong, 2014). In contrast, in human tissues, Tβ4 has only two other highly conserved variants (thymosin β10 and thymosin β15) (Bayard, Barrows, & Zetter, 2009). Thymosin β10, like Tβ4, is present ubiquitously in most cells, albeit at a much lower level, whereas thymosin β15 is rarely expressed in normal tissues. Other than in embryonic brain, thymosin β15 is present in highly metastatic prostate cancers but not in normal prostate tissue (Bao et al., 1996). It is not known if either of these other thymosin βs affects dermal wound repair. These differences distinguish Tβ4 both structurally and functionally from the families of growth factors.

Most of the pharmacological agents that have been tested to date in patients with chronic dermal wounds have been growth factors and their limited actions may be the reason for their general lack of efficacy in patient wounds (Collins & Seraj, 2010). In fact, molecular analysis of chronic wounds indicates that there is no shortage of growth factors in the wound site, but rather an overabundance of inflammatory cytokines and chemokines due to the dysregulation of the normal cycle of inflammation, proliferation, and remodeling that must occur if a wound is to heal.

5. PRECLINICAL ANIMAL STUDIES WITH Tβ4

Tβ4 has multiple, well-defined biological activities that contribute significantly to reduce tissue damage following injury and to promote tissue repair and regeneration (Goldstein & Kleinman, 2015). The activity of Tβ4 in various preclinical models of dermal injury has been extensively investigated (Table 1). Tβ4 acts equally well when administered topically or by intraperitoneal injection in rats (Malinda et al., 1999). It promotes full-thickness dermal healing in normal rats and mice as well as in compromised animal models, such as diabetic rats, aged mice, and steroid-treated rats (Table 1). Both recombinant and synthetic Tβ4 work well. Synthetic Tβ4-derived peptides containing the minimal sequence LKKTETQ (amino acids 17–23, which is the actin binding site plus Q) have also been found to have repair activity when used topically in wounds in aged animals (Fig. 3; Philp, Badamchian, et al., 2003; Philp et al., 2004). Recently, it has been shown to increase the survival of skin flaps in rats and also to promote dermal healing after burn injury in animal models (Kim & Kwon, 2014, 2015; Lin et al., 2015). A dimeric Tβ4 prepared recombinantly acts faster than the monomer in promoting dermal healing (Xu et al., 2013). These studies provide a solid rationale for use in humans for various skin injuries. It should also be noted that no adverse effects, such as weight loss, tumor development, and inertia, were noted in any of the animal studies. The animals did in many cases have increased hair growth around the wound edges that came in contact with Tβ4, and Tβ4 alone on shaved, unwounded skin also promoted increased hair growth (Philp, Badamchian, et al., 2003). Finally, when dermal punch wound healing was studied in the Tβ4 overexpressing mouse, no increase in healing was observed over non overexpressing controls (Cha et al., 2010). The addition of Tβ4 also did not increase healing of these Tβ4-overexpressing mice. This is likely due to the fact that Tβ4 is a biological response modifier and when too much factor is present, activity is lost. The response is the so-called bell-shaped curve of activity (Occleston et al., 2011). Indeed, in an injured brain stroke model, Tβ4 was found to show maximal activity at 30 mg/kg but higher levels resulted in decreased activity (Morris et al., 2014). Similar bell-shaped curves of Tβ4-treated responding tissues were observed in patients with either stasis or pressure ulcers and with epidermolysis bullosa lesions where the mid dose out of three doses showed the greatest activity (Treadwell et al., 2012).

6. HUMAN DERMAL STUDIES WITH Tβ4

6.1 Phase 1 Topical Safety Trial

A phase 1 safety trial was carried out with 15 normal volunteers treated sequentially with varying doses of topical Tβ4 or placebo for 28 days (Goldstein & Kleinman, 2015). There were a total of 1 control and 4 treated volunteers in the dosage groups consisting of 250, 500, and 1000 μg/5 mL that was applied to the skin and covered with a clear bandage. The study was carried out at one center in the United States. No significant adverse events were found, and the treatment was well tolerated. In addition, analysis of the blood for Tβ4 levels showed no measurable increase in Tβ4, suggesting that there was little systemic exposure of the peptide. Since this study demonstrated the topical safety of Tβ4, the FDA allowed phase 2 trials to be considered.

6.2 Phase 2 Trials on Patients with Pressure Ulcers, Stasis Ulcers, and Epidermolysis Bullosa Lesions

6.2.1 Pressure Ulcers (NCT00382174)

A phase 2-randomized, placebo-controlled phase 2 trial on pressure ulcers has been completed (Table 2) (Treadwell et al., 2012). There were 72 patients in the trial with wounds that had been stable and present for at least a month and were 5–70 cm^2, ie, stages III and IV. The patients were divided into 4 groups of 18 patients each: placebo, 0.01%, 0.02%, and 0.1% Tβ4 (w/w) in a gel solution that was given every day for 84 days. The study was carried out in the United States at various locations. No adverse events were observed in any of the patients, and the treatment was well tolerated. No differences were seen in healing at any of the doses over placebo at the conclusion of the study (day 84). However, there was a trend toward faster healing in the 0.02% by week 4 that persisted through week 9 at which point the placebo caught up.

6.2.2 Venous Stasis Ulcers (NCT00832091)

A phase 2-randomized, placebo-controlled phase 2 trial on venous stasis ulcers has been completed (Table 2; Treadwell et al., 2012). There were 72 patients in the trial with leg wounds that had been stable and present for at least 6 weeks before enrollment and were 3–30 cm^2. The patients were divided into 4 groups of 18 patients each: placebo, 0.01%, 0.03%, and 0.1% Tβ4 (w/w) in a gel solution that was given every day for 84 days. The study was carried out at various medical centers in both Italy and Poland. No adverse events were observed in any of the patients, and the treatment was well tolerated. No differences were seen in healing at any of the doses over placebo at the conclusion of the study (day 84). However, there was a trend toward faster healing in the 0.03% by week 4 that persisted through week 8.

6.2.3 Epidermolysis Bullosa Wounds (NCT00311766)

Epidermolysis bullosa has an inherited defect in laminin-322 production (Fine, 2007). Since Tβ4 increases laminin-322 production, it was tested for its effects on both junctional and dystrophic epidermolysis bullosa patients (Sosne, Xu, et al., 2004). This disease is rare and debilitating as the patients' wounds do not heal and often become infected and eventually convert to malignancy. Epidermolysis bullosa has been given orphan drug status by the FDA, so fewer patients are required for clinical trials.

A phase 2-randomized, placebo-controlled phase 2 trial on epidermolysis bullosa wounds has been completed (Table 2; Treadwell et al., 2012). There were 28 patients in the trial with leg wounds that had been stable and present for at least 14–60 days before enrollment and were 5–50 cm^2. The patients were divided into 4 treatment groups: placebo, 0.01%, 0.03%, and 0.1% Tβ4 (w/w) in a gel solution that was given every day for 56 days. The study was carried out at various medical centers in the United States. No adverse events were observed in any of the patients in this at high-risk/fragile treatment group, and the treatment was well tolerated. No differences were seen in healing at any of the doses over placebo at the conclusion of the study (day 56). However, there was a trend toward faster healing in the 0.03% treated group.

In summary, Tβ4 has been shown to promote an increase in the rate of dermal healing in various patient populations with 0.02–0.03% (w/w) gel. Tβ4 was safe and well tolerated by the patients. Higher or lower doses were less active/inactive. Future trials should focus on the 0.02–0.03% range and perhaps better define the optimal concentrations for each wound type.

7. THYMOSIN β4 SAFETY

Cancers have been described as wounds that do not heal since the abnormal cell behaviors of chronic wounds share many properties with cancer (Riss et al., 2006). Tβ4 cellular levels are upregulated in metastatic tumor cells (Cha et al., 2003; Clark, Golub, Lander, & Hynes, 2000; Huang et al., 2011, Ribenja, Wongkham, Wongkham, Yao, & Chen, 2013; Tang et al., 2011; Wang, Carretero, et al., 2004). Furthermore, when tumor cells overexpress Tβ4 in vivo by targeted gene expression, they become more metastatic (Cha et al., 2003; Kobayashi et al., 2002). It should be noted that this increase in metastasis is also observed with many other important cellular molecules vital for normal growth, such as growth hormone, estrogen, testosterone, and growth factors (such as VEGF), which attain high levels in metastatic tumors and can increase malignant activity also when overexpressed in tumor cells (Korc & Friesel, 2012; Lee, Jeong, Han, & Baek, 2015; Munoz, Wheler, & Kurzrock, 2014; Wei, Wang, Markeo, & Li, 2013). Tumor cells can use normal cellular proteins, such as Tβ4 and VEGF, to stimulate their own growth and survival. Interestingly, Tβ4 increases VEGF expression so it has dual mechanisms for promoting tumor growth and metastasis (Cha et al., 2003; Jo et al., 2010; Yoon, Lee, Park, Kim, et al., 2011). Tβ4 has not shown any tumor-promoting activity when applied topically to the skin in various animal models and to the skin in phase 1 and 2 human clinical trials. In a phase 2 trial, patients with epidermolysis bullosa were treated with Tβ4, and no cancers were found in these highly cancer-prone patients (Goldstein & Kleinman, 2015). Furthermore, experiments in mice, where this gene is overexpressed by 50-fold in the skin and in some organs, have shown that there is no increase in cancer rates even after many generations (Cha et al., 2010). Finally, the use of cancer-promoting agents on the skin of these Tβ4-overexpressing mice did not increase tumor formation over that seen with controls (Cha et al., 2010). In contrast to studies that indicate Tβ4 plays a role in metastasis, a recent study of murine myeloma, where cells have been manipulated to overexpress Tβ4, showed a significantly decreased tumor cell proliferation and an increased sensitivity to different antimyeloma agents compared to control myeloma cells (Caers et al., 2010). Most importantly, when Tβ4-overexpressing murine myeloma cells were injected into mice, there was a significant decrease in tumor formation and prolonged survival compared to mice injected with myeloma cells that did not overexpress Tβ4. Also, studies in patients with multiple myeloma point to a potential positive role for Tβ4 in the diagnosis and perhaps in the treatment of this cancer (Caers et al., 2010). A recent study of 171 patients with multiple myeloma found that patients with the highest levels of Tβ4 had a greater chance of not having a recurrence of the disease following therapy and longer overall survival than patients with lower levels of Tβ4. Thus, for multiple myeloma, Tβ4 may be a tumor suppressor. These studies reinforce an earlier report that Tβ4 is capable of downregulating the expression of WT1 (Wilms' tumor 1 gene), a universal tumor-associated antigen, expressed in Molt-4 cells derived from a patient with lymphoblastic leukemia and that Tβ4 is capable of inducing phenotypic changes in these cells representing the early stages in maturation of the T-cell lineage (Kokkinopoulos, Perez, & Papmichall, 1985). Thus, while elevated in some cancer cells lines and in tumors, current evidence suggests that Tβ4 will not cause cancer and may in fact provide a benefit to reduce cancer in certain forms of the disease.

Thus, topical Tβ4 is safe and well tolerated in phase 1 and 2 dermal trials (a total of 200 patients) (Treadwell et al., 2012). Furthermore, human subjects have been treated with single intravenous pharmacological doses (42, 140, 420, and 1260 mg qd) of Tβ4 and multiple doses for 2 weeks, and there were no findings of Tβ4 being associated with the appearance of any adverse events. These human studies document the safety and tolerability of Tβ4.

8. CONCLUSIONS AND FUTURE STUDIES WITH Tβ4 IN DERMAL WOUND REPAIR

8.1 What We Know

Unlike many pharmacological treatments for various pathologies, much is known about the mechanisms of action of Tβ4. The active site(s), receptor, signaling pathways, and biological activities are known. The safety profile is excellent, and no preclinical toxicology has been found. Efficacy trends have been observed in phase 2 clinical trials where the healing time was reduced by almost one half. Also a near optimal dose (0.02–0.03%, w/w) has been identified in dosing studies for dermal healing. Finally, no adverse reactions have been observed with topical or injected Tβ4 in humans and no autoantibodies have been detected. Tβ4 is a naturally occurring molecule that can be synthesized at a relatively low cost. The cost is important given that other treatments are not in wide use due to the high cost and limited efficacy.

8.2 What We Do Not Know

While a lot has been done in the heart in terms of determining Tβ4 efficacy and mechanisms of action, less has been done in dermal healing preclinical models. For example, the optimal dosing and dosing schedule for burns have not yet been determined. The functional receptor for each of the activities of Tβ4 (such as angiogenesis, migration, antiapoptosis, antiinflammation, cytoprotection, etc.) identified in dermal repair is not known. More clinical studies are needed to further define the efficacy, dosing amount, and dosing schedule for each type of wound (aged patients, diabetics, burns, stasis ulcers, pressure ulcers, etc.). Based on studies in other tissues, there may be differences in the optimal amounts and dosing schedule dependent on the type and severity of the injury and general cause of the injury. It is likely that Tβ4 treatment may be able to be stopped when healing has progressed to a certain point and is not needed for full wound closure, ie, jump start healing and then let the natural process take over. Finally, additional clinical trials are needed using Tβ4 alone and in combination with other potential factors/treatments.

REFERENCES

Badamchian, M., Damavandy, A. A., Damavandy, H., Wadhwa, S. D., Katz, B., & Goldstein, A. L. (2007). Identification and quantification of thymosin beta4 in human saliva and tears. *Annals of the New York Academy of Sciences*, *1112*, 458–465.

Badamchian, M., Fagarasan, M. O., Danner, R. L., Suffredini, A. F., Damavandy, H., & Goldstein, A. L. (2003). Thymosin beta(4) reduces lethality and down-regulates inflammatory mediators in endotoxin-induced septic shock. *International Immunology*, *5*, 1317–1329.

Bao, L., et al. (1996). Thymosin beta 15: A novel regulator of tumor cell motility upregulated in metastatic prostate cancer. *Nature Medicine*, *2*, 1322–1328.

Bayard, J., Barrows, C., & Zetter, B. R. (2009). Differential regulation of thymosin beta 15 isoforms by transforming growth factor beta 1. *Genes, Chromosomes and Cancer*, *48*, 502–509.

Bicer, A., Karakurt, O., Akdemir, R., Erden, G., et al. (2011). Thymosin beta 4 is associated with collateral vessel development in coronary artery disease. *Scandinavian Journal of Clinical & Laboratory Investigation*, 7, 625–630.
Bock-Marquette, I., Saxena, A., White, M. D., Dimaio, J. M., & Srivastava, D. (2004). Thymosin beta4 activates integrin-linked kinase and promotes cardiac cell migration, survival and cardiac repair. *Nature*, *432*, 466–472.
Bock-Marquette, I., Shrivastava, S., Pipes, G. C., Thatcher, J. E., Blystone, A., Shelton, J. M., et al. (2009). Thymosin beta4 mediated PKC activation is essential to initiate the embryonic coronary developmental program and epicardial progenitor cell activation in adult mice in vivo. *Journal of Molecular and Cellular Cardiology*, *46*, 728–738.
Bollini, S., Riley, P. R., & Smart, N. (2015). Thymosin beta 4: Multiple functions in protection, repair, and regeneration of the mammalian heart. *Expert Opinion on Biological Therapy*, *22*, 1–12.
Caers, J., Hose, D., Kuipers, I., Bos, T. J., Valckenborgh, E. V., Menu, E., et al. (2010). Thymosin beta 4 has tumor suppressive effects and its decrease expression results in poor prognosis and decreased survival in multiple myeloma. *Haematologica*, *95*, 163–167.
Castoldi, G., Di Gioia, C. R., Bombardi, C., Preziuso, C., Lepizzi, M., Maestroni, S., et al. (2013). Renal antifibrotic effect of N-acetyl-seryl-aspartyl-lysyl-proline in diabetic rats. *American Journal of Nephrology*, *37*, 65–73.
Cavasin, M. A. (2006). Therapeutic potential of thymosin beta 4 and its derivative N-acetyl-seryl-aspartyl-lysyl-proline (Ac-SDKP) in cardiac healing after infarction. *American Journal of Cardiovascular Drugs*, *6*, 305–311.
Cavasin, M. A., Liao, T. D., Yang, X. P., Yang, J. J., & Carretero, O. A. (2007). Decreased endogenous levels of Ac-SDKP promotes organ fibrosis. *Hypertension*, *50*, 130–136.
Cha, H. J., Jeong, M. J., & Kleinman, H. K. (2003). Role of thymosin beta 4 in tumor metastasis and angiogenesis. *Journal of the National Cancer Institute*, *95*, 1674–1680.
Cha, H. J., Philp, D., Lee, S. H., Moon, H. S., Kleinman, H. K., & Nakamura, T. (2010). Over-expression of thymosin beta 4 promotes abnormal tooth development and stimulation of hair growth. *International Journal of Developmental Biology*, *54*, 135–140.
Cheng, P., Kuang, F., Zhang, H., Ju, G., & Wang, J. (2014). Beneficial effects of thymosin beta 4 on spinal cord injury in the rat. *Neuropharmacology*, *85*, 408–416.
Chopp, M., & Zhang, Z. G. (2015). Thymosin beta 4 as a restorative/regenerative therapy for neurological injury and neurodegenerative disease. *Expert Opinion on Biological Therapy*, *22*, 1–4.
Clark, E. A., Golub, T. R., Lander, E. S., & Hynes, R. O. (2000). Genomic analysis of metastases reveals an essential role for Rho C. *Nature*, *406*, 532–535.
Collins, L., & Seraj, S. (2010). Diagnosis and treatment of venous stasis ulcers. *American Family Physician*, *81*, 9.
De Santis, M., Inzitari, R., Bosello, S. L., Peluso, G., et al. (2011). Beta thymosins and interstitial lung disease: A study of a scleroderma cohort with a one year follow up. *Respiratory Research*, *11*, 12–22.
Dube, K. N., Bollini, S., Smart, N., & Riley, P. R. (2012). Thymosin beta 4 protein therapy for cardiac repair. *Current Pharmaceutical Design*, *18*, 799–806.
Ehrlich, H. P., & Hazard, S. W. (2010). Thymosin beta 4 enhances repair by organizing connective tissue and preventing the appearance of myofibroblasts. *Annals of the New York Academy of Sciences*, *1194*, 118–124.
Ehrlich, H. P., & Hazard, S. W. (2012). Thymosin β4 affecting the cytoskeleton organization of the myofibroblasts. *Annals of the New York Academy of Sciences*, *1269*, 74–78.
Fine, J. D. (2007). Epidermolysis bullosa: A genetic disease of altered cell adhesion and wound healing, and the possible clinical utility of topically applied thymosin beta4. *Annals of the New York Academy of Sciences*, *1112*, 396–406.

Freeman, K. W., Bowman, B. R., & Zetter, B. R. (2011). Regenerative protein thymosin beta 4 is a novel regulator of purinergic signaling. *The FASEB Journal, 25*, 907–915.

Fromm, M., Gunne, H., Bergman, A. C., et al. (1996). Biochemical and antibacterial analysis of human wound and blister fluid. *European Journal of Biochemistry, 237*, 86–92.

Gallina, C., Turunetto, V., & Giachino, C. (2015). A new paradigm in cardiac regeneration: The mesenchymal stem cell secretome. *Stem Cells International, 2015*, 765846.

Gao, X., Liang, H., Hou, F., Zhang, S., Nuo, M., Guo, X., et al. (2015). Thymosin beta 4-induced mouse hair growth. *PLoS One, 10*(6), e0130040.

Goldstein, A. L., Hannappel, E., Sosne, G., & Kleinman, H. K. (2012). Thymosin beta 4: A multi-functional regenerative peptide. Basic properties and clinical applications. *Expert Opinion on Biological Therapy, 12*, 37–51.

Goldstein, A. L., & Kleinman, H. K. (2015). Advances in the basic and clinical applications of thymosin beta 4. *Expert Opinion on Biological Therapy, 15*, S139–S145.

Gong, S. G. (2014). Isoforms of receptors for fibroblast growth factors. *Journal of Cellular Physiology, 229*, 1887–1895.

Gonzalez, G. E., Rhaleb, N. E., Nakagawa, P., Lioa, T. D., Liu, Y., Leung, P., et al. (2014). N-Acetyl-seryl-aspartyl-lysyl-proline reduces cardiac collagen cross-linking and inflammation in angiotensin II-induced hypertensive rats. *Clinical Science, 126*, 85–94.

Gould, L., Abadir, P., Brem, H., et al. (2015). Chronic wound repair and healing in older adults: Current status and future research. *Wound Repair and Regeneration, 23*, 1–13.

Grant, D. S., Kinsella, J. L., Kibbey, M. C., LaFlamme, S., Burbelo, P. D., Goldstein, A. L., et al. (1995). A novel role for thymosin beta 4: A Matrigel-induced gene involved in endothelial cell differentiation and angiogenesis. *Journal of Cell Science, 108*, 3685–3694.

Hannappel, E., & van Kampen, M. (1987). Determination of thymosin beta 4 in human blood cells and serum. *Journal of Chromatography, 26*, 279–285.

Hinkel, R., Ballm, H. L., DiMiado, J. M., Srivastava, S., et al. (2015). C-terminal variable AGES domain of thymosin beta 4: The molecule's primary contribution in support of post-ischemic cardiac function and repair. *Journal of Molecular and Cellular Cardiology, 87*, 113–125.

Hinkel, R., El-Aouni, C., Olson, T., Horstkotte, J., et al. (2008). Thymosin beta4 is an essential paracrine factor of embryonic endothelial progenitor cell-mediated cardioprotection. *Circulation, 117*, 2232–2240.

Ho, J. H., Chuang, C. H., Ho, C. Y., Shih, Y. R., Lee, O. K., & Su, Y. (2007). Internalization is essential for the antiapoptotic effects of exogenous thymosin beta-4 on human corneal epithelial cells. *Investigative Ophthalmology & Visual Science, 48*, 27–33.

Ho, J. H., Tseng, K. C., Ma, W. H., Chen, K. H., Lee, O. K., & Su, Y. (2008). Thymosin beta-4 upregulates anti-oxidative enzymes and protects corneal epithelial cells against oxidative damage. *The British Journal of Ophthalmology, 92*, 992–997.

Huang, C. M., Wang, C. C., Barnes, S., & Elmets, C. A. (2006). In vivo detection of secreted proteins from wounded skin using capillary ultrafiltration probes and mass spectrometric proteomics. *Proteomics, 6*, 5805–5814.

Huang, L., Zheng, M., Zon, C. M., Zhang, M. Y., Jia, W. H., Yun, J. P., et al. (2011). Identification of a gene expression signature for predicting lymph node metastases in patients with early stage cervical cancer. *Cancer, 117*, 3363–3376.

Jo, J. O., Kim, S. R., Bae, M. K., Kang, Y. J., Ock, M. S., Kleinman, H. K., et al. (2010). Thymosin beta 4 induces the expression of vascular endothelial growth factors (VEGF) in a hypoxia-inducible factor (HIF)-1 alpha-dependent manner. *Biochimica et Biophysica Acta, 1803*, 1244–1251.

Kaur, H., & Mutus, B. (2012). Platelet function and thymosin beta 4. *Biological Chemistry, 393*, 595–598.

Kim, S., & Kwon, J. (2014). Thymosin beta4 improves dermal burn wound healing via downregulation of receptor of advanced glycation end products in db/db mice. *Biochimica et Biophysica Acta, 1840*, 3452–3459.
Kim, S., & Kwon, J. (2015). Thymosin beta 4 has a major role in dermal burn wound healing that involves actin cytoskeletal remodeling via heat shock protein 70. *Journal of Tissue Engineering, 1840*, 3452–3459.
Kobayashi, T., Okada, F., Fujii, N., Tomita, N., Ito, H., Tazawa, H., et al. (2002). Thymosin beta 4 regulates motility and metastasis of malignant mouse fibrosarcoma cells. *American Journal of Pathology, 160*, 869–882.
Kokkinopoulos, D., Perez, S., & Papmichall, M. (1985). Thymosin beta 4 induced phenotypic changes in Molt-4 leukemic cell line. *Blut, 50*, 341–348.
Korc, M., & Friesel, R. E. (2012). The role of fibroblast growth factors in tumor growth. *Current Cancer Drug Targets, 9*, 639–651.
Kumar, S., & Gupta, S. (2011). Thymosin beta 4 prevents oxidative stress by targeting anti--oxidant and anti-apoptotic genes in cardiac fibrosis. *PLoS One, 6*(10), e26912.
Kupatt, C., Bosk-Marquette, I., & Bockstegers, P. (2008). Embryonic endothelial progenitor cell-mediated cardioprotection requires thymosin beta 4. *Trends in Cardiovascular Medicine, 18*, 205–210.
Lee, S. H., Jeong, D., Han, Y. S., & Baek, M. J. (2015). Pivotal role of vascular endothelial cell growth factor pathway in tumor angiogenesis. *Annals of Surgical Treatment and Research, 89*, 1–8.
Li, X., Zheng, L., Peng, F., Qi, C., Zhang, X., Zhou, A., et al. (2007). Recombinant thymosin beta4 can promote full-thickness cutaneous wound healing. *Protein Expression and Purification, 56*, 229–236.
Lin, Y., Lin, B., Lin, D., Huang, G., & Cao, B. (2015). Effect of thymosin beta 4 on the survival of skin flaps in rats. *Journal of Reconstructive Microsurgery, 31*, 464–470.
Malinda, K. M., Goldstein, A. L., & Kleinman, H. K. (1997). Thymosin beta 4 stimulates directional migration of human umbilical vein endothelial cells. *The FASEB Journal, 11*, 474–481.
Malinda, K. M., Sidhu, G. S., Mani, H., Banaudha, K., Maheshwari, R. K., Goldstein, A. L., et al. (1999). Thymosin beta 4 accelerates wound healing. *Journal of Investigative Dermatology, 113*, 364–368.
Mohring, T., Kellmann, M., Jurgens, M., & Schrader, M. (2005). Top-down identification of endogenous peptides up to 9 kDa in cerebrospinal fluid and brain tissues by nonelectrospray quadrupole time-of-flight tandem mass spectrometry. *Journal of Mass Spectrometry, 40*, 214–226.
Morris, D. C., Chopp, M., Zhang, L., Lu, M., & Zhang, Z. G. (2010). Thymosin beta 4 improves functional neurological outcomes in a rat model of embolic stroke. *Neuroscience, 169*, 674–682.
Morris, D. C., Cui, Y., Cheung, W. L., Lu, M., Zhang, L., Zhang, Z. G., et al. (2014). A dose–response study of thymosin β4 for the treatment of acute stroke. *Journal of Neurological Science, 345*(1–2), 61–67.
Munoz, J., Wheler, J., & Kurzrock, R. (2014). Expression of estrogen and progesterone receptors across human malignancies: New therapeutic opportunities. *Cancer Metastasis Reviews, 34*(4), 547–561.
Occleston, N. L., O'Kane, S. O., Laverty, H. G., Cooper, M., Fairlamb, D., Mason, T., et al. (2011). Discovery and development of avoterim (recombinant human transforming growth factor beta 3): A new class of prophylactic therapeutic for the improvement of scarring. *Wound Repair and Regeneration, 19*, S38–S48.
Parihar, A., Parihar, M. S., Miner, S., & Bhat, S. (2008). Oxidative stress and anti-oxidative mobilization in burn injury. *Burns, 34*, 6–17.

Philp, D., Badamchian, M., Scheremeta, B., Nguyen, M., Goldstein, A. L., & Kleinman, H. K. (2003). Thymosin beta 4 and a synthetic peptide containing its actin-binding domain promote dermal wound repair in db/db diabetic mice and in aged mice. *Wound Repair and Regeneration, 11*, 19–24.

Philp, D., Huff, T., Gho, Y. S., Hannappel, E., & Kleinman, H. K. (2003). The actin binding site on thymosin beta 4 promotes angiogenesis. *The FASEB Journal, 17*, 2103–2105.

Philp, D., & Kleinman, H. K. (2010). Animal studies with thymosin beta 4: A multifunctional tissue repair and regeneration peptide. *Annals of the New York Academy of Sciences, 1194*, 81–86.

Philp, D., Nguyen, M., Scheremeta, B., St-Surin, S., Villa, A. M., Orgel, A., et al. (2004). Thymosin increases hair growth by the activation of clonogenic hair follicle stem cells. *The FASEB Journal, 18*, 385–387.

Philp, D., Scheremeta, B., Sibliss, K., Zhou, M., Fine, E. L., Nguyen, M., et al. (2006). Thymosin beta 4 promotes matrix metalloproteinases expression during wound healing. *Journal of Cellular Physiology, 208*, 195–200.

Popoli, P., Pepponi, R., Martire, A., Armida, M., Pèzzola, A., Galluzzo, M., et al. (2007). Neuroprotective effects of thymosin beta4 in experimental models of excitotoxicity. *Annals of the New York Academy of Sciences, 1112*, 219–224.

Postrach, J., Schmidt, M., Thormann, M., Thein, E., Burdorf, L., Reichart, B., et al. (2014). Adeno-associated viral vector 2.9 thymosin β4 application attenuates rejection after heart transplantation: Results of a preclinical study in the pig. *Transplantation, 98*, 835–843.

Reiger, S., Zhao, H., Martin, P., Abe, K., & Lisse, T. S. (2015). The role of nuclear hormone receptors in cutaneous wound repair. *Cell Biochemistry and Function, 33*, 1–13.

Ribenja, S., Wongkham, S., Wongkham, C., Yao, Q., & Chen, C. (2013). Roles and mechanisms of beta thymosins in cell migration and cancer metastasis: An update. *Cancer Investigation, 31*, 103–110.

Riss, J., et al. (2006). Cancers as wounds that do not heal: Differences and similarities between renal regeneration/repair and renal cell carcinoma. *Cancer Research, 66*, 7216–7224.

Roy, P., Raifur, Z., Jones, D., Marriott, G., Loew, L., & Jacobson, K. (2001). Local photorelease of caged thymosin beta 4 in locomoting keratocytes causes cell turning. *Journal of Cell Biology, 153*, 1035–1048.

Singer, A. J., & Clark, R. A. F. (1999). Cutaneous wound healing. *New England Journal of Medicine, 341*, 738–746.

Smart, N., Riseboro, C. A., Melville, A. A., Moses, K., Schwartz, R. J., Chien, K. R., et al. (2007). Thymosin beta 4 induces adult cardiac progenitor mobilization and neovascularization. *Nature, 445*, 177–182.

Sopko, N., Qin, Y., Finan, A., Dadabayev, A., Chigurupati, S., Qin, J., et al. (2011). Significance of thymosin β4 and implication of PINCH-1-ILK-α-Parvin (PIP) complex in human dilated cardiomyopathy. *PLoS One, 6*(5), e20184.

Sosne, G., Chrispopherson, P. L., Barrett, R. P., & Fridman, R. (2005). Thymosin beta 4 modulates matrix metalloproteinase levels and polymorphonuclear cell infiltration after alkali injury. *Investigative Ophthalmology & Visual Science, 46*, 2388–2395.

Sosne, G., Qiu, P., Christopherson, P. L., & Wheater, M. K. (2007). Thymosin beta 4 suppression of corneal NFkappaB: A potential anti-inflammatory pathway. *Experimental Eye Research, 84*, 663–669.

Sosne, G., Qiu, P., Goldstein, A. L., & Wheater, M. (2010). Biological activities of thymosin beta4 defined by active sites in short peptide sequences. *The FASEB Journal, 24*, 2144–2151.

Sosne, G., Siddiqi, A., & Kurpakus-Wheater, M. (2004). Thymosin-beta4 inhibits corneal epithelial cell apoptosis after ethanol exposure in vitro. *Investigative Ophthalmology and Visual Science, 45*, 1095–1100.

Sosne, G., Szliter, E. A., Barrett, R., Kernacki, K. A., Kleinman, H., & Hazlett, L. D. (2002). Thymosin beta 4 promotes corneal wound healing and decreases inflammation in vivo following alkali injury. *Experimental Eye Research, 74*, 293–299.

Sosne, G., Xu, L., Prach, L., Mrock, L. K., Kleinman, H. K., Letterio, J. J., et al. (2004). Thymosin beta 4 stimulates laminin-5 production independent of TGF-beta. *Experimental Cell Research, 293*, 175–183.

Tang, M. C., Chen, L. C., Yeh, Y. C., Chon, T. Y., Wang, W. S., & Su, Y. (2011). Thymosin beta 4 induces colon cancer, cell migration and clinical metastases via enhancing ILK/IQGAP1/Rac1 signal transduction pathway. *Cancer Letters, 308*, 162–171.

Ti, D., Hao, H., Xia, L., Tong, C., Liu, J., Dong, L., et al. (2015). Controlled release of thymosin beta 4 using a collagen-chitosan sponge scaffold augments cutaneous wound healing and increases angiogenesis in diabetic rats with hindlimb ischemia. *Tissue Engineering Part A, 21*, 541–549.

Treadwell, T., Kleinman, H. K., Crockford, D., Hardy, M. A., Guarnera, G. T., & Goldstein, A. L. (2012). The regenerative peptide thymosin β4 accelerates the rate of dermal healing in preclinical animal models and in patients. *Annals of the New York Academy Science, 1270*, 37–44.

Trenkwalder, T., Deindl, E., Bongiovanni, D., Lei, S., Schunkert, H., Kupatt, C., et al. (2015). Thymosin beta 4-mediated therapeutic neovascularization: The role of the PI3/AKT pathway. *Expert Opinion on Biological Therapy, 4*, 1–11.

Wang, D., Carretero, O. A., Yang, X. Y., Rhaleb, N. E., et al. (2004). N-Acetyl-seryl-aspartyl-lysyl-proline stimulates angiogenesis in vitro and in vivo. *American Journal of Physiology. Heart Circulation and Physiology, 287*, H2099–H2105.

Wang, L., Chopp, M., Szalad, A., Liu, Z., Lu, M., Zhang, L., et al. (2012). Thymosin β4 promotes the recovery of peripheral neuropathy in type II diabetic mice. *Neurobiology of Disease, 48*, 546–555.

Wang, M., Liu, R., Jia, X., Mu, S., & Xie, S. (2010). N-Acetyl-seryl-aspartyl-lysyl-proline attenuates renal inflammation and tubulointerstitial fibrosis in rats. *International Journal of Molecular Medicine, 26*, 795–801.

Wei, C., Kim, I. K., Li, L., Wu, L., & Gupta, S. (2014). Thymosin beta 4 protects mice from monocrotaline-induced pulmonary hypertension and right ventricular hypertrophy. *PLoS One, 9*(11), e110598.

Wei, C., Kumar, S., Kim, I. K., & Gupta, S. (2012). Thymosin beta 4 protects cardiomyocytes from oxidative stress by targeting anti-oxidative enzymes and anti-apoptotic genes. *PLoS One, 7*(8), e42586.

Wei, J., Wang, Z., Markeo, D., & Li, X. (2013). Current treatments and novel therapeutic targets for castration resistant prostate cancer with bone metastases. *American Journal of Clinical and Experimental Urology, 1*, 30–38.

Weinstein, A. L., Lalezarzadeh, F. D., Soares, M. S., Saddeh, B., & Ceradini, D. J. (2015). Normalizing dysfunctional purine metabolism accelerates diabetes wound healing. *Wound Repair and Regeneration, 23*, 14–21.

Woodley, D. T., Wysong, A., DeClerck, B., Chen, M., & Li, W. (2015). Keratinocyte migration and a hypothetical new role for extracellular heat shock protein 90 alpha in orchestrating skin wound healing. *Advances in Wound Care, 4*, 203–212.

Wyczoikowska, J., Walczak-Drzewiecka, A., Wagner, W., & Dastych, J. (2007). Thymosin beta 4 and thymosin beta 4-derived peptides induce mast cell exocytosis. *Peptides, 28*, 752–759.

Xiong, Y., Mahmood, A., Meng, Y., et al. (2011). Treatment of traumatic brain injury with thymosin beta4 in rats. *Journal of Neurosurgery, 114*, 102–115.

Xu, T. J., Wang, Q., Ma, X. W., Zhang, Z., Zhang, W., Xue, X. C., et al. (2013). A novel dimeric thymosin beta 4 with enhanced activities accelerates the rate of wound healing. *Drug Design, Development and Therapy*, 7, 1075–1088.
Yoon, S. Y., Lee, H. R., Park, Y., Kim, J. H., et al. (2011). Thymosin beta 4 expression correlates with lymph nodes metastases thru hypoxia inducible factor-alpha induction in breast cancer. *Oncology Reports*, *25*, 23–31.
Young, J. D., Lawrence, A. J., Mac Lean, A. G., Leung, B. P., MacLunes, I. B., Canas, B., et al. (1999). Thymosin beta 4 sulfoxide is an anti-inflammatory agent generated by monocytes in the presence of glucocorticoids. *Nature Medicine*, *5*, 1424–1427.
Zhang, J., Zhang, Z. G., Morris, D., et al. (2009). Neurological functional recovery after thymosin beta 4 treatment in mice with experimental auto encephalomyelitis. *Neuroscience*, *164*, 1887–1893.
Zuo, Y., Chun, B., Potthoff, S. A., Kazi, N., Brolin, T. J., Orhan, D., et al. (2013). Thymosin β4 and its degradation product, Ac-SDKP, are novel reparative factors in renal fibrosis. *Kidney International*, *84*, 1166–1175.

CHAPTER TWELVE

Thymosin Beta 4: A Potential Novel Therapy for Neurotrophic Keratopathy, Dry Eye, and Ocular Surface Diseases

G. Sosne[*,1], **D. Rimmer**[*,†], **H.K. Kleinman**[‡,§], **G. Ousler**[*,†]
[*]Kresge Eye Institute, Wayne State University School of Medicine, Detroit, MI, United States
[†]ORA Inc., Andover, MA, United States
[‡]George Washington University, Washington, DC, United States
[§]NIDCR/NIH, Bethesda, MD, United States
[1]Corresponding author: e-mail address: gsosne@med.wayne.edu

Contents

Abstract

Chronic ocular surface diseases such as dry eye, blepharitis, and neurotrophic keratopathies represent a significant and a growing therapeutic challenge. The basis of this expanding prevalence is multifactorial and may due to issues such as an aging population, an increasing use of video display terminals, and increases in frequency of refractive surgeries. The growing incidence of diseases such as diabetes may also be a contributing factor. Current treatments for ocular surface disease include artificial tears, lubricants, tear duct plugs, steroids, antibiotics, cyclosporine, scleral lenses, and serum tears. Treatment choices depend on the type and severity of the disease, but in general positive outcomes are limited because many of these treatments do not fully address the underlying disease process or promote mechanisms that facilitate long-term wound

Vitamins and Hormones, Volume 102
ISSN 0083-6729
http://dx.doi.org/10.1016/bs.vh.2016.04.012

repair. From minor corneal injuries to more severe inflammatory-mediated pathologies, clinicians need agents that promote corneal healing and reduce the inflammatory response to prevent visual disturbances and improve quality of life. A focus on treatments that reduce the inflammatory responses while accelerating corneal epithelial growth would represent a major step forward from current treatment options. Increasing evidence suggests that thymosin beta 4 (Tβ4), a naturally occurring polypeptide, can elicit this spectrum of therapeutic responses: a rapid corneal reepithelialization and a reduction in corneal inflammation. This chapter serves as a review of standard therapies as well as recent advancements in the development of newer therapies that includes the use of Tβ4 that is proving to be an exciting new agent for the management of ocular surface disease.

1. INTRODUCTION

Thymosins include three families of polypeptides termed α, β, and γ. Of the 16 known family members, thymosin β4 (Tβ4), thymosin β10 (Tβ10), and thymosin β15 (Tβ15) are found in man. Tβ4 is biochemically distinct from and is the most abundant and biologically active member of the family in most mammalian cells, comprising about 70–80% of total β-thymosin content (Huff, Muller, Otto, Netzker, & Hannappel, 2001). First characterized from a thymus gland extract in the calf, Tβ4 is a low-molecular weight, 43-amino acid protein that is critical to cell survival that has unique and broad-ranging wound healing and antiinflammatory activities that are important at different stages of tissue repair (Sosne, Qiu, Kurpakus-Wheater, & Matthew, 2010) (Fig. 1). Tβ4 is a G-actin-sequestering peptide that regulates the cellular actin cytoskeleton, migration, and cellular inflammation. Tβ4 is present in high concentrations (up to 0.4–2.1 μg/mL in human serum) in all tissue types except red blood cells, with highest levels occurring in platelets, white blood cells, plasma, and wound fluid (Hannappel & van Kampen, 1987; Huff et al., 2001; Sosne, Hafeez, Greenberry, & Kurpakus-Wheater, 2002). It has been shown both in vivo (corneal scrape and alkali injury) and in vitro (human corneal epithelial cells and fibroblasts) that Tβ4 treatment decreases cytokines (including TNF-α) and chemokines (including MIP-2 and KC), nuclear factor kappa B (NF-kB) activation, nuclear translocation, and stress fiber density (Sosne, Szliter, et al., 2002). Based on its multifunctional activities during tissue regeneration, Tβ4 has the potential for clinical application in a wide range of pathological conditions including ocular surface diseases.

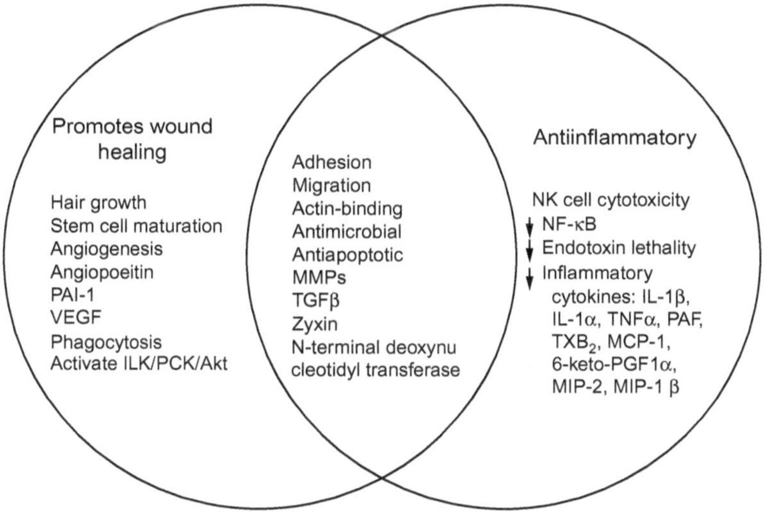

Fig. 1 Biological activities of Tβ4. Many of the activities are distinct and important in wound healing and in inflammation, while other activities are important to both processes and are shown in the overlapping area of the two circles. Some of these activities might depend on the processing and/or degradation of intact Tβ4; such processing may occur inside the cell or tissue and be difficult to determine. *Copyright Permission: Sosne, G., Qiu, P., Goldstein, A. L., & Wheater, M. (2010). Biological activities of thymosin beta4 defined by active sites in short peptide sequences.* The FASEB Journal, *24(7), 2144–2151. doi:10.1096/fj.09-142307. Used with permission from © Federation of American Societies for Experimental Biology.*

2. Tβ4 IN WOUND HEALING

Given its importance in cellular dynamics and ubiquity in all mammalian tissues, Tβ4 is thought to function as a critical element in cell migration, repair, and regeneration of damaged tissues of any origin. Tβ4 has been shown to be a major determinant of cell protection and survival in a number of organ systems. Wound healing involves a dynamic process of stem cell migration and differentiation that is needed to rapidly reestablish the epithelial barrier. In many tissues including the heart and central nervous system, Tβ4 has been shown to reduce inflammation, increase stem cell recruitment, increase cell regeneration, and decrease scar formation. Several clinical studies have shown that higher levels of Tβ4 in serum or tissue fluid are predictive of recovery or survival in cases of myocardial infarction, stroke, heart failure, liver failure, neurological disease, skin and ocular wounds, ulcerations, burns, and infections (Bock-Marquette, Saxena, White, Dimaio, & Srivastava, 2004; Han et al., 2010; Lv, Cheng, Xu, Wang, & Xu, 2011; Morris et al., 2012; Santra et al., 2012; Sopko et al., 2011; Sosne, Szliter, et al., 2002; Vartiainen, Pyykonen, Hokfelt, & Koistinaho, 1996; Xiong et al., 2012) (Fig. 2). In wounds, platelets and polymorphonuclear leukocytes (PMNs) contain Tβ4 in high concentrations that vary from 500 to 600 μM (Hannappel & van Kampen, 1987); therefore, it is not surprising that they are the first cells to arrive at the site of injury. In the case of dermal wounds, Tβ4 promotes full-thickness wound healing, with antiinflammatory properties that promote both endothelial and keratinocyte migration due to its chemotactic and haptotactic activity (Al-Nedawi et al., 2004; Philp et al., 2006). The ability of the epithelium to migrate and reepithelialize a wound is an essential step in the cascade of cellular events in the healing process.

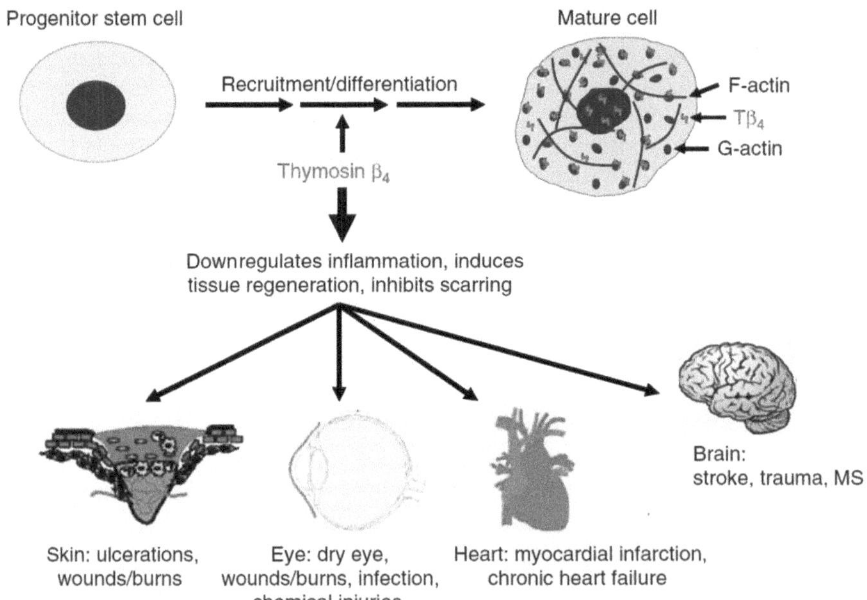

Fig. 2 Known and newly emerging clinical applications of Tβ4. Shown schematically are various organs and the injuries that have been found to be repaired or affected by Tβ4. *Copyright Permission: Goldstein, A. L., Hannappel, E., Sosne, G., & Kleinman, H. K. (2012). Thymosin β4: A multifunctional regenerative peptide. Basic properties and clinical applications.* Expert Opinion on Biological Therapy, 12*(1), 37–51. doi:10.1517/14712598.2012.634793. Used with permission: © Taylor & Francis LLC. http://www.tandfonline.com. Previous work for noncommercial use (doi:10.1517/14712598.2012.634793).* (See the color plate.)

Tβ4 is a G-actin-binding protein and serves as the major actin sequestering molecule in mammalian cells to promote cell migration (Crockford, Turjman, Allan, & Angel, 2010). This interaction with G-actin, forming a 1:1 complex, is the process by which Tβ4 maintains a large pool of actin monomers, controlling the assembly and disassembly of filaments that regulate the dynamics of the actin cytoskeleton. The mechanism by which Tβ4 influences cell migration, differentiation, and survival is thought to be linked to its ability to maintain equilibrium between G-actin and F-actin that is critical to the reorganization and stability of the cytoskeleton, particularly after cell injury (Safer, Elzinga, & Nachmias, 1991; Shelton, Poole, Reese, & Bader, 2013; Smart, Rossdeutsch, & Riley, 2007). At the leading edge of the migrating endothelial cells, Tβ4 interaction with G-actin is reduced and allows actin to be available for filament assembly.

Despite the diversity of these pleiotropic effects of Tβ4, the exact mechanisms by which healing occurs remains unclear. Studies of recombinant and synthetic Tβ4 molecules have defined bioactivity of amino acid domains within the Tβ4 molecule. At least three sites have been identified: (1) peptide 1–4 "SDKP" appears to play specific roles in inflammation and angiogenesis; (2) peptide 17–22, "LKKTETQ" is involved in actin binding and in wound healing; and (3) a larger domain including SDKP amino acids 1–15 may function in cytotoxicity protection (Sosne, Qiu, Goldstein, & Wheater, 2010). Much less is known about cell surface binding molecules that mediate Tβ4's effects. Purinergic signaling pathways have been reported to be involved in the molecule's activity, although until recently data have been scarce regarding the nature of potential receptors. A 2011 study examining the role of Tβ4 in human umbilical vein endothelial cells (HUVECs) migration sheds light on this issue (Freeman, Bowman, & Zetter, 2011). When exposed to Tβ4, these cells respond with an increase in extracellular ATP and with an increase in cell migration. This study demonstrated that Tβ4 binding to the β-subunit of cell-membrane ATP synthase appears to act as an inhibitor of extracellular ATP hydrolysis. This is thought to initiate signaling by increasing levels of extracellular ATP and triggering an ATP-responsive P2X4 receptor required for Tβ4-mediated HUVEC migration. The two major classes of purinergic receptors are the ligand-gated ionotropic P2X receptors and the G protein-coupled P2Y receptors. It has been speculated that since Tβ4 regulates the ligand ATP, it may act through additional ATP-responsive purinergic receptors in both the P2X and P2Y classes (Freeman et al., 2011).

Tβ4 also appears to modulate a number of downstream biological factors that are important in cell differentiation, migration, and inflammation. Tβ4 participates in extracellular matrix remodeling to reestablish the corneal epithelial barrier (Lu, Reinach, & Kao, 2001). Evidence exists that matrix metalloproteinases (MMPs) which are expressed in the cornea, also play an important role in epithelial cell migration (Sivak et al., 2000). When MMP activity is blocked, Tβ4-mediated epithelial cell migration is inhibited (Qiu, Kurpakus-Wheater, & Sosne, 2007). However, overexpression of MMPs can lead to exacerbation of epithelial defects (Matsubara, Zieske, & Fini, 1991). Thus, part of Tβ4's role in wound healing lies in its ability to maintain this balance by transiently increasing MMP activity via its central actin domain (Philp, Badamchian, et al., 2003). Tβ4 can alter the expression of cytokines during corneal reepithelialization after injury and increases conjunctival cell deposition of laminin-332 (LM-332, formerly LM-5), either directly or indirectly through modulation of TGFβ-1 after wounding (Sosne, Xu, et al., 2004). LM-332 is a component of anchoring filaments in the basement membrane of many tissues and provides structural integrity of the epithelial–subepithelial junction. It serves as a ligand for epithelial cell adhesion, migration, and collagen deposition (Ebihara et al., 2000). LM-332 is degraded by proteases into a smaller fragment that promotes cell migration by anchoring cells to each other and to the substratum, thus allowing a sheet of epithelium to migrate over the corneal wound. This ability of LM-332 to maintain cell–cell and cell–matrix interaction may be aided by Tβ4's antiapoptotic activity (Ho et al., 2007; Kainulainen et al., 1998; Rousselle & Beck, 2013).

3. Tβ4 AND INFLAMMATION

TNF-α has been shown to bind to its receptor (TRFN1) resulting in a signaling cascade leading to activation of NF-kB, a transcriptional factor that is normally maintained in an inactivated state by a family inhibitory proteins (IkBs). IkB can be phosphorylated following cellular stimulation, resulting in its degradation and release from the NF-kB dimer. Activated NF-kB translocates to the nucleus where it binds to a range of proinflammatory genes. Because of its ability to regulate the expression of inflammatory genes, NF-kB is believed to play a major role in the inflammatory process (Hayden & Ghosh, 2004). Moreover, translocation of TNFR1 to membrane rafts, essential for recruitment of membrane receptors, adaptor molecules, signaling propagators and is important for TNF-α-mediated NF-kB activation and inflammation (Legler, Micheau, Doucey, Tschopp, & Bron, 2003; Yamamoto, Yamamoto, Petroll, Cavanagh, & Jester, 2005). Sosne and colleagues demonstrated that in human epithelial corneal cells stimulated with TNF-α, Tβ4 significantly decreased nuclear NF-kB activation, p65 subunit phosphorylation, and nuclear translocation (Sosne, Qiu, Christopherson, & Wheater, 2007). However, the regulatory effects of Tβ4 on TNF-α-mediated signaling cascade are less known.

Additionally, Tβ4 blocks TNF-α-induced NF-kB, blocks RelA/p65 translocation and the sensitizing effects of its binding partners, LIM-domain adaptor protein (PINCH), and integrin-linked kinase. By a combination of these mechanisms, Tβ4 reduces further injury and facilitates the migration of epithelial cells and reduces inflammation both of which are necessary for wound repair (Sosne et al., 2007).

4. NEUROTROPHIC KERATOPATHY

Neurotrophic keratopathy (NK) is a rare degenerative corneal disease that is prevalent in less than 5/10,000 individuals, and therefore has been classified as an orphan disease (Sacchetti et al., 2015). It is characterized by decreased or absence of corneal sensitivity, degradation of the corneal epithelium, altered tear composition, decreased blink reflex, blurred vision, and impaired corneal wound healing. In advanced cases, pathological changes to the corneal epithelium may result in the development of corneal ulceration, stromal melting, and corneal perforation. Recent evidence from in vivo corneal confocal microscopy suggests that changes to keratocytes are involved as well (Lambiase, Sacchetti, Mastropasqua, & Bonini, 2013). Although there are various systemic, central nervous system, or ocular surface conditions that can cause NK such as diabetes mellitus, herpes zoster, chemical burns, surgery and long-term use of contact lenses, the final common insult is impairment of the trigeminal nerve (fifth cranial nerve), or one of its branches (Bonini, Rama, Olzi, & Lambiase, 2003; Semeraro et al., 2014).

The ophthalmic branch of the trigeminal nerve provides critical corneal sensory information as well as support to the ocular surface by releasing a variety of neurotrophic factors (Tseng & Tsubota, 1997). Historically, trigeminal nerve ablation studies demonstrated a loss of corneal transparency; a condition termed "neuroparalytic keratitis," believed to have arisen from a trophic disturbance. However, Flemming (1898) in a brief review of the literature suggested an alternative hypothesis based largely on the work that foreign bodies may lodge on an insensitive cornea that "not being felt" may cause irritation and ulceration (Flemming, 1898). More recently, it has been shown that the trigeminal nerve also serves as the afferent limb of two reflex arcs via the facial nerve. A motor arc drives the blink response and an autonomic arc that regulates the secretion of goblet cells, lacrimal, and meibomian glands. Taken together, this exquisitely integrated neuroanatomic mechanism serves as the basis of the ocular surface defense system. The relative importance any individual component of this feedback loop as a regulator of the corneal epithelium remains unclear. However, it is likely that peptides, neurotrophins, and neurotransmitters released from sensory corneal neurons (eg, nerve growth factor (NGF), substance P, calcitonin gene-related peptide, IFG-1, and acetylcholine), as well as the volume and composition of fluid from the lacrimal apparatus and substances released from metabolically active epithelial cells, work synergistically to provide trophic support to the cornea and in the case of ocular injury, promote corneal reepithelialization and wound healing (Aloe, Tirassa, & Lambiase, 2008; Ko et al., 2014; Muller, Marfurt, Kruse, & Tervo, 2003). Proper functioning of this neural loop and lacrimal functional unit is essential for maintaining a healthy ocular surface. Impairment or lesioning that may occur with any aspect of the neural pathways associated with the trigeminal nerve can lead to loss of corneal sensitivity, disruption of the reflex neural arc, diminution of neurotrophic factors, and ultimately the development of corneal ulceration (Marfurt, Cox, Deek, & Dvorscak, 2010; Muller et al., 2003; Semeraro et al., 2014; Zander & Weddell, 1951).

5. TREATMENT FOR NK

Treatment of NK depends on the stage and severity of the disease. The goal is to prevent the progression of corneal damage and to promote corneal healing. Recently, promising new therapies are being investigated for the treatment of NK that take advantage of the biologically active molecules found in serum and tear fluids.

Amniotic membrane transplantation (AMT), a surgical technique used to cover corneal ulcers, is a treatment option for epithelial defects that are otherwise refractory (Tseng, Prabhasawat, & Lee, 1997). AMT is effective in supporting corneal reepithelialization by providing a basement membrane matrix, epithelial donor cells, providing growth factors and cytokines, reducing vascularization, and reducing ocular surface inflammation.

However, Khokhar and colleagues have showed that the efficacy of AMT was analogous to conventional management (tarsorrhaphy vs bandage contact lens) in eyes with refractory neurotrophic ulcers (Khokhar et al., 2005).

Autologous serum-based artificial tears are also being developed (Tsubota, Goto, Shimmura, & Shimazaki, 1999; Young et al., 2004). These preparations contain neurotrophic factors such as epidermal growth factor (EGF), transforming growth factor-β (TGF-β), substance P, insulin-like growth factor (IGF-1), platelet-derived growth factor, and NGF, among others that may facilitate the proliferation, migration, and differentiation of ocular surface epithelium. However, there are few masked, placebocontrolled studies demonstrating the efficacy of these preparations (Aloe et al., 2008; Philp, Huff, Gho, Hannappel, & Kleinman, 2003). Moreover, it is not known what the optimal concentration of autologous serum should be, there is no standardized method for developing the tears from whole blood, and there are a myriad of regulatory restrictions on the manufacturing of autologous tears.

Some patients may derive more benefit from alternative therapies such as umbilical cord blood serum or platelet-rich fibrin tears. Vajpayee and colleagues studied the comparative efficacy of autologous serum vs umbilical cord serum in 59 patients with persistent corneal epithelial defects. Their results demonstrated that the cord serum group had faster healing times and a greater number of defects with complete reepithelialization (Vajpayee et al., 2003). The superiority of cord blood was confirmed by a subsequent study that demonstrated that cord blood was more efficacious in treating ocular chemical injury than artificial tears or autologous serum (Lambiase, Rama, Bonini, Caprioglio, & Aloe, 1998; Sharma et al., 2011).

NGF is a soluble protein belonging to a family of factors called neurotrophins that are responsible for the growth, migration, and survival of neurons. NGF has been reported to be effective in human ocular diseases such as NK and immune corneal ulcers. NGF is stored in human and rat corneal epithelial cells that express its TrkA receptor (Kawamoto & Matsuda, 2004). Lambiase and colleagues observed a transient increase of corneal NGF levels in the rat after corneal epithelial injury, suggesting that it may play a key role in the regeneration of injured neurons and target tissues (Lambiase et al., 2000). Bonini and colleagues evaluated the efficacy of NGF in 45 eyes of NK patients unresponsive to other nonsurgical therapies. The patients received murine NGF in eyedrops every 2 h for 2 days followed by one drop six times daily until the ulcer healed. During the treatment, the size and depth of the ulcer, corneal sensitivity, and best corrected visual acuity were examined. All the patients showed complete remission after 12 days to 6 weeks of treatment with NGF and visual function such as corneal sensitivity and visual acuity were improved (Bonini, Lambiase, Rama, Caprioglio, & Aloe, 2000).

6. WOUND HEALING AND ANTIINFLAMMATORY EFFECTS OF Tβ4

Tβ4 reduces inflammatory mediators in a variety of cells in in vitro and in vivo studies in many tissues, including the cornea, by modulating the infiltration of proinflammatory leukocytes and cytokines to reduce swelling and further tissue damage (Badamchian et al., 2003; Cavasin, 2006; Goldstein, Hannappel, Sosne, & Kleinman, 2012; Qiu et al., 2007; Sosne, Hafeez, et al., 2002). Although some degree of inflammation is necessary for healing to occur, the severity of the inflammatory response must be balanced. Tβ4 may modulate this response by decreasing the expression of IL-1, MIP-1α, MCP-1, and MIP-2 mRNA. The reduction in cytokine and chemokine production, in turn, stems the massive infiltration of PMNs and mononuclear leukocytes to the site of injury.

The efficacy of Tβ4 for corneal healing in vivo was first shown in a rat debridement model (Sosne et al., 2001). After treatment with Tβ4, reepithelialization of wounds was increased, along with cytokine mRNA transcripts for IL1-β and IL-6, indicating an antiinflammatory activity of Tβ4. In a subsequent murine model (Sosne, Szliter, et al., 2002), corneas of 129 mice were treated with either Tβ4 or PBS twice daily after alkali injury. Treatment with Tβ4 promoted reepithelialization and reduced mRNA transcripts for proinflammatory and chemokine proteins, including PMNs and chemoattractants MIP-2, KC, and IL-8 (a critical chemokine for PMN influx to the wound site). TNF-α is known to be one of the most potent proinflammatory cytokines and is increased in tears during infection and inflammation of the cornea (Bitko et al., 2004; Kumar, Zhang, & Yu, 2004). To further explore the role of TNF-α and the mechanisms by which Tβ4 regulates corneal inflammation, Sosne and colleagues stimulated cultured human corneal epithelial cells and adult human epidermal keratinocytes with TNF-α either in the presence or absence of Tβ4. Results showed that Tβ4 treatment inhibited injury-induced proinflammatory cytokine production and suppressed TNF-α-mediated NF-kB activation (Sosne et al., 2007).

TNF-α mainly binds to its membrane receptor, TNFR1, that results in a signaling cascade involving the intermediates TRADD–TRAFs–RIP–TIK–IKKs leading to NF-kB activation. In turn, activated NF-kB regulates the transcription of numerous proinflammatory cytokine and chemokine genes. Tβ4 inhibits corneal NF-kB activation and nuclear translocation following TNF-α stimulation and suppresses IL-8 secretion. However, the regulatory effects of Tβ4 on the TNF signaling cascade involving the intermediates that lead to proinflammatory cytokine and chemokine gene transcription have yet to be fully elucidated (Sosne et al., 2007).

The translocation of TNFR1 to membrane rafts (lipid rafts or detergent resistant membranes) is essential for TNF-α-mediated NF-kB activation (Legler et al., 2003). Membrane rafts are biological bilayers containing a large variety of molecules and are enriched in cholesterol, glycolsphingolipids, and phospholipids with saturated acyl chains. There are also specific proteins that localize preferentially to membrane rafts. Such localization of proteins is normally observed in resting cells and prevents unnecessary signaling. Upon stimulation, the small dynamic rafts aggregate into bigger platforms sustaining signal transduction (Grzybek, Kozubek, Dubielecka, & Sikorski, 2005). In this way, membrane rafts are essential for recruiting and concentrating membrane receptors, adaptor molecules, and signaling propagators and are critically linked to ocular surface inflammation and infection (Helms & Zurzolo, 2004; Yamamoto et al., 2005). The transcription factor NF-kB is a central mediator of the human immune response and it plays a crucial role in many inflammatory diseases, including those in the cornea (Pahl, 1999). NF-kB is composed of transcription factors from the NF-kB/Rel family. In the NF-kB pathway, the p50:RelA (p65) heterodimer functions in conjunction with the signaling axis of TRADD–TRAF2/5–RIP1–TAK1/MEKK3–IKKα/β/γ (IKK complex) (Perkins, 2007). NF-kB is maintained in an inactive state by a family of inhibitory proteins (IkBs). The recruitment of signaling adaptors initiates the activation of the IKK that phosphorylates IkB. Following degradation of IkB proteins, NF-kB is released from the binding of IkB, thereby promoting its translocation to the nucleus allowing it to target promoter regions of proinflammatory genes (Karin & Delhase, 2000). Tβ4 has a binding affinity for intracellular proteins, including focal adhesion proteins (LIM protein PINCH-1 and the Ankyrin repeat protein ILK, hMLH1, Ku80, and stabilin-2). Focal adhesion complex proteins contribute directly to TNF-α signaling transduction. Additional intracellular NF-kB activators or inhibitors anchored into both the nuclear matrix and cytoplasmic stress fibers have been shown to physically interact with NF-kB (Bednarek et al., 2008;

Bock-Marquette et al., 2004; Brieger, Plotz, Zeuzem, & Trojan, 2007; Fazal, Minhajuddin, Bijli, McGrath, & Rahman, 2007; Lee, So, Park, & Kim, 2008; Zhang et al., 2006). Tβ4 is a major intracellular monomeric G-actin-sequestering molecule. Qui and colleagues have shown that the antiinflammatory properties of Tβ4 are related to its association with actin and these intracellular binding partners (Qiu, Wheater, Qiu, & Sosne, 2011). Their results suggest that Tβ4 inhibits TNF-α-mediated NF-kB activation by directly targeting the NF-kB subunit RelA/p65. Furthermore, Tβ4 inhibits the sensitizing effects of its intracellular binding partners, PINCH-1 and ILK, in an actin-independent manner.

Taken together, evidence suggests that the pleiotropic molecule Tβ4 represents a novel wound-healing agent in a variety of tissues and disorders. Its biological actions present a unique linkage between the extracellular matrix and inflammatory mediators. Studies that further explore the interactions among Tβ4, G-actin, RelA/p65, and PINCH-ILK would help elucidate the mechanisms connecting the NF-kB transcriptional regulatory network to the cytoskeleton and integrin adhesion complex. Information gleaned from these experiments would facilitate translation of Tβ4's mechanisms of action into effective therapeutic regimens for the treatment of inflammatory-mediated disorders associated with ocular surface diseases in humans.

7. CLINICAL EFFICACY OF Tβ4 IN WOUND HEALING DUE TO NK

Results from key clinical trials have demonstrated the promising role for Tβ4 as a novel wound-healing agent in the management of nonhealing neurotrophic corneal epithelial disease. Tβ4 has been shown to accelerate cell migration and reepithelialization, reduce the number of proinflammatory cytokines, suppress NF-kB, regulate the delicate balance of MMPs (eg, MMP-1, MMP-2, MMP-9), and inhibit cell apoptosis (Sosne, Siddiqi, & Kurpakus-Wheater, 2004).

The first published data in humans came from a compassionate use study that utilized topical Tβ4 to treat a patient with a persistent corneal defect resulting from severe diabetic NK (Sosne, Qiu, et al., 2010). Conventional therapies including vitrectomy/retinal surgery and corneal scraping failed to heal the corneal defect for 23 days prior to Tβ4 treatment. Topical Tβ4 (5 μg) was administered four times daily. By the fourth day of treatment, a decrease in ocular surface inflammation and an improvement in the corneal epithelial defect were observed (Fig. 3). Examination on the 11th day of treatment revealed that the cornea was 95% reepithelialized and inflammation was almost entirely abated without significant side effects.

In another compassionate use study, nine patients with NK that had epithelial defects that were present from 6 weeks to more than 10 years were treated with Tβ4 (Dunn et al., 2010). Patients were divided into two groups based on slit lamp appearance of their defect: Group 1 was comprised of six patients with geographical defects; Group 2 was comprised of three patients with punctate erosions (Table 1). Group 1 was subdivided into four patients

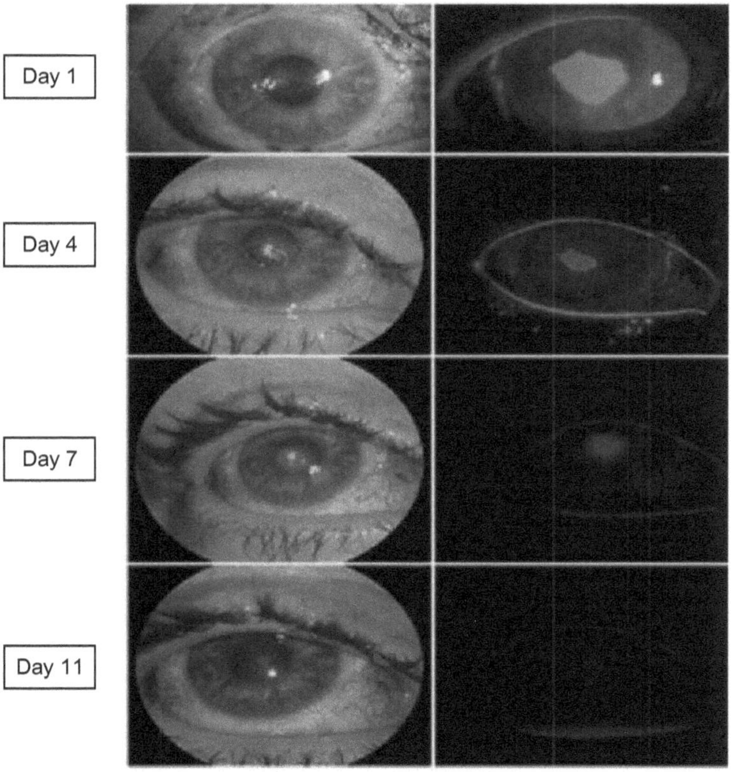

Fig. 3 Slit lamp photographs (*left column*) show the clinical appearance of the right eye at days 1, 4, 7, and 11 after Tβ4 therapy was initiated. The central corneal epithelial defect can be seen (*arrows*) as it decreases in size with the Tβ4 treatment. The fluorescein pictures (*right column*) show areas stained that represent areas absent of epithelium. The defect decreases to almost full closure by day 11 of Tβ4 treatment. *Copyright Permission: Originally published in Sosne, G., Qiu, P., Kurpakus-Wheater, M., & Matthew, H. (2010). Thymosin beta4 and corneal wound healing: Visions of the future.* The Annals of the New York Academy Sciences, 1194, *190–198. © The New York Academy of Sciences.* (See the color plate.)

Table 1 Geographical defects (Group 1) patients demographics and treatment response using thymosin beta 4[a]

Case	Age/Sex	Eye	Sensation	Diagnosis	Initial Defect	Tx End	End of f/u
#1	81/M	R	Decrease	HZ Ophthalmicus diabetes, dry eye	2.0 × 1.0	0.0 × 0.0	0.0 × 0.0
#2	47/F	R	Decrease	Erosion diabetes × 23 years	3.5 × 3.0	1.0 × 0.25	Sliver-like defect
#3	84/M	L	Absent	HZ ophthalmicus dry eye	1.0 × 0.5	1.1 × 0.5	Healing line 3 irregular areas < 0.1 mm
#4	57/M	L	Absent	HZ ophthalmicus diabetes × 10 years	4.5 × 3.6	0.3 × 0.8	0.0 × 0.0
#5	64//F	R	Absent	Diabetes × 20 years	4.1 × 3.6	6.0 × 6.0	0.0 × 0.0[b]
#6	73/F	R	Absent	HZ ophthalmicus	3.3 × 3.8	3.5 × 2.8	0.0 × 0.0

[a]Cases 1–4, 28-day treatment; Cases 5 and 6, 49-day treatment.
[b]Bandage contact lens inserted.

who were treated for 28 days and two patients who were treated for 49 days. All Group 2 patients were treated for 49 days. Patients were followed for an additional 30-days posttreatment. Four of six patients in Group 1 and all patients in Group 2 lacked corneal sensation.

One patient (Group 1) healed completely during the 28- and 49-day treatment period. All other patients showed improvement after discontinuation of treatment compared to baseline. All six patients in Group 1 showed dramatic healing by the end of the posttreatment follow-up period with four patients healing completely (Fig. 4). Three patients in Group 2 with punctate defects did not have a demonstrable change in their clinical findings. Importantly, extension of the treatment period from 28 to 49 days did not facilitate healing of the corneal defect. Major changes in healing were observed after treatment was discontinued. This observation has two important implications: (1) that a shorter treatment period may be more efficacious and (2) that Tβ4 may be "priming the pump" of the wound repair mechanism during the early stages of healing.

The results of the preclinical studies, in addition to data from compassionate use studies in patients with nonhealing corneal surface defects, suggest that Tβ4 has significant potential to be a novel, safe, and effective therapeutic treatment for neurotrophic corneal defects. Ocular surface

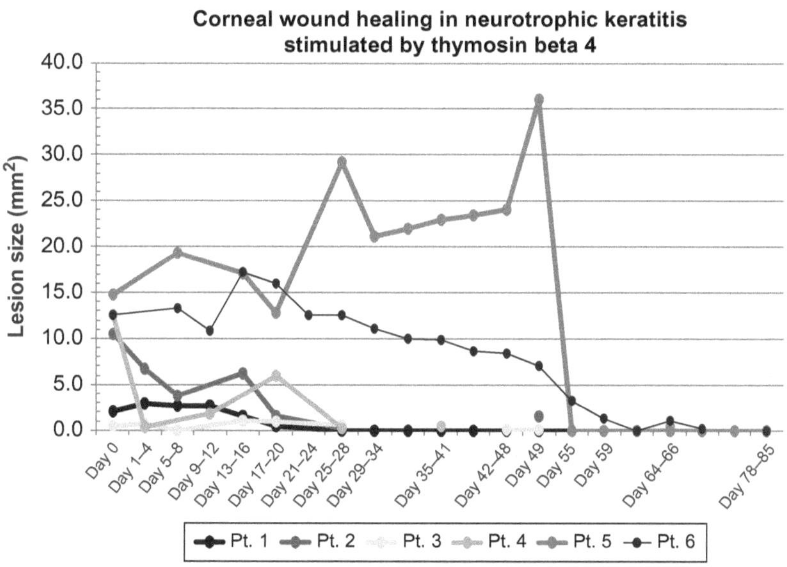

Fig. 4 Corneal wound healing in neurotrophic keratitis stimulated by Tβ4. *Copyright Permission: Originally published in Dunn, S. P., Heidemann, D. G., Chow, C. Y., Crockford, D., Turjman, N., Angel, J., …, Sosne, G. (2010). Treatment of chronic nonhealing neurotrophic corneal epithelial defects with thymosin beta4.* The Annals of the New York Academy of Sciences, 1194, *199–206. © The New York Academy of Sciences.* (See the color plate.)

inflammation plays a major role in corneal epithelial and sensory nerve disease and causes the patient discomfort, pain, and visual disturbances. It is likely that the potential of Tβ4 to improve patient's quality of life is mainly due to its antiinflammatory and wound-healing properties.

8. DRY EYE DISEASE

Dry eye syndrome (DES) is a common disorder affecting an estimated 25–30 million people in the United States, with prevalence estimates varying widely from 7.8% to almost 58%. Dry eye incidence rises sharply with age, and women are more affected than men, purportedly due to the pathophysiological role of androgens and the complex nexus of the endocrine-immunological systems. The Dry Eye Workshop (DEWS) has redefined dry eye as "a multifactorial disease of the tears and ocular surface that results in symptoms of discomfort, visual disturbance, and tear film instability with potential damage to the ocular surface, accompanied by increased osmolarity of the tear film and inflammation of the ocular surface" (Anonymous, 2007). The human tear film is typically described as having aqueous, lipid, and mucin components. However, it is much more complex and also contains electrolytes, antioxidants (eg, vitamin C, glutathione), antiinfectives (eg, lactoferrin, lysosome), immune and inflammatory proteins (eg, sIgA, IgG, INF-γ, IL-2, IL-6, IL-8, IL-17, TNF-α), proteins involved in tear film-maintenance (eg, lipocalins, mucins), proteins involved in wound healing (eg, Tβ4, fibronectin, growth factors, insulin), neuropeptides (eg, calcitonin, substance-P), and lipid components (eg, wax and cholesterol esters). These components originate from a variety of sources including the lacrimal glands, ocular surface cells, nerve endings, and Meibomian glands (Ableson, Doane, & Ousler, 2008; Butovich, Millar, & Ham, 2008; Gipson, 2007). Furthermore, the tear film is constantly being wiped from the ocular surface and replenished with every blink (Doane, 1980; Ousler et al., 2014). Thus, the concentrations of these tear film components can vary drastically from one tear film to another and deficiencies in an individual's quality or quantity of the tear film can lead to the development of dry eye disease of varying etiologies. Given this, the diagnosis and treatment of dry eye are complicated by the existence of subtypes of the disease, largely defined as aqueous deficient and evaporative dry eye, the latter also defined as lipid or mucin deficient dry eye, although further subsets undoubtedly exist, such as Sjögren's and non-Sjögren's aqueous deficiency, and meibomian gland deficient evaporative disease vs ocular surface irregularities

(McGinnigle, Naroo, & Eperjesi, 2012). The clinical diagnosis of dry eye is based on a constellation of signs and symptoms, most nonexclusive to the disease, in conjunction with a battery of functional tests such as tear film break-up time, which measures the time to corneal epithelium exposure with tear film evaporation; the Schirmer's test of aqueous tear production, as well as various vital stains such as fluorescein and lissamine green that help to visualize the keratitis that results from epithelial exposure. Underlying intrinsic variables can cause subtle differences in outcomes on objective tests and/or in signs and symptoms, while extrinsic variables such as environment (increased air flow, decreased humidity) and lifestyle (smoking, contact lens wear) undoubtedly affect one type of dry eye more than another. This great variability in presentation, as well as the grouping of different subgroups in the same study cohort, has greatly hampered the successful identification of both etiology and treatment.

Chronic dry eye disease is hypothesized to be a Th17-mediated disease (Chen, Chauhan, Lee, Saban, & Dana, 2014; Kodati et al., 2014). Inflammation of the lacrimal gland, the conjunctiva, and the cornea occurs as an amalgam of cytokine-mediated primary and secondary insults to the ocular surface tissue (Enriquez-de-Salamanca et al., 2010). Hyperosmolar stress activates key inflammatory pathways in epithelial and inflammatory cells, including NF-kB and IL-8 (Png et al., 2011). NF-kB signaling plays a central role in generating a host of proinflammatory cytokine, chemokine, and adhesion factor responses that mediate tissue responses in inflammation. Reports indicate that Sjogren's syndrome patients showed decreased tear fluid EGF and significantly increased levels of IL-1, IL-1α, IL-6, IL-8, TNF-α, and TGF-β1 in their conjunctival epithelium compared to control patients (de Paiva & Pflugfelder, 2008; Lam et al., 2009; Lisi, Sisto, Scagliusi, Mitolo, & D'Amore, 2007; Massingale et al., 2009). Recent evidence suggests that elevated levels of IL-1 may impair the secretory function of the lacrimal gland and inhibit neurally mediated lacrimal gland secretion (Zoukhri, Hodges, Byon, & Kublin, 2002). Innervation is known to play a role, not only at the origin of the pain and discomfort brought on by corneal exposure, but by the modifications and dysregulation in neurotransmitter-receptor activity driven by excessive T cell-derived cytokines (Tashiro, Okamoto, Chang, & Bereiter, 2010).

Approaches to treatment vary. However, disease modification usually has targeted the inflammatory aspects of the disease, and the only currently approved therapy is cyclosporin ophthalmic emulsion (Restasis). Cyclosporin A (CSA) is a fungal-derived peptide that prevents activation and nuclear translocation of cytoplasmic transcription factors that are required for T-cell activation and inflammatory cytokine production (Matsuda & Koyasu, 2000). It is an established immunomodulatory used to prevent transplant rejection, and as a treatment for autoimmune diseases. However, 85% of CSA-treated patients did not experience this significant increase in tear production (Sall, Stevenson, Mundorf, & Reis, 2000). Considering that ultimately the clinical manifestation of dry eye is chronic corneal epithelial damage, Tβ4 as an antiinflammatory agent and promoter of wound healing has been an appropriate target of study for its treatment.

9. EFFICACY OF Tβ4 IN ANIMAL MODELS OF DRY EYE

In a recent exploratory study, researchers demonstrated in a murine model of dry eye disease using a controlled adverse environment that Tβ4 eye drops successfully decreased corneal fluorescein staining more than saline (negative) controls, and as well as Restasis and doxycycline (positive) controls. Tβ4 also improved corneal epithelial intercellular adhesions following injury in animal models of dry eye. Tβ4 reduced corneal staining significantly more than positive and negative controls for dry eye (Sosne, Kim, & Kleinman, 2015).

10. CLINICAL EFFICACY OF Tβ4 IN DRY EYE

Few clinical trials have explored the use of Tβ4 in dry eye. One small ($N=9$), single-centered pilot study investigated the efficacy of Tβ4 (0.1%) in severe dry eye, including those with dry eye associated with graft vs host disease. The 12 eyes treated with Tβ4 had a 35% reduction of ocular discomfort compared to vehicle control (6 eyes) ($p=0.0141$), and a 59% reduction of total corneal fluorescein staining compared to vehicle control ($p=0.0108$) (Sosne, Dunn, & Kim, 2015).

In a subsequent study in 72 subjects, a 1-month treatment with Tβ4 or vehicle was evaluated. Subjects were entered who had a diagnosis of dry eye and moderately severe signs of corneal staining and discomfort symptoms at baseline, after a run-in period of placebo, and after a controlled adverse environment (CAE™) exposure. This assured a population of dry eye subjects who had moderately severe, yet modifiable disease, who did not respond solely to tear substitution therapy, thereby minimizing high placebo responders. In this study, after 28 days of Tβ4 therapy, subjects were again exposed to a CAE™, and signs and symptoms were assessed 2 and 4 days after, in order to assess if Tβ4 facilitated corneal surface healing after CAE-elicited damage. While this proof-of-concept study failed to meet its primary endpoints of ocular discomfort and inferior corneal staining 2 days after the Day-28 CAE™ exposure, Tβ4 did demonstrate improvements in secondary endpoints, lowering scores by 27% in ocular discomfort during the 90-min CAE exposure ($p=0.0244$), and improving superior ($p=0.0075$) and peripheral ($p=0.0210$) corneal staining. These results suggest that Tβ4 acts to improve recovery from insult, possibly by its wound-healing effects. The study revealed no safety signals with Tβ4 use for 30 days, with no clinically significant changes in safety assessments (slit lamp biomicroscopy, visual acuity, intraocular pressure, fundoscopy, corneal sensitivity) and no treatment-emergent adverse events (Sosne & Ousler, 2015).

11. CLINICAL SAFETY EVALUATIONS OF Tβ4

One Phase 1 study double-blind, randomized, placebocontrolled, dose–response clinical trial (TB4-01-02) was performed to evaluate the safety and tolerability of Tβ4 when administered topically to 15 healthy volunteers. There were three unexpected AEs that were related to study drug observed in two subjects: one drug-related unexpected AE, pruritus, was observed in the 250 μg Tβ4 dose group. In the 1000 μg Tβ4 dose group, there were two potentially drug-related unexpected AEs reported in one subject: headache and dizziness. Following 28 consecutive days of dermal dosing up to 1 mg Tβ4/day, it was concluded that Tβ4 is safe and well tolerated in humans under the conditions of use for this study.

Another Phase 1 study evaluated the safety, tolerability, and pharmacokinetics (PK) of intravenously administered Tβ4 after single (Phase 1A) and multiple (Phase 1B) doses in healthy volunteers in a randomized, double-blind, placebocontrolled fashion. Four cohorts with 10 healthy subjects each were given a single intravenous dose of Tβ4 or placebo (Ruff, Crockford, Girardi, & Zhang, 2010) in ascending doses of 42, 140, 420, or 1260 mg (Phase 1A). Following review of these data, subjects were given the same dosage regimen daily for 14 consecutive days (Phase 1B). Safety evaluations, incidence of unexpected adverse events, and PK parameters were evaluated. Adverse events were infrequent, mild, and moderate in intensity. There were no deaths, no dose-limiting toxicities, or serious adverse events. The PK profile showed a dose proportional response, increasing half-life with increasing dose and minimal drug accumulation. In Phase 1A, a subject in the 140 mg-dose group experienced dizziness that could have been related to Tβ4. No other subjects in any of the treatment groups experienced any unexpected AEs. In Phase 1B, two subjects treated with Tβ4 experienced two unexpected adverse events that were reported as possibly related to study drug: in the 140 mg-dose group, one subject experienced pyrexia and another subject reported a headache. No other subjects in any of the treatment groups experienced any unexpected adverse events. The study concluded that Tβ4 was well tolerated and no safety issues were identified. Based on predefined criteria, the assessment of C_{max} and AUC indicated approximate dose proportionality of Tβ4 over the studied dose range following both single and multiple intravenous dose administration. Following 14 days of multiple intravenous administration of Tβ4 once daily, accumulation of Tβ4 was minimal. Clearance remained constant in all dose groups and was similar following single and multiple intravenous administrations, indicating the dose independency of total clearance. Apparent mean $T_{1/2}$ increased with dose over the studied dose range. Apparent V_d increased with dose over the studied dose range. Tβ4 did not appear to induce an antibody response in this study.

12. POTENTIAL INDICATIONS FOR Tβ4

Our understanding of role of Tβ4 in health and disease has increased over the past 35 years. However, management of patients with NK remains challenging because current treatment regimens are limited. Therefore, it is paramount that the pathophysiology underlying the disease and the mechanisms of action of treatment modalities continue to be examined. As a first line of treatment, preventing the progression of corneal damage and promoting corneal healing through the eye's innate reparative ability is imperative. However, the development of new therapies that regulate specific steps in ocular surface healing process has been slow to occur. Tβ4 has been demonstrated to promote corneal reepithelialization, reduce inflammation, and inhibit apoptosis; properties which form the basis for future use as a corneal wound-healing molecule. Although the literature suggesting that Tβ4 decreases inflammation is growing, the antiinflammatory mechanisms of Tβ4 are still being elucidated. Well-controlled studies exploring the potential effects of Tβ4 on TNF-α-mediated NF-kB activation and proinflammatory protein synthesis at multiple levels are needed. These include: (1) the initiation of TNF-α signaling pathways by TNFR receptor clustering on membrane rafts, (2) signal intermediate complex formation in the cytoplasm, and (3) signaling transduction into nuclei via NF-kB. Findings from these experiments could expand our knowledge of the manner in which Tβ4 regulates the corneal responses to injury-related inflammation and uncover new molecular pathways involved in these processes that could feasibly be targeted for therapy. This strategy represents a comprehensive approach for studying the consequences of the effects of Tβ4 on corneal inflammation. Defining Tβ4-modulated pathways that regulate corneal inflammation will facilitate translation of basic scientific findings into safe and effective corneal therapeutic regimens for the treatment of corneal inflammatory-mediated disorders.

To this end, Sosne and colleagues are developing a Tβ4 drug delivery system to the ocular surface through a Tβ4-hydrogel bandage contact lens. Bandage lenses are a group of soft, thin, highly oxygen permeable, hydrophilic lenses that are fit over the cornea. They are used to protect the eye from further insult and evidence suggests that they improve the efficacy of certain topical drugs by keeping the drug in contact with the ocular surface for longer periods of time. Only about 5% of a topically administered drug enters the anterior segment, and concentrations remain low even when drops are applied six or more times a day (Guarnera, DeRosa, Camerini, & 8 European sites, 2010). However, there is significant evidence that bandage contact lenses improve the results of concomitant topical medications. Studies have demonstrated that lenses enhance the effects of antivirals (idoxuridine) on rabbit corneas infected with herpes virus and help keep various therapeutic agents in contact with the ocular surface for longer periods of time (Aquavella, 1976; Kaufman, Ellison, & Townsend, 1970; Matoba & McCulley, 1985). Innovations such as the bandage contact lens that soak the lens in a drug not only protects the ocular surface from mechanical forces such a blinking but also acts as a reservoir, facilitating drug delivery into the anterior segment (Podos, Becker, Asseff, & Hartstein, 1972).

Based on the defined mechanism of action of Tβ4 in the eye, it is likely that it will promote repair in variety of other eye disorders involving corneal injury such as blepharitis. Blepharitis is an inflammatory condition where the meibomian glands in the lids fail to function properly, resulting in inflamed, irritated, itchy eyelids, and abnormal tear film. This can cause dry eye, scarring in the eyelid and can result in injury to the corneal surface. Recent findings demonstrated that Tβ4 is the sixth highest expressed gene in human meibomian gland disease samples (Liu, Richards, et al., 2011). Since keratinization plays an important role in meibomian gland disease, Tβ4 may be an excellent candidate drug to ameliorate this condition as it reduces inflammation, scarring and promotes corneal repair.

Strategies that manipulate biological modulators such MicroRNA (miRNAs) levels may provide novel therapeutic tools for managing corneal wound healing. miRNAs are natural and powerful gene expression regulators and key mediators of basic biological processes in invertebrate and vertebrate organisms including in the human cornea. Since their discovery in *Caenorhabditis elegans*, miRNAs have been shown to regulate gene expression via translational inhibition, mRNA degradation, or a combination of both mechanisms (Lagos-Quintana, Rauhut, Lendeckel, & Tuschl, 2001; O'Carroll & Schaefer, 2013). The emerging evidence suggests the miRNAs are important in many phases of corneal epithelial wound healing with some miRNAs promoting certain signaling pathways and genes while inhibiting others. For example, during migration and proliferation of corneal epithelial cells at the wound edge, upregulated miR-205 targets the Akt-that signaling pathway that promotes the survival and growth of cells and enhances wound sealing likely through modification of F-actin organization (Yu et al., 2010). It has been further shown that miR-205 facilitates wound-healing process through inhibition of the KCNJ10 channel in human corneal epithelial cells (Lin et al., 2013). Certain miRNAs such as miRNA-146a and miR-424 have been identified as regulators of corneal epithelial wound healing by modulating the innate immune and inflammatory responses to injury and downregulating proinflammatory genes within the wound. Other miRNAs have shown to regulate multiple mechanisms during the recovery from neurological injury via neurogenesis and axonal and dendritic plasticity (Funari et al., 2013; Roy & Sen, 2011). Studying the molecular mechanism that underlying innate neuroprotection can help elucidate the manner in which the nervous system is protected against damaging insults such as that which occurs to the trigeminal nerve in NK, and other ocular surface diseases.

The subventricular zone produces neuroprogenitor cells (NPCs) and oligoprogenitor cells (OPCs) (Arvidsson, Collin, Kirik, Kokaia, & Lindvall, 2002; Morris et al., 2012; Zhang et al., 2009). Increasing evidence in animal data suggests that microRNAs regulate the differentiation of NPCs into neurons and OPCs into mature oligodendrocytes that remyelinate injured axons. Among the superfamily of miRNAs, the miRNA-200 family (eg, miRNA-200a) members appear to show the most prominent neuroprotective, neuroregenerative, and apoptotic effects (Choi et al., 2008; Lee et al., 2010; Liu, Chopp, et al., 2011; Rink & Khanna, 2011; Wang, Ji, Cheng, Chen, & Bai, 2014; Zhang et al., 2013). Thus, agents that promote differentiation, proliferation, and survival of these cells may represent new therapies for the treatment of NK. To this end, Santra and colleagues demonstrated that Tβ4 upregulated miR-200a, -200b, and -429, all of which belong to miR-200b-a-429 cluster gene. Their studies (in vivo and in vitro) showed that upregulation of miRNA-200a by Tβ4 treatment, markedly upregulated AKT which is known to phosphorylate more than 30 downstream targets that regulate cell survival, growth, differentiation, migration (Somanath, Razorenova, Chen, & Byzova, 2006). The results from their studies suggest that activation of AKT increased myelin basic protein synthesis and POC differentiation into mature myelinating oligodendrocytes. Treatment with Tβ4 also reduced the p53 transcription factor (an inducer of a proapoptotic gene) expression by phosphorylation of mouse double minute 2 homolog. Through a combination of these effects, Tβ4 may promote NPG and OPC survival through a novel miR-200a-dependent pathway (Morris, Chopp, Zhang, Lu, & Zhang, 2010; Santra et al., 2012).

These data provide insight into novel miR-200-mediated mechanisms of action for the neuroprotective and neurorestorative effects of Tβ4 and its wound-healing capabilities. Further investigations examining the use of Tβ4 for therapeutic modulation of miRNA levels are needed to in order to develop new tools for managing ocular diseases such as NK and Dry Eye Disease.

13. CONCLUSION

In summary, although there has been progress in understanding the manner in which the cornea heals, current therapies that promote epithelial healing are limited. Thus, there is an increasing need for better treatment of corneal defects and other ocular disorders. The overwhelming evidence presented thus far suggests that Tβ4 has important future clinical implications as a wound-healing molecule in several tissues including the human cornea. The mechanisms of action for wound healing are only beginning to come to light, therefore, additional clinical trials are needed to explore the potential therapeutic indications for Tβ4 in the eye to improve patients quality of life.

REFERENCES

Ableson, M. B., Doane, M., & Ousler, G. (2008). Tear film and blink dynamics. In F. Jakobiec & D. M. Albert (Eds.), *Principles and practice of ophthalmology*. Canada: Elsevier.

Al-Nedawi, K. N., Czyz, M., Bednarek, R., Szemraj, J., Swiatkowska, M., Cierniewska-Cieslak, A., … Cierniewski, C. S. (2004). Thymosin beta 4 induces the synthesis of plasminogen activator inhibitor 1 in cultured endothelial cells and increases its extracellular expression. *Blood, 103*(4), 1319–1324. http://dx.doi.org/10.1182/blood-2003-04-1015.

Aloe, L., Tirassa, P., & Lambiase, A. (2008). The topical application of nerve growth factor as a pharmacological tool for human corneal and skin ulcers. *Pharmacological Research, 57*(4), 253–258. http://dx.doi.org/10.1016/j.phrs.2008.01.010.

Anonymous (2007). The definition and classification of dry eye disease: Report of the Definition and Classification Subcommittee of the International Dry Eye WorkShop. *The Ocular Surface, 5*(2), 75–92.

Aquavella, J. V. (1976). New aspects of contact lenses in ophthalmology. *Advances in Ophthalmology, 32*, 2–34.

Arvidsson, A., Collin, T., Kirik, D., Kokaia, Z., & Lindvall, O. (2002). Neuronal replacement from endogenous precursors in the adult brain after stroke. *Nature Medicine, 8*(9), 963–970. http://dx.doi.org/10.1038/nm747.

Badamchian, M., Fagarasan, M. O., Danner, R. L., Suffredini, A. F., Damavandy, H., & Goldstein, A. L. (2003). Thymosin beta(4) reduces lethality and down-regulates inflammatory mediators in endotoxin-induced septic shock. *International Immunopharmacology, 3*(8), 1225–1233. http://dx.doi.org/10.1016/S1567-5769(03)00024-9.

Bednarek, R., Boncela, J., Smolarczyk, K., Cierniewska-Cieslak, A., Wyroba, E., & Cierniewski, C. S. (2008). Ku80 as a novel receptor for thymosin beta4 that mediates its intracellular activity different from G-actin sequestering. *The Journal of Biological Chemistry, 283*(3), 1534–1544. http://dx.doi.org/10.1074/jbc.M707539200.

Bitko, V., Garmon, N. E., Cao, T., Estrada, B., Oakes, J. E., Lausch, R. N., & Barik, S. (2004). Activation of cytokines and NF-kappa B in corneal epithelial cells infected by respiratory syncytial virus: Potential relevance in ocular inflammation and respiratory infection. *BMC Microbiology, 4*, 28. http://dx.doi.org/10.1186/1471-2180-4-28.

Bock-Marquette, I., Saxena, A., White, M. D., Dimaio, J. M., & Srivastava, D. (2004). Thymosin beta4 activates integrin-linked kinase and promotes cardiac cell migration, survival and cardiac repair. *Nature, 432*(7016), 466–472. http://dx.doi.org/10.1038/nature03000.

Bonini, S., Lambiase, A., Rama, P., Caprioglio, G., & Aloe, L. (2000). Topical treatment with nerve growth factor for neurotrophic keratitis. *Ophthalmology, 107*(7), 1347–1351. discussion 1351–1342.

Bonini, S., Rama, P., Olzi, D., & Lambiase, A. (2003). Neurotrophic keratitis. *Eye (London, England), 17*(8), 989–995. http://dx.doi.org/10.1038/sj.eye.6700616.

Brieger, A., Plotz, G., Zeuzem, S., & Trojan, J. (2007). Thymosin beta 4 expression and nuclear transport are regulated by hMLH1. *Biochemical and Biophysical Research Communications, 364*(4), 731–736. http://dx.doi.org/10.1016/j.bbrc.2007.10.010.

Butovich, I. A., Millar, T. J., & Ham, B. M. (2008). Understanding and analyzing meibomian lipids—A review. *Current Eye Research, 33*(5), 405–420. http://dx.doi.org/10.1080/02713680802018419.

Cavasin, M. A. (2006). Therapeutic potential of thymosin-beta4 and its derivative N-acetyl-seryl-aspartyl-lysyl-proline (Ac-SDKP) in cardiac healing after infarction. *American Journal of Cardiovascular Drugs, 6*(5), 305–311.

Chen, Y., Chauhan, S. K., Lee, H. S., Saban, D. R., & Dana, R. (2014). Chronic dry eye disease is principally mediated by effector memory Th17 cells. *Mucosal Immunology*, 7(1), 38–45. http://dx.doi.org/10.1038/mi.2013.20.

Choi, P. S., Zakhary, L., Choi, W. Y., Caron, S., Alvarez-Saavedra, E., Miska, E. A., … Dulac, C. (2008). Members of the miRNA-200 family regulate olfactory neurogenesis. *Neuron, 57*(1), 41–55. http://dx.doi.org/10.1016/j.neuron.2007.11.018.

Crockford, D., Turjman, N., Allan, C., & Angel, J. (2010). Thymosin beta4: Structure, function, and biological properties supporting current and future clinical applications. *The Annals of the New York Academy of Sciences, 1194*, 179–189. http://dx.doi.org/10.1111/j.1749-6632.2010.05492.x.

de Paiva, C. S., & Pflugfelder, S. C. (2008). Rationale for anti-inflammatory therapy in dry eye syndrome. *Arquivos Brasileiros de Oftalmologia, 71*(6 Suppl.), 89–95.
Doane, M. G. (1980). Interactions of eyelids and tears in corneal wetting and the dynamics of the normal human eyeblink. *American Journal of Ophthalmology, 89*(4), 507–516.
Dunn, S. P., Heidemann, D. G., Chow, C. Y., Crockford, D., Turjman, N., Angel, J., … Sosne, G. (2010). Treatment of chronic nonhealing neurotrophic corneal epithelial defects with thymosin beta4. *The Annals of the New York Academy of Sciences, 1194*, 199–206. http://dx.doi.org/10.1111/j.1749-6632.2010.05471.x.
Ebihara, N., Mizushima, H., Miyazaki, K., Watanabe, Y., Ikawa, S., Nakayasu, K., & Kanai, A. (2000). The functions of exogenous and endogenous laminin-5 on corneal epithelial cells. *Experimental Eye Research, 71*(1), 69–79. http://dx.doi.org/10.1006/exer.2000.0857.
Enriquez-de-Salamanca, A., Castellanos, E., Stern, M. E., Fernandez, I., Carreno, E., Garcia-Vazquez, C., … Calonge, M. (2010). Tear cytokine and chemokine analysis and clinical correlations in evaporative-type dry eye disease. *Molecular Vision, 16*, 862–873.
Fazal, F., Minhajuddin, M., Bijli, K. M., McGrath, J. L., & Rahman, A. (2007). Evidence for actin cytoskeleton-dependent and -independent pathways for RelA/p65 nuclear translocation in endothelial cells. *The Journal of Biological Chemistry, 282*(6), 3940–3950. http://dx.doi.org/10.1074/jbc.M608074200.
Flemming, P. (1898). The pathology of neuropathic keratitis. *The Lancet*, 18–20. II(July–September).
Freeman, K. W., Bowman, B. R., & Zetter, B. R. (2011). Regenerative protein thymosin beta-4 is a novel regulator of purinergic signaling. *The FASEB Journal, 25*(3), 907–915. http://dx.doi.org/10.1096/fj.10-169417.
Funari, V. A., Winkler, M., Brown, J., Dimitrijevich, S. D., Ljubimov, A. V., & Saghizadeh, M. (2013). Differentially expressed wound healing-related microRNAs in the human diabetic cornea. *PLoS One, 8*(12). e84425. http://dx.doi.org/10.1371/journal.pone.0084425.
Gipson, I. K. (2007). The ocular surface: The challenge to enable and protect vision: The Friedenwald lecture. *Investigative Ophthalmology & Visual Science, 48*(10), 4390. http://dx.doi.org/10.1167/iovs.07-0770. 4391–4398.
Goldstein, A. L., Hannappel, E., Sosne, G., & Kleinman, H. K. (2012). Thymosin beta4: A multi-functional regenerative peptide. Basic properties and clinical applications. *Expert Opinion on Biological Therapy, 12*(1), 37–51. http://dx.doi.org/10.1517/14712598.2012.634793.
Grzybek, M., Kozubek, A., Dubielecka, P., & Sikorski, A. F. (2005). Rafts—The current picture. *Folia Histochemica et Cytobiologica, 43*(1), 3–10.
Guarnera, G., DeRosa, A., Camerini, R., & 8 European sites (2010). The effect of thymosin treatment of venous ulcers. *The Annals of the New York Academy of Sciences, 1194*, 207–212. http://dx.doi.org/10.1111/j.1749-6632.2010.05490.x.
Han, T., Liu, Y., Liu, H., Zhu, Z. Y., Li, Y., Xiao, S. X., … Zhao, Z. G. (2010). Serum thymosin beta4 levels in patients with hepatitis B virus-related liver failure. *World Journal of Gastroenterology, 16*(5), 625–630.
Hannappel, E., & van Kampen, M. (1987). Determination of thymosin beta 4 in human blood cells and serum. *Journal of Chromatography, 397*, 279–285.
Hayden, M. S., & Ghosh, S. (2004). Signaling to NF-kappaB. *Genes & Development, 18*(18), 2195–2224. http://dx.doi.org/10.1101/gad.1228704.
Helms, J. B., & Zurzolo, C. (2004). Lipids as targeting signals: Lipid rafts and intracellular trafficking. *Traffic, 5*(4), 247–254. http://dx.doi.org/10.1111/j.1600-0854.2004.0181.x.
Ho, J. H., Chuang, C. H., Ho, C. Y., Shih, Y. R., Lee, O. K., & Su, Y. (2007). Internalization is essential for the antiapoptotic effects of exogenous thymosin beta-4 on human corneal epithelial cells. *Investigative Ophthalmology & Visual Science, 48*(1), 27–33. http://dx.doi.org/10.1167/iovs.06-0826.

Huff, T., Muller, C. S., Otto, A. M., Netzker, R., & Hannappel, E. (2001). beta-Thymosins, small acidic peptides with multiple functions. *The International Journal of Biochemistry & Cell Biology*, *33*(3), 205–220.

Kainulainen, T., Hakkinen, L., Hamidi, S., Larjava, K., Kallioinen, M., Peltonen, J., … Oikarinen, A. (1998). Laminin-5 expression is independent of the injury and the microenvironment during reepithelialization of wounds. *The Journal of Histochemistry and Cytochemistry*, *46*(3), 353–360.

Karin, M., & Delhase, M. (2000). The I kappa B kinase (IKK) and NF-kappa B: Key elements of proinflammatory signalling. *Seminars in Immunology*, *12*(1), 85–98. http://dx.doi.org/10.1006/smim.2000.0210.

Kaufman, H. E., Ellison, E. D., & Townsend, W. M. (1970). The chemotherapy of herpes iritis with adenine arabinoside and cytarabine. *Archives of Ophthalmology*, *84*(6), 783–787.

Kawamoto, K., & Matsuda, H. (2004). Nerve growth factor and wound healing. *Progress in Brain Research*, *146*, 369–384. http://dx.doi.org/10.1016/S0079-6123(03)46023-8.

Khokhar, S., Natung, T., Sony, P., Sharma, N., Agarwal, N., & Vajpayee, R. B. (2005). Amniotic membrane transplantation in refractory neurotrophic corneal ulcers: A randomized, controlled clinical trial. *Cornea*, *24*(6), 654–660.

Ko, J. A., Mizuno, Y., Ohki, C., Chikama, T., Sonoda, K. H., & Kiuchi, Y. (2014). Neuropeptides released from trigeminal neurons promote the stratification of human corneal epithelial cells. *Investigative Ophthalmology & Visual Science*, *55*(1), 125–133. http://dx.doi.org/10.1167/iovs.13-12642.

Kodati, S., Chauhan, S. K., Chen, Y., Dohlman, T. H., Karimian, P., Saban, D., & Dana, R. (2014). CCR7 is critical for the induction and maintenance of Th17 immunity in dry eye disease. *Investigative Ophthalmology & Visual Science*, *55*(9), 5871–5877. http://dx.doi.org/10.1167/iovs.14-14481.

Kumar, A., Zhang, J., & Yu, F. S. (2004). Innate immune response of corneal epithelial cells to Staphylococcus aureus infection: Role of peptidoglycan in stimulating proinflammatory cytokine secretion. *Investigative Ophthalmology & Visual Science*, *45*(10), 3513–3522. http://dx.doi.org/10.1167/iovs.04-0467.

Lagos-Quintana, M., Rauhut, R., Lendeckel, W., & Tuschl, T. (2001). Identification of novel genes coding for small expressed RNAs. *Science*, *294*(5543), 853–858. http://dx.doi.org/10.1126/science.1064921.

Lam, H., Bleiden, L., de Paiva, C. S., Farley, W., Stern, M. E., & Pflugfelder, S. C. (2009). Tear cytokine profiles in dysfunctional tear syndrome. *American Journal of Ophthalmology*, *147*(2), 198–205. http://dx.doi.org/10.1016/j.ajo.2008.08.032. e191.

Lambiase, A., Manni, L., Bonini, S., Rama, P., Micera, A., & Aloe, L. (2000). Nerve growth factor promotes corneal healing: Structural, biochemical, and molecular analyses of rat and human corneas. *Investigative Ophthalmology & Visual Science*, *41*(5), 1063–1069.

Lambiase, A., Rama, P., Bonini, S., Caprioglio, G., & Aloe, L. (1998). Topical treatment with nerve growth factor for corneal neurotrophic ulcers. *The New England Journal of Medicine*, *338*(17), 1174–1180. http://dx.doi.org/10.1056/NEJM199804233381702.

Lambiase, A., Sacchetti, M., Mastropasqua, A., & Bonini, S. (2013). Corneal changes in neurosurgically induced neurotrophic keratitis. *JAMA Ophthalmology*, *131*(12), 1547–1553. http://dx.doi.org/10.1001/jamaophthalmol.2013.5064.

Lee, S. T., Chu, K., Jung, K. H., Yoon, H. J., Jeon, D., Kang, K. M., … Roh, J. K. (2010). MicroRNAs induced during ischemic preconditioning. *Stroke*, *41*(8), 1646–1651. http://dx.doi.org/10.1161/STROKEAHA.110.579649.

Lee, S. J., So, I. S., Park, S. Y., & Kim, I. S. (2008). Thymosin beta4 is involved in stabilin-2-mediated apoptotic cell engulfment. *FEBS Letters*, *582*(15), 2161–2166. http://dx.doi.org/10.1016/j.febslet.2008.03.058.

Legler, D. F., Micheau, O., Doucey, M. A., Tschopp, J., & Bron, C. (2003). Recruitment of TNF receptor 1 to lipid rafts is essential for TNFalpha-mediated NF-kappaB activation. *Immunity*, *18*(5), 655–664.

Lin, D., Halilovic, A., Yue, P., Bellner, L., Wang, K., Wang, L., & Zhang, C. (2013). Inhibition of miR-205 impairs the wound-healing process in human corneal epithelial cells by targeting KIR4.1 (KCNJ10). *Investigative Ophthalmology & Visual Science*, *54*(9), 6167–6178. http://dx.doi.org/10.1167/iovs.12-11577.

Lisi, S., Sisto, M., Scagliusi, P., Mitolo, V., & D'Amore, M. (2007). Siogren's syndrome: Anti-Ro and anti-La autoantibodies trigger apoptotic mechanism in the human salivary gland cell line, A-253. *Panminerva Medica*, *49*(3), 103–108.

Liu, X. S., Chopp, M., Zhang, R. L., Tao, T., Wang, X. L., Kassis, H., … Zhang, Z. G. (2011). MicroRNA profiling in subventricular zone after stroke: MiR-124a regulates proliferation of neural progenitor cells through Notch signaling pathway. *PLoS One*, *6*(8), e23461. http://dx.doi.org/10.1371/journal.pone.0023461.

Liu, S., Richards, S. M., Lo, K., Hatton, M., Fay, A., & Sullivan, D. A. (2011). Changes in gene expression in human meibomian gland dysfunction. *Investigative Ophthalmology & Visual Science*, *52*(5), 2727–2740. http://dx.doi.org/10.1167/iovs.10-6482.

Lu, L., Reinach, P. S., & Kao, W. W. (2001). Corneal epithelial wound healing. *Experimental Biology and Medicine (Maywood, NJ)*, *226*(7), 653–664.

Lv, S., Cheng, G., Xu, Y., Wang, Y., & Xu, G. (2011). Relationship between serum thymosin beta4 levels and coronary collateral development. *Coronary Artery Disease*, *22*(6), 401–404. http://dx.doi.org/10.1097/MCA.0b013e3283487d68.

Marfurt, C. F., Cox, J., Deek, S., & Dvorscak, L. (2010). Anatomy of the human corneal innervation. *Experimental Eye Research*, *90*(4), 478–492. http://dx.doi.org/10.1016/j.exer.2009.12.010.

Massingale, M. L., Li, X., Vallabhajosyula, M., Chen, D., Wei, Y., & Asbell, P. A. (2009). Analysis of inflammatory cytokines in the tears of dry eye patients. *Cornea*, *28*(9), 1023–1027. http://dx.doi.org/10.1097/ICO.0b013e3181a16578.

Matoba, A. Y., & McCulley, J. P. (1985). The effect of therapeutic soft contact lenses on antibiotic delivery to the cornea. *Ophthalmology*, *92*(1), 97–99.

Matsubara, M., Zieske, J. D., & Fini, M. E. (1991). Mechanism of basement membrane dissolution preceding corneal ulceration. *Investigative Ophthalmology & Visual Science*, *32*(13), 3221–3237.

Matsuda, S., & Koyasu, S. (2000). Mechanisms of action of cyclosporine. *Immunopharmacology*, *47*(2–3), 119–125.

McGinnigle, S., Naroo, S. A., & Eperjesi, F. (2012). Evaluation of dry eye. *Survey of Ophthalmology*, *57*(4), 293–316. http://dx.doi.org/10.1016/j.survophthal.2011.11.003.

Morris, D. C., Chopp, M., Zhang, L., Lu, M., & Zhang, Z. G. (2010). Thymosin beta4 improves functional neurological outcome in a rat model of embolic stroke. *Neuroscience*, *169*(2), 674–682. http://dx.doi.org/10.1016/j.neuroscience.2010.05.017.

Morris, D. C., Zhang, Z. G., Zhang, J., Xiong, Y., Zhang, L., & Chopp, M. (2012). Treatment of neurological injury with thymosin beta4. *The Annals of the New York Academy of Sciences*, *1269*, 110–116. http://dx.doi.org/10.1111/j.1749-6632.2012.06651.x.

Muller, L. J., Marfurt, C. F., Kruse, F., & Tervo, T. M. (2003). Corneal nerves: Structure, contents and function. *Experimental Eye Research*, *76*(5), 521–542.

O'Carroll, D., & Schaefer, A. (2013). General principals of miRNA biogenesis and regulation in the brain. *Neuropsychopharmacology*, *38*(1), 39–54. http://dx.doi.org/10.1038/npp.2012.87.

Ousler, G. W., 3rd, Abelson, M. B., Johnston, P. R., Rodriguez, J., Lane, K., & Smith, L. M. (2014). Blink patterns and lid-contact times in dry-eye and normal subjects. *Clinical Ophthalmology*, *8*, 869–874. http://dx.doi.org/10.2147/OPTH.S56783.

Pahl, H. L. (1999). Activators and target genes of Rel/NF-kappaB transcription factors. *Oncogene*, *18*(49), 6853–6866. http://dx.doi.org/10.1038/sj.onc.1203239.

Perkins, N. D. (2007). Integrating cell-signalling pathways with NF-kappaB and IKK function. *Nature Reviews. Molecular Cell Biology*, *8*(1), 49–62. http://dx.doi.org/10.1038/nrm2083.

Philp, D., Badamchian, M., Scheremeta, B., Nguyen, M., Goldstein, A. L., & Kleinman, H. K. (2003). Thymosin beta 4 and a synthetic peptide containing its actin-binding domain promote dermal wound repair in db/db diabetic mice and in aged mice. *Wound Repair and Regeneration, 11*(1), 19–24.

Philp, D., Huff, T., Gho, Y. S., Hannappel, E., & Kleinman, H. K. (2003). The actin binding site on thymosin beta4 promotes angiogenesis. *The FASEB Journal, 17*(14), 2103–2105. http://dx.doi.org/10.1096/fj.03-0121fje.

Philp, D., Scheremeta, B., Sibliss, K., Zhou, M., Fine, E. L., Nguyen, M., … Kleinman, H. K. (2006). Thymosin beta4 promotes matrix metalloproteinase expression during wound repair. *Journal of Cellular Physiology, 208*(1), 195–200. http://dx.doi.org/10.1002/jcp.20650.

Png, E., Samivelu, G. K., Yeo, S. H., Chew, J., Chaurasia, S. S., & Tong, L. (2011). Hyperosmolarity-mediated mitochondrial dysfunction requires Transglutaminase-2 in human corneal epithelial cells. *Journal of Cellular Physiology, 226*(3), 693–699. http://dx.doi.org/10.1002/jcp.22389.

Podos, S. M., Becker, B., Asseff, C., & Hartstein, J. (1972). Pilocarpine therapy with soft contact lenses. *American Journal of Ophthalmology, 73*(3), 336–341.

Qiu, P., Kurpakus-Wheater, M., & Sosne, G. (2007). Matrix metalloproteinase activity is necessary for thymosin beta 4 promotion of epithelial cell migration. *Journal of Cellular Physiology, 212*(1), 165–173. http://dx.doi.org/10.1002/jcp.21012.

Qiu, P., Wheater, M. K., Qiu, Y., & Sosne, G. (2011). Thymosin beta4 inhibits TNF-alpha-induced NF-kappaB activation, IL-8 expression, and the sensitizing effects by its partners PINCH-1 and ILK. *The FASEB Journal, 25*(6), 1815–1826. http://dx.doi.org/10.1096/fj.10-167940.

Rink, C., & Khanna, S. (2011). MicroRNA in ischemic stroke etiology and pathology. *Physiological Genomics, 43*(10), 521–528. http://dx.doi.org/10.1152/physiolgenomics.00158.2010.

Rousselle, P., & Beck, K. (2013). Laminin 332 processing impacts cellular behavior. *Cell Adhesion & Migration*, 7(1), 122–134. http://dx.doi.org/10.4161/cam.23132.

Roy, S., & Sen, C. K. (2011). MiRNA in innate immune responses: Novel players in wound inflammation. *Physiological Genomics, 43*(10), 557–565. http://dx.doi.org/10.1152/physiolgenomics.00160.2010.

Ruff, D., Crockford, D., Girardi, G., & Zhang, Y. (2010). A randomized, placebo-controlled, single and multiple dose study of intravenous thymosin beta4 in healthy volunteers. *The Annals of the New York Academy of Sciences, 1194*, 223–229. http://dx.doi.org/10.1111/j.1749-6632.2010.05474.x.

Sacchetti, M., Mantelli, F., Marenco, M., Macchi, I., Ambrosio, O., & Rama, P. (2015). Diagnosis and Management of Iridocorneal Endothelial Syndrome. *Biomed Research International, 2015*, 763093. http://dx.doi.org/10.1155/2015/763093.

Safer, D., Elzinga, M., & Nachmias, V. T. (1991). Thymosin beta 4 and Fx, an actin-sequestering peptide, are indistinguishable. *The Journal of Biological Chemistry, 266*(7), 4029–4032.

Sall, K., Stevenson, O. D., Mundorf, T. K., & Reis, B. L. (2000). Two multicenter, randomized studies of the efficacy and safety of cyclosporine ophthalmic emulsion in moderate to severe dry eye disease. CsA Phase 3 Study Group. *Ophthalmology, 107*(4), 631–639.

Santra, M., Chopp, M., Zhang, Z. G., Lu, M., Santra, S., Nalani, A., … Morris, D. C. (2012). Thymosin beta 4 mediates oligodendrocyte differentiation by upregulating p38 MAPK. *Glia, 60*(12), 1826–1838. http://dx.doi.org/10.1002/glia.22400.

Semeraro, F., Forbice, E., Romano, V., Angi, M., Romano, M. R., Filippelli, M. E., … Costagliola, C. (2014). Neurotrophic keratitis. *Ophthalmologica, 231*(4), 191–197. http://dx.doi.org/10.1159/000354380.

Sharma, N., Goel, M., Velpandian, T., Titiyal, J. S., Tandon, R., & Vajpayee, R. B. (2011). Evaluation of umbilical cord serum therapy in acute ocular chemical burns. *Investigative*

Ophthalmology & Visual Science, *52*(2), 1087–1092. http://dx.doi.org/10.1167/iovs.09-4170.
Shelton, E. L., Poole, S. D., Reese, J., & Bader, D. M. (2013). Omental grafting: A cell-based therapy for blood vessel repair. *Journal of Tissue Engineering and Regenerative Medicine*, 7(6), 421–433. http://dx.doi.org/10.1002/term.528.
Sivak, J. M., Mohan, R., Rinehart, W. B., Xu, P. X., Maas, R. L., & Fini, M. E. (2000). Pax-6 expression and activity are induced in the reepithelializing cornea and control activity of the transcriptional promoter for matrix metalloproteinase gelatinase B. *Developmental Biology*, *222*(1), 41–54.
Smart, N., Rossdeutsch, A., & Riley, P. R. (2007). Thymosin beta4 and angiogenesis: Modes of action and therapeutic potential. *Angiogenesis*, *10*(4), 229–241. http://dx.doi.org/10.1007/s10456-007-9077-x.
Somanath, P. R., Razorenova, O. V., Chen, J., & Byzova, T. V. (2006). Akt1 in endothelial cell and angiogenesis. *Cell Cycle*, *5*(5), 512–518.
Sopko, N., Qin, Y., Finan, A., Dadabayev, A., Chigurupati, S., Qin, J., … Gupta, S. (2011). Significance of thymosin beta4 and implication of PINCH-1-ILK-alpha-parvin (PIP) complex in human dilated cardiomyopathy. *PLoS One*, *6*(5). e20184. http://dx.doi.org/10.1371/journal.pone.0020184.
Sosne, G., Chan, C. C., Thai, K., Kennedy, M., Szliter, E. A., Hazlett, L. D., & Kleinman, H. K. (2001). Thymosin beta 4 promotes corneal wound healing and modulates inflammatory mediators in vivo. *Experimental Eye Research*, *72*(5), 605–608. http://dx.doi.org/10.1006/exer.2000.0985.
Sosne, G., Dunn, S. P., & Kim, C. (2015). Thymosin beta4 significantly improves signs and symptoms of severe dry eye in a phase 2 randomized trial. *Cornea*, *34*(5), 491–496. http://dx.doi.org/10.1097/ICO.0000000000000379.
Sosne, G., Hafeez, S., Greenberry, A. L., 2nd, & Kurpakus-Wheater, M. (2002). Thymosin beta4 promotes human conjunctival epithelial cell migration. *Current Eye Research*, *24*(4), 268–273.
Sosne, G., Kim, C., & Kleinman, H. K. (2015). Thymosin beta4 significantly reduces the signs of dryness in a murine controlled adverse environment model of experimental dry eye. *Expert Opinion on Biological Therapy*, *15*(Suppl. 1), S155–S161. http://dx.doi.org/10.1517/14712598.2015.1019858.
Sosne, G., & Ousler, G. W. (2015). Thymosin beta 4 ophthalmic solution for dry eye: A randomized, placebo-controlled, Phase II clinical trial conducted using the controlled adverse environment (CAE) model. *Clinical Ophthalmology*, *9*, 877–884. http://dx.doi.org/10.2147/OPTH.S80954.
Sosne, G., Qiu, P., Christopherson, P. L., & Wheater, M. K. (2007). Thymosin beta 4 suppression of corneal NFkappaB: A potential anti-inflammatory pathway. *Experimental Eye Research*, *84*(4), 663–669. http://dx.doi.org/10.1016/j.exer.2006.12.004.
Sosne, G., Qiu, P., Goldstein, A. L., & Wheater, M. (2010a). Biological activities of thymosin beta4 defined by active sites in short peptide sequences. *The FASEB Journal*, *24*(7), 2144–2151. http://dx.doi.org/10.1096/fj.09-142307.
Sosne, G., Qiu, P., Kurpakus-Wheater, M., & Matthew, H. (2010b). Thymosin beta4 and corneal wound healing: Visions of the future. *The Annals of the New York Academy of Sciences*, *1194*, 190–198. http://dx.doi.org/10.1111/j.1749-6632.2010.05472.x.
Sosne, G., Siddiqi, A., & Kurpakus-Wheater, M. (2004). Thymosin-beta4 inhibits corneal epithelial cell apoptosis after ethanol exposure in vitro. *Investigative Ophthalmology & Visual Science*, *45*(4), 1095–1100.
Sosne, G., Szliter, E. A., Barrett, R., Kernacki, K. A., Kleinman, H., & Hazlett, L. D. (2002). Thymosin beta 4 promotes corneal wound healing and decreases inflammation in vivo following alkali injury. *Experimental Eye Research*, *74*(2), 293–299. http://dx.doi.org/10.1006/exer.2001.1125.

Sosne, G., Xu, L., Prach, L., Mrock, L. K., Kleinman, H. K., Letterio, J. J., ... Kurpakus-Wheater, M. (2004). Thymosin beta 4 stimulates laminin-5 production independent of TGF-beta. *Experimental Cell Research*, *293*(1), 175–183.

Tashiro, A., Okamoto, K., Chang, Z., & Bereiter, D. A. (2010). Behavioral and neurophysiological correlates of nociception in an animal model of photokeratitis. *Neuroscience*, *169*(1), 455–462. http://dx.doi.org/10.1016/j.neuroscience.2010.04.034.

Tseng, S. C., Prabhasawat, P., & Lee, S. H. (1997). Amniotic membrane transplantation for conjunctival surface reconstruction. *American Journal of Ophthalmology*, *124*(6), 765–774. PMID: 9402822.

Tseng, S. C. G., & Tsubota, K. (1997). Important concepts for treating ocular surface and tear disorders. *American Journal of Ophthalmology*, *124*(6), 825–835.

Tsubota, K., Goto, E., Shimmura, S., & Shimazaki, J. (1999). Treatment of persistent corneal epithelial defect by autologous serum application. *Ophthalmology*, *106*(10), 1984–1989. http://dx.doi.org/10.1016/S0161-6420(99)90412-8.

Vajpayee, R. B., Mukerji, N., Tandon, R., Sharma, N., Pandey, R. M., Biswas, N. R., ... Melki, S. A. (2003). Evaluation of umbilical cord serum therapy for persistent corneal epithelial defects. *The British Journal of Ophthalmology*, *87*(11), 1312–1316.

Vartiainen, N., Pyykonen, I., Hokfelt, T., & Koistinaho, J. (1996). Induction of thymosin beta(4) mRNA following focal brain ischemia. *Neuroreport*, 7(10), 1613–1616.

Wang, C., Ji, B., Cheng, B., Chen, J., & Bai, B. (2014). Neuroprotection of microRNA in neurological disorders (Review). *Biomedical Reports*, *2*(5), 611–619. http://dx.doi.org/10.3892/br.2014.297.

Xiong, Y., Zhang, Y., Mahmood, A., Meng, Y., Zhang, Z. G., Morris, D. C., & Chopp, M. (2012). Neuroprotective and neurorestorative effects of thymosin beta4 treatment initiated 6 hours after traumatic brain injury in rats. *Journal of Neurosurgery*, *116*(5), 1081–1092. http://dx.doi.org/10.3171/2012.1.JNS111729.

Yamamoto, N., Yamamoto, N., Petroll, M. W., Cavanagh, H. D., & Jester, J. V. (2005). Internalization of Pseudomonas aeruginosa is mediated by lipid rafts in contact lens-wearing rabbit and cultured human corneal epithelial cells. *Investigative Ophthalmology & Visual Science*, *46*(4), 1348–1355. http://dx.doi.org/10.1167/iovs.04-0542.

Young, A. L., Cheng, A. C., Ng, H. K., Cheng, L. L., Leung, G. Y., & Lam, D. S. (2004). The use of autologous serum tears in persistent corneal epithelial defects. *Eye (London, England)*, *18*(6), 609–614. http://dx.doi.org/10.1038/sj.eye.6700721.

Yu, J., Peng, H., Ruan, Q., Fatima, A., Getsios, S., & Lavker, R. M. (2010). MicroRNA-205 promotes keratinocyte migration via the lipid phosphatase SHIP2. *The FASEB Journal*, *24*(10), 3950–3959. http://dx.doi.org/10.1096/fj.10-157404.

Zander, E., & Weddell, G. (1951). Observations on the innervation of the cornea. *Journal of Anatomy*, *85*(1), 68–99.

Zhang, H. M., Keledjian, K. M., Rao, J. N., Zou, T., Liu, L., Marasa, B. S., ... Wang, J. Y. (2006). Induced focal adhesion kinase expression suppresses apoptosis by activating NF-kappaB signaling in intestinal epithelial cells. *American Journal of Physiology. Cell Physiology*, *290*(5), C1310–C1320. http://dx.doi.org/10.1152/ajpcell.00450.2005.

Zhang, Y., Ueno, Y., Liu, X. S., Buller, B., Wang, X., Chopp, M., & Zhang, Z. G. (2013). The MicroRNA-17-92 cluster enhances axonal outgrowth in embryonic cortical neurons. *The Journal of Neuroscience*, *33*(16), 6885–6894. http://dx.doi.org/10.1523/JNEUROSCI.5180-12.2013.

Zhang, J., Zhang, Z. G., Morris, D., Li, Y., Roberts, C., Elias, S. B., & Chopp, M. (2009). Neurological functional recovery after thymosin beta4 treatment in mice with experimental auto encephalomyelitis. *Neuroscience*, *164*(4), 1887–1893. http://dx.doi.org/10.1016/j.neuroscience.2009.09.054.

Zoukhri, D., Hodges, R. R., Byon, D., & Kublin, C. L. (2002). Role of proinflammatory cytokines in the impaired lacrimation associated with autoimmune xerophthalmia. *Investigative Ophthalmology & Visual Science*, *43*(5), 1429–1436.

INDEX

Note: Page numbers followed by "*f*" indicate figures, and "*t*" indicate tables.

Q

R

S

T

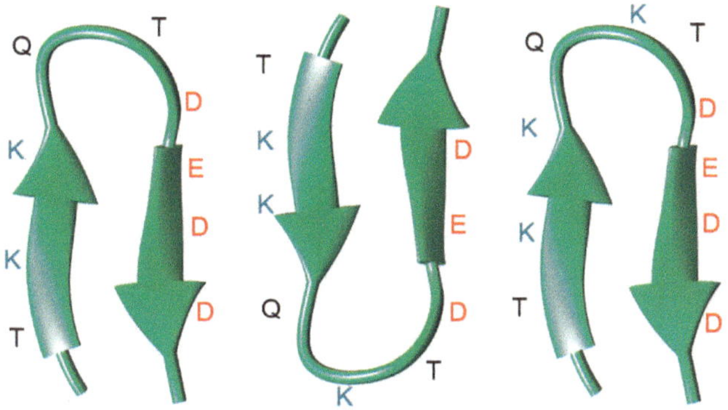

K. Hoch and D.E. Volk, Fig. 1 One of the models proposed for the nuclear localization signal after proteolitic cleavage from thymosin α1 (Skopeliti et al., 2006) to account for CD measurements most consistent with an antiparallel β-pleated sheet. An alternating pattern of linear extended beta structures would also neutralize the charges. Images were created using Chimera (Petterson et al., 2004).

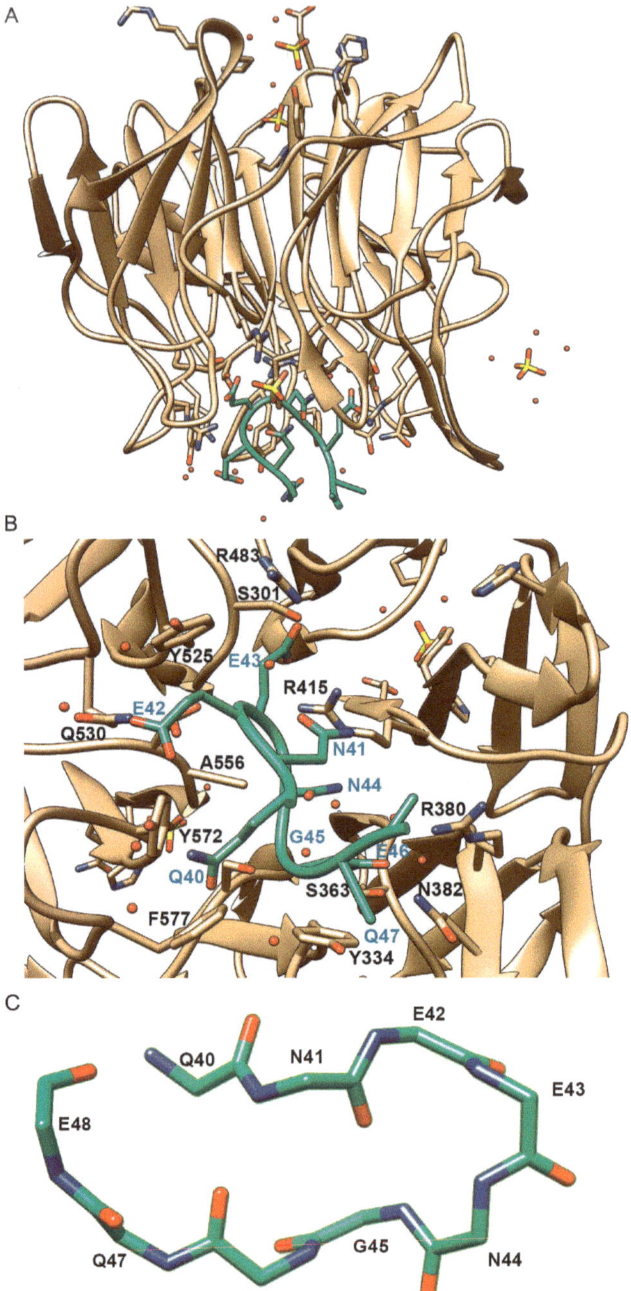

K. Hoch and D.E. Volk, Fig. 2 (A) Crystal structure of ProTα bound to the Kelch domain of Keap1. (B) Structure of ProTα when bound to the Kelch domain. (C) Structure of the ProTα hairpin that interacts with the Kelch domain. Atomic coordinates were obtained from the Protein Data Bank (Berman et al., 2000; PDB ID 2Z32; Padmanabhan et al., 2008).

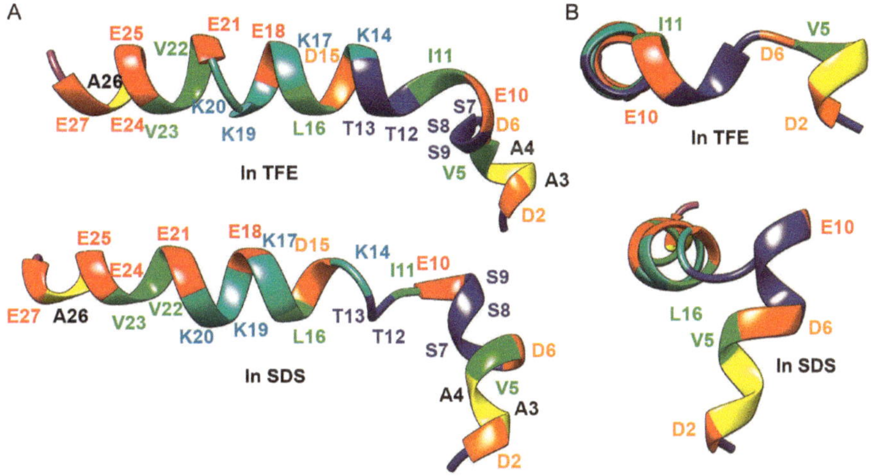

K. Hoch and D.E. Volk, Fig. 3 (A) NMR-derived structures of thymosin α1 protein in 40% TFE solution and in a membrane-like environment with SDS micelles. (B) Same structures rotated 90 degrees. Atomic coordinates were obtained from the Protein Data Bank for thymosin α1 in 40%TFE (PDB ID 2L9I; Elizondo-Riojas et al., 2011) and in SDS (PDB ID 2MNQ; Nepravishta et al., 2015). Images were created using Chimera (Petterson et al., 2004).

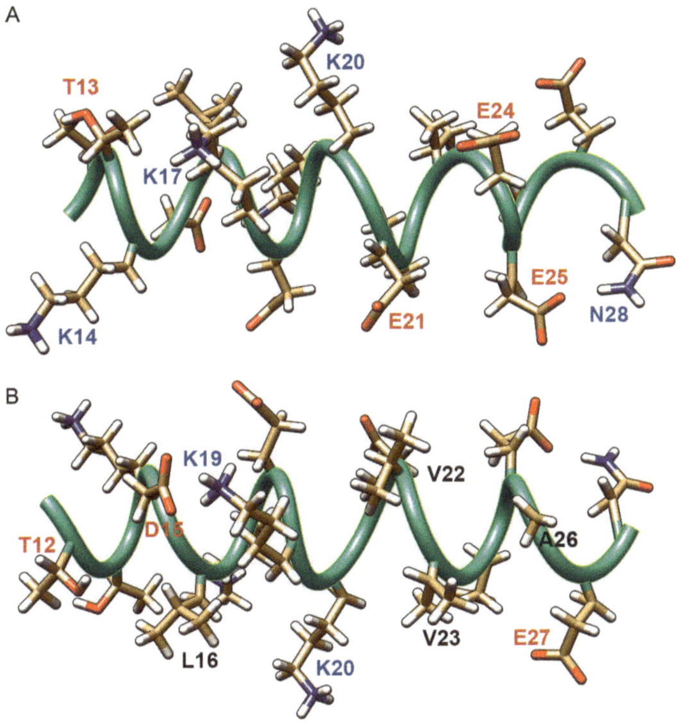

K. Hoch and D.E. Volk, Fig. 4 *Ribbon* and *stick* diagrams of the C-terminal helix of thymosin α1 spanning residues T12–N28. Aliphatic residues are indicated by *black* labels, while positive and negative side chains are indicated by *blue* or *red* labels, respectively. (A) Looking down on the hydrophilic side of the helix. (B) Looking down at the opposite side of the helix. Images were created using Chimera (Petterson et al., 2004), and coordinates were obtained from the Protein Data Bank (PDB ID 2L9I; Elizondo-Riojas et al., 2011).

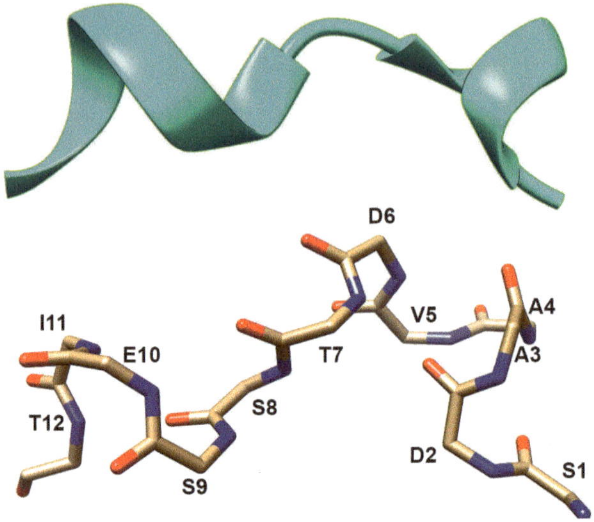

K. Hoch and D.E. Volk, Fig. 5 The N-terminal domain of thymosin α1 in 40% TFE. The N-terminal domain of Tα1 in TFE contains two double beta-turns and nearly forms a complete helix. However, an unusual kink along residues 6 and 7 disrupts the helical structure. Atomic coordinates were obtained from the Protein Data Bank (PDB ID 2L9I; Elizondo-Riojas et al., 2011).

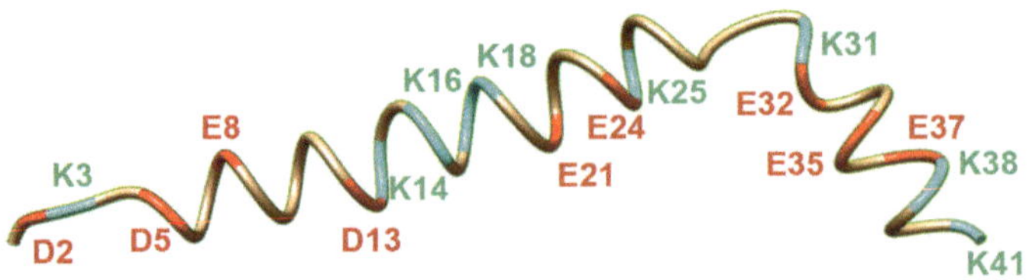

K. Hoch and D.E. Volk, Fig. 6 NMR-derived structure of thymosin β9 in hexafluoropropanol. Atomic coordinates were obtained from the Protein Data Bank (PDB ID 1HJ0; Stoll et al., 1997).

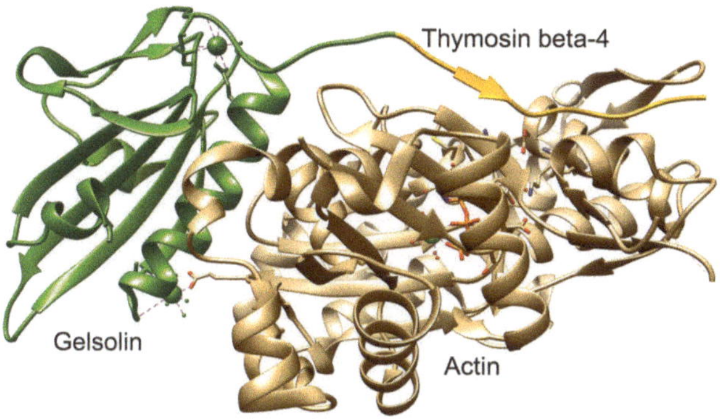

K. Hoch and D.E. Volk, Fig. 7 Crystal structure of alpha actin bound to a gelsolin–thymosin beta-4 C-terminal helix chimera. The helix is the minus-end capping helix, and it competes with DNAse I binding at this site. Atomic coordinates were obtained from the Protein Data Bank (PDB ID 1T44; Irobi et al., 2004).

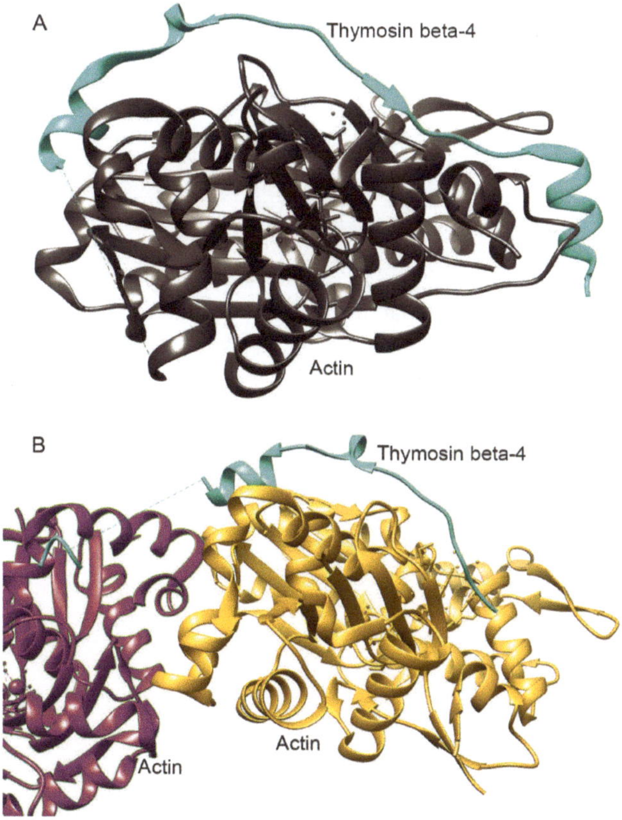

K. Hoch and D.E. Volk, Fig. 8 (A) Crystal structure of an actin thymosin beta-4 hybrid (PDB ID 4PL7). (B) Crystal structure (PDB ID 4PL8) of actin bound to a peptide construct containing tandem thymosin β4 units. The two actin proteins are shown in *purple* or *goldenrod*, and the tandem thymosin β4 peptide is colored *cyan*. While much of the second thymosin β4 unit (D5–E24) was detected binding to the actin unit on the *right*, the final C-terminal helix is disrupted.

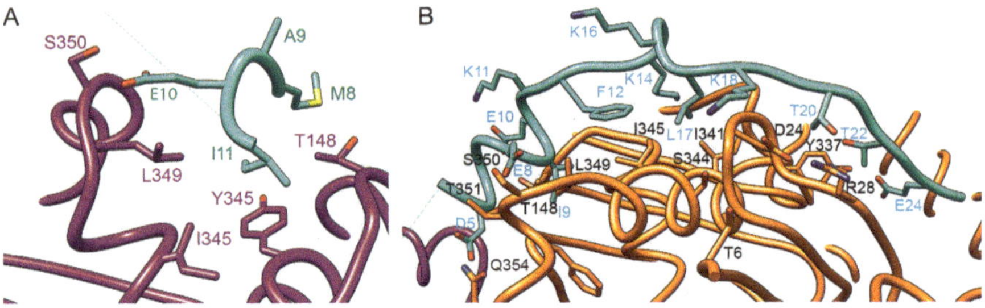

K. Hoch and D.E. Volk, Fig. 9 Expanded views of the actin dimer bound to a peptide containing tandem thymosin β4 units. (A) The interactions of the first thymosin β4 unit. (B) The interactions of the second thymosin β4 unit. The peptides have individually been renumbered to match the thymosin β4 sequence for convenience. Atomic coordinates were obtained from the Protein Data Bank (PDB ID 4PL8; Xue et al., 2014).

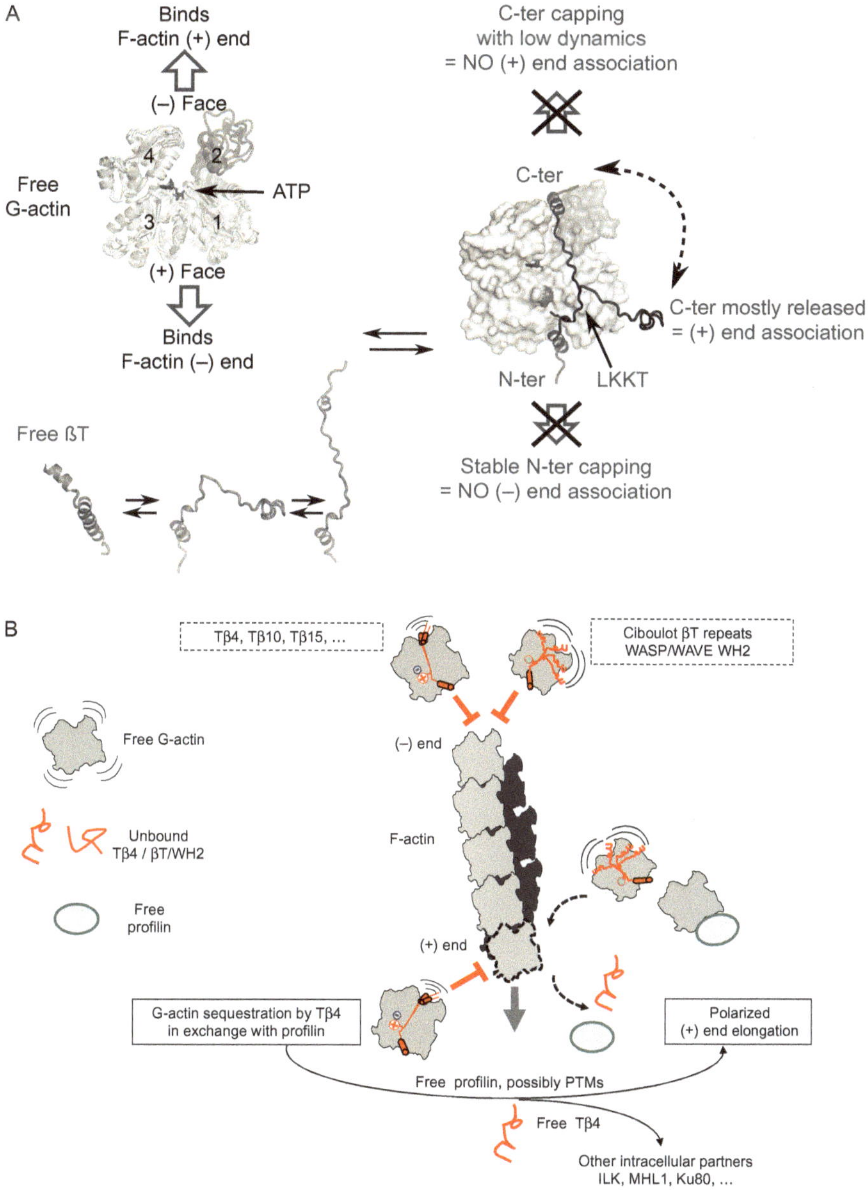

L. Renault, Fig. 4 Structural basis for the regulations of G-actin by Tβ4 and other βT/WH2 domains in actin assembly. (A) Conformational variability of free G-actin, and free and actin-bound βT. Actin subdomains 1 and 3 (G-actin barbed face), and 2 and 4 (G-actin pointed face) are indicated on free G-actin. Its intrinsic conformational flexibility is shown by four overlaid G-actin conformations (PDB: 1HLU, 2VCP, 2PDB or SAXS model (Didry et al., 2012)), and the flexibility of free βT by the solution structure of free

(Continued)

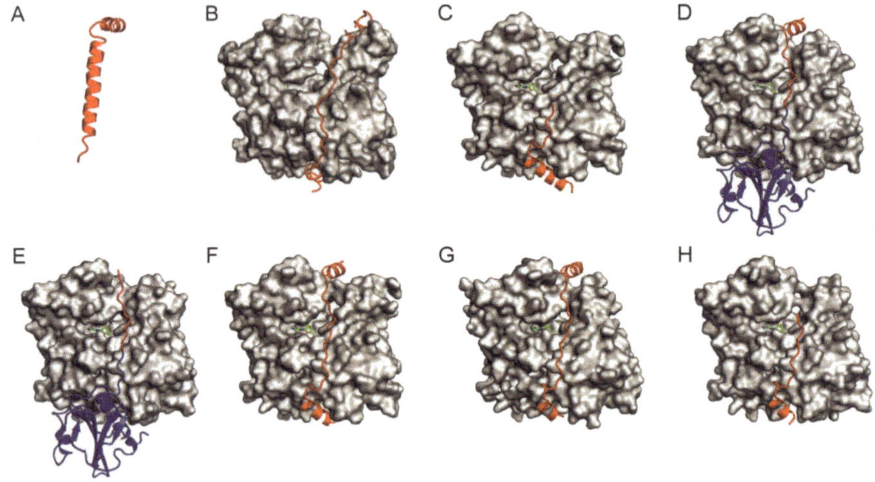

B. Xue and R.C. Robinson, Fig. 1 Structures of beta-thymosins. The structures were solved by X-ray crystallography unless stated otherwise. Beta-thymosins are shown as *red* cartoons; actin is shown as a grey surface, with its bound nucleotide shown as *green* sticks; gelsolin domain 1 (G1) in two of the structures is shown as a *blue* cartoon. PDB IDs are given following the abbreviations of the structures, whose meanings can be found in the main text. (A) Solution NMR structure of Tβ10 (1HJ0). (B) NMR-based model of Tβ4:actin (1UY5). (C) Cib1N:actin (1SQK). (D) G1-Tβ4:actin (1T44). (E) G1-Cib23:actin (2FF6). (F) Structure-based model of Tβ4:actin. (G) Tβ4FL:actin (4PL7). (H) Tβ4N:actin (4PL8). This and the subsequent structural figures were prepared with the PyMOL Molecular Graphics System (DeLano Scientific LLC).

Fig. 4—Cont'd Tβ9, which adopts a mostly helical folding in 40% 1,1,1,3,3,3-hexafluoro-2-propanol (HFP) (Stoll et al., 1997), and different SAXS conformations of bound CibD1 βT, shown here without G-actin (Didry et al., 2012). The right panel overlay two full SAXS models of G-actin bound to CibD1 βT obtained either at low (sequestering function like Tβ4) or near physiological (profilin-like function) ionic strength (Didry et al., 2012). Tβ4 adopts the same sequestering conformation with G-actin (Didry et al., 2012; Xue et al., 2014). (B) βT/WH2 domains interact with actin monomers to either sequester G-actin, or promote specifically F-actin barbed-end elongation like profilin. They display always a stable N-terminal binding and capping over G-actin barbed face that prevents G-actin association at (−) ends. In contrast, their C-terminal interactions over G-actin pointed face display different dynamics between the different sequences of βT/WH2 domains: with low dynamics like in Tβ4 they sequester, with high dynamics in most other βT/WH2 they allow association at (+) ends and thus may control different (+) end growth rates (Didry et al., 2012). The sequestration of G-actin by Tβ4 is in rapid exchange with free profilin. Other intracellular interactions and regulations may be thus stimulated by massive and spatially restricted profilin-dependent dissociation of G-actin:Tβ4, or possibly by PTMs that may also switch or weaken the interactions with G-actin.

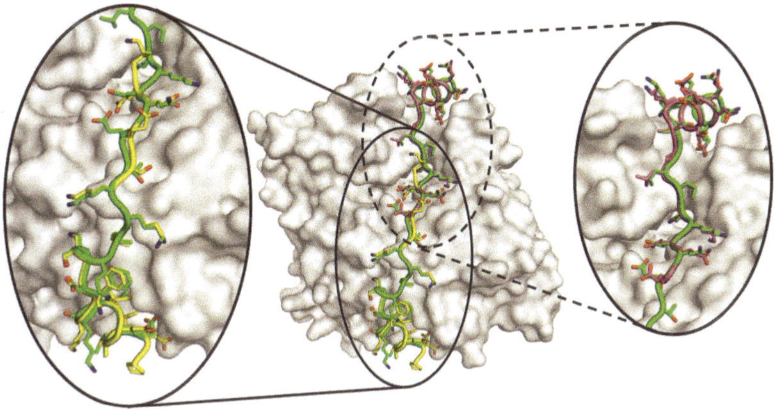

B. Xue and R.C. Robinson, Fig. 2 Superimposition of the three known Tβ4:actin structures. Tβ4s in Tβ4FL:actin, Tβ4N:actin, and G1-Tβ4:actin are colored in *green*, *yellow*, and *violet*, respectively. Actin from Tβ4FL:actin is shown as a gray surface.

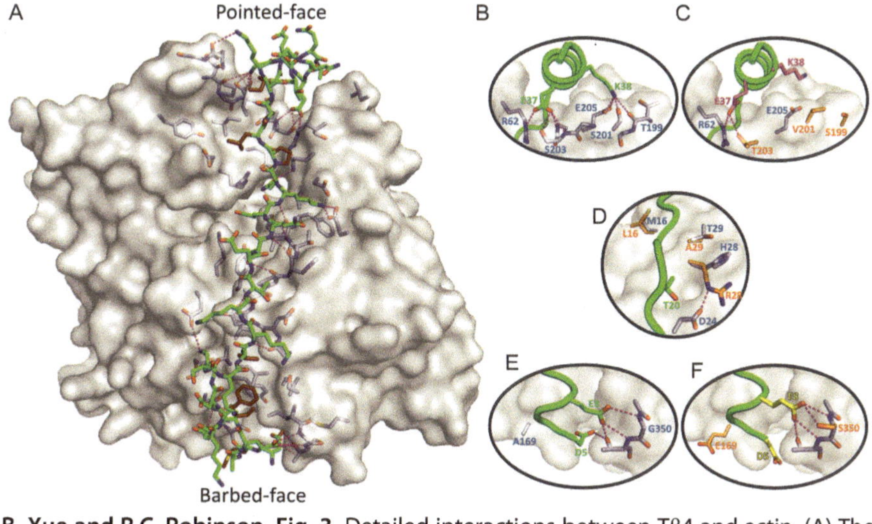

B. Xue and R.C. Robinson, Fig. 3 Detailed interactions between Tβ4 and actin. (A) The Tβ4:actin interface based on the Tβ4FL:actin structure. Tβ4 is shown as *green* sticks except for its buried hydrophobic residues, which are colored in *chocolate*. Actin is shown as a gray surface, with its interface residues shown as *light blue* sticks. *Dashed lines* in *magenta* represent hydrogen bonds and/or salt bridges. (B–E) Differences between the *Pichia* and rabbit skeletal muscle isoforms on the pointed-face (B vs C), side (D), and the barbed-face (E vs F) of actin. Tβ4 residues from Tβ4FL, Tβ4N, and G1-Tβ4 are colored in *green*, *yellow*, and *violet*, respectively. Nonidentical actin residues from the *Pichia* and the muscle isoforms are colored in *light blue* and *orange*, respectively.

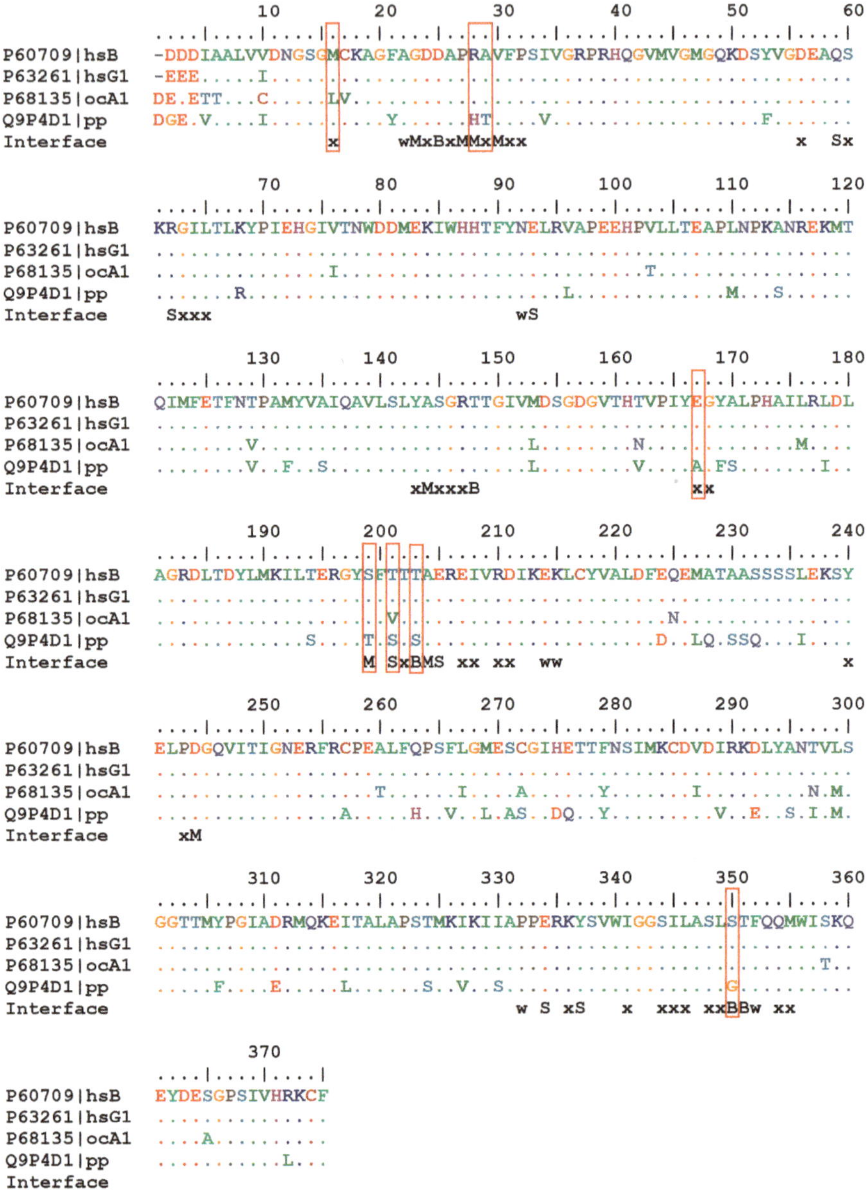

B. Xue and R.C. Robinson, Fig. 4 Sequence alignment of actin isoforms. hsB and hsG1: *Homo sapiens* (human) β- and γ1-actins; ocA1: *Oryctolagus cuniculus* (rabbit) skeletal muscle actin; pp: *Pichia pastoris* (yeast) actin. Residues lying on the Tβ4:actin interface are generally denoted by an (x), except for those with less than 1.6 $Å^2$ in buried surface area (w) according to PISA (Krissinel & Henrick, 2007) and for those that are involved in hydrogen bonding with Tβ4 through their main chain (M), side chain (S), or both (B). *Red boxes* highlight nonidentical interface residues.

A

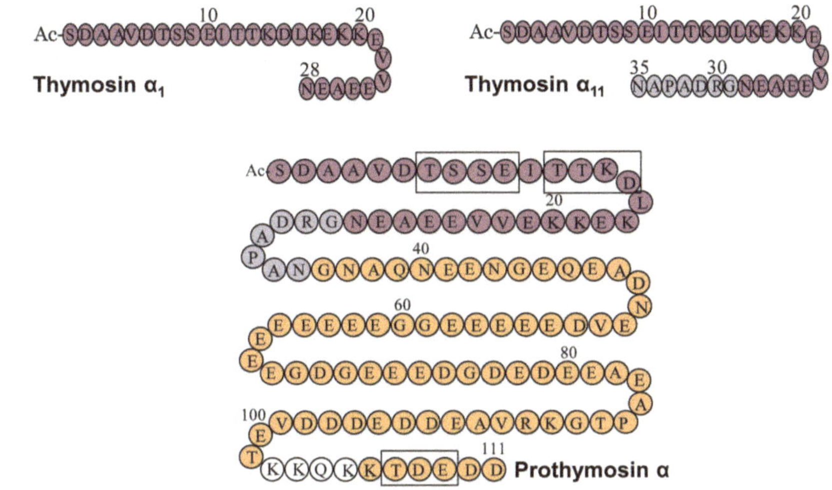

B

```
                         20                  40
Human: SDAAVDTSSEITTKDLKEKKEVVEEAENGRDAPANGNA-NEENGEQEADNEVDEEEEEG: 58
Rat  : SDAAVDTSSEITTKDLKEKKEVVEEAENGRDAPANGNAQNEENGEQEADNEVDEEEEEG: 59
Mouse: SDAAVDTSSEITTKDLKEKKEVVEEAENGRDAPANGNAQNEENGEQEADNEVDEEEEEG: 59
Calf : SDAAVDTSSEITTKDLKEKKEVVEEAENGREAPANGNA-NEENGEQEADNEVDEEEEEG: 58
Pig  : SDAAVDTSSEITTKDLKXXKEVVEEAENGREAPANGNA-NEENGEQEADNEVDEEEEEG: 58
Goat : SBAAVBTSSZITTKBLKEKKEVVEEAENGREAPANGNA-NEENGEQEABBZVBZZZZZG: 58

       60                  80                  100
Human: GEEEEEEEEGDGEEEDGDEDEEAESATGKRAAEDDEDDDVDTKKQK-TDEDD: 109
Rat  : GEEEEEEEEGDGEEEDGDEDEEAEAPTGKRVAEDDEDDDVETKKQKKTDEDD: 111
Mouse: GEEEEEEEEGDGEEEDGDEDEEAEAPTGKRVAEDDEDDDVDTKKQK-TEEDD: 110
Calf : GEEEEEEEEGDGEEEDGDEDEEAEAATGKRAAEDDEDDDVDTKKQK-TDEDD: 109
Pig  : GEEEEEEEEGDGEEEDGDEDEEAEAATGKRAAEDDEDDDVDTKKQK-TDEDD: 109
Goat : GZZZZZZZZGBGZZZBGBZBZZAZAPTGKRAAEDDEDDDVDTKKQK-TDEDD: 109
```

M. Freire *et al.*, Fig. 1 (A) Amino acid sequence of thymosin α_1 (Goldstein et al., 1977), thymosin α_{11} (Caldarella et al., 1983), and prothymosin α (Haritos, Blacher, Stein, Caldarella, & Horecker, 1985). *White circles* correspond to the NLS and *boxes* contain the CK-2 phosphorylation sites. (B) Alignment of the amino acid sequences of prothymosin α from human spleen (Goodall, Domínguez, & Horecker, 1986), rat spleen (Frangou-Lazaridis, Clinton, Goodall, & Horecker, 1988), mouse Ehrlich ascites cells (Schmidt & Werner, 1991), bovine thymus (Panneerselvam, Wellner, & Horecker, 1988), pig intestine (Wintero, Fredholm, & Davies, 1996), and goat spleen (Frillingos et al., 1991).

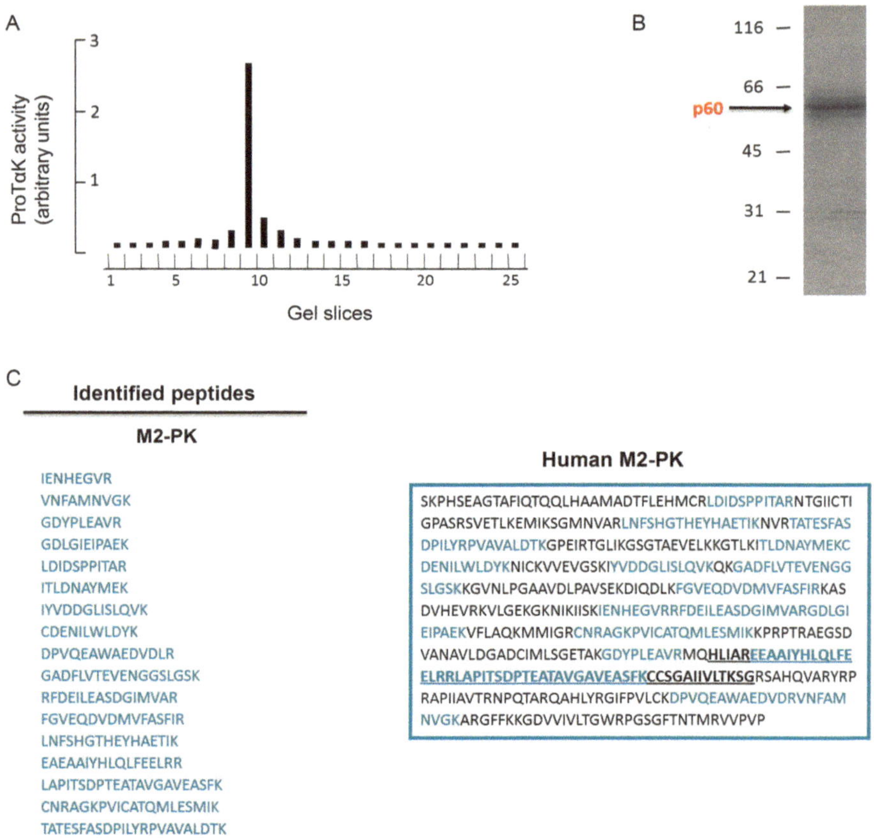

M. Freire *et al.*, Fig. 5 Structural characterization of the protein with ProTαK activity obtained from NC37 cells. (A) ProTαK activity in the slices from the renatured SDS-PAGE gels in which chromatographically purified ProTαK was separated. (B) SDS-PAGE analysis of the components in the slice with maximal ProTαK activity. (C) Peptides identified in the mass spectrometry analysis of the p60 protein in (B). The identified peptides are displayed in the human M2 sequence showed in the *inset*, and the 56 amino acids from exon 10, which differentiate the M2 from the M1 isoenzyme, are *underlined*. *Díaz-Jullien, C., Moreira, D., Sarandeses, C. S., Covelo, G., Barbeito, P., & Freire, M. (2011). The M2-type isoenzyme of pyruvate kinase phosphorylates prothymosin α in proliferating lymphocytes.* Biochimica et Biophysica Acta, 1814, *355–365. Copyright (2011) Elsevier.*

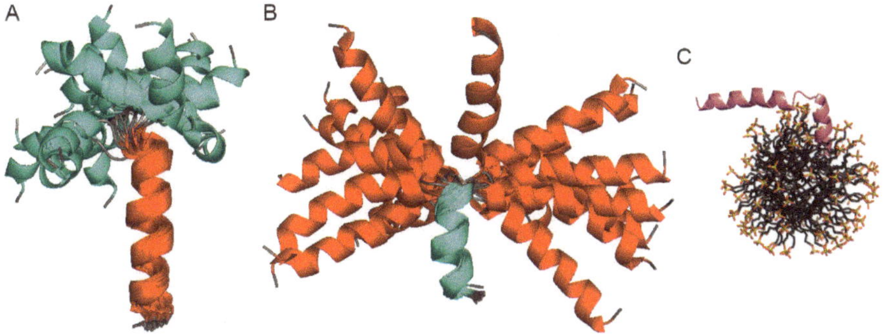

R. Nepravishta *et al.*, Fig. 1 (A, B, and C) Thymosinα1 structured upon binding to SDS micelles: (A) the Thymosinα1 in SDS micelles aligned with the N-terminus tract, (B) the same as in (A) aligning the C-terminus tract, and (C) representing its insertion on the SDS micelle.

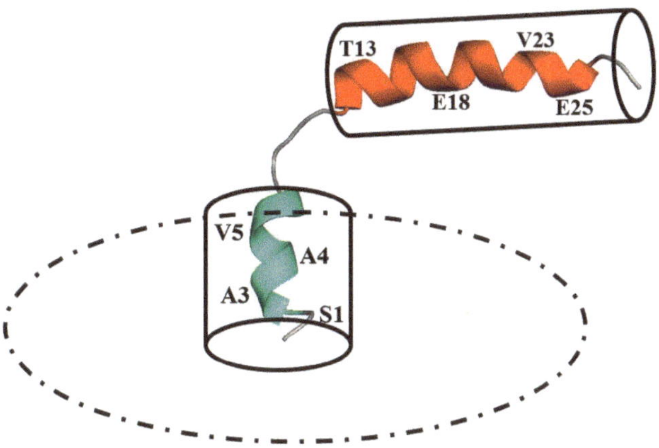

R. Nepravishta *et al.*, Fig. 2 Thymosinα1 inserted in mixed micelles of DPC and perdeuterated SDS.

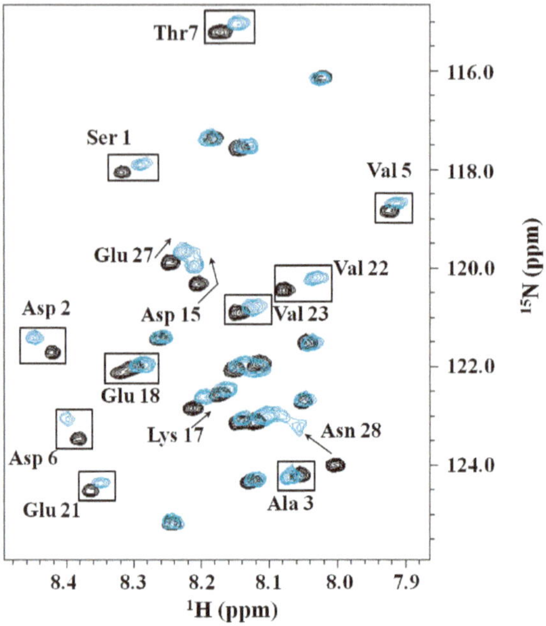

R. Nepravishta *et al.*, Fig. 4 Assignment of the resonances of natural abundance ^{15}N HSQC NMR of Thymosinα1 in the presence of 20 mM of PC–PS (4:1) (*black*) and in the presence of increasing PC/PS 4:1 *gray*. *Black* PC–PS 5 mM; *gray* PC–PS 20 mM.

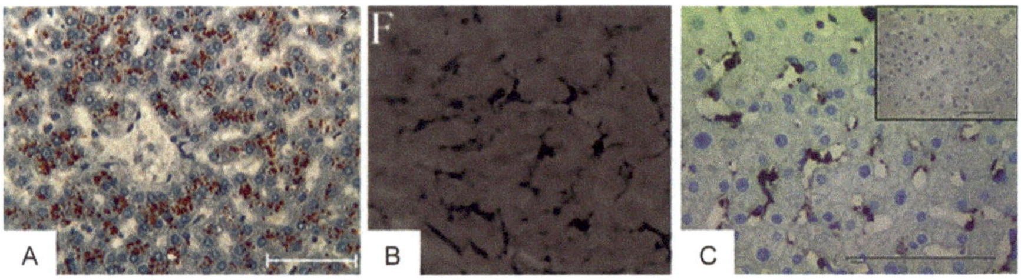

J. Kim and Y. Jung, Fig. 1 Immunohistochemistry analysis of Tβ4 in the liver. (A) A Tβ4-immunoreactivity was observed in the cytoplasm of hepatocyte in healthy human liver. Scale bars represent 400 μm. (B) Kupffer-like cells are positive for Tβ4 in healthy liver of mice. These cells have small cell bodies with long branched processes. (C) Hepatic expression of Tβ4 was observed in Kupffer cells in liver of HCV-infected patients. The inserted image shows a negative control omitting primary antibody. Scale bars represent 100 μm. *Panel A: Figure from Nemolato, S., Van Eyken, P., Cabras, T., Cau, F., Fanari, M., Locci, A., …, Castagnola, M. (2011). Expression pattern of thymosin beta 4 in the adult human liver.* European Journal of Histochemistry, 55*(3), e25, PagePress, Italy, ©2011 Nemolato et al. Panel B: Figure from Paulussen, M., Landuyt, B., Schoofs, L., Luyten, W., & Arckens, L. (2009). Thymosin beta 4 mRNA and peptide expression in phagocytic cells of different mouse tissues.* Peptides, 30*(10), 1822–1832 ©2009 Elsevier Inc. Panel C: Figure from Henkel, C., Schwamborn, K., Zimmermann, H. W., Tacke, F., Kuhnen, E., Odenthal, M., …, Weiskirchen, R. (2011). From proteomic multimarker profiling to interesting proteins: Thymosin-beta(4) and kininogen-1 as new potential biomarkers for inflammatory hepatic lesions.* Journal of Cellular and Molecular Medicine, 15*(10), 2176–2188, ©2011 John Wiley and Sons Inc.*

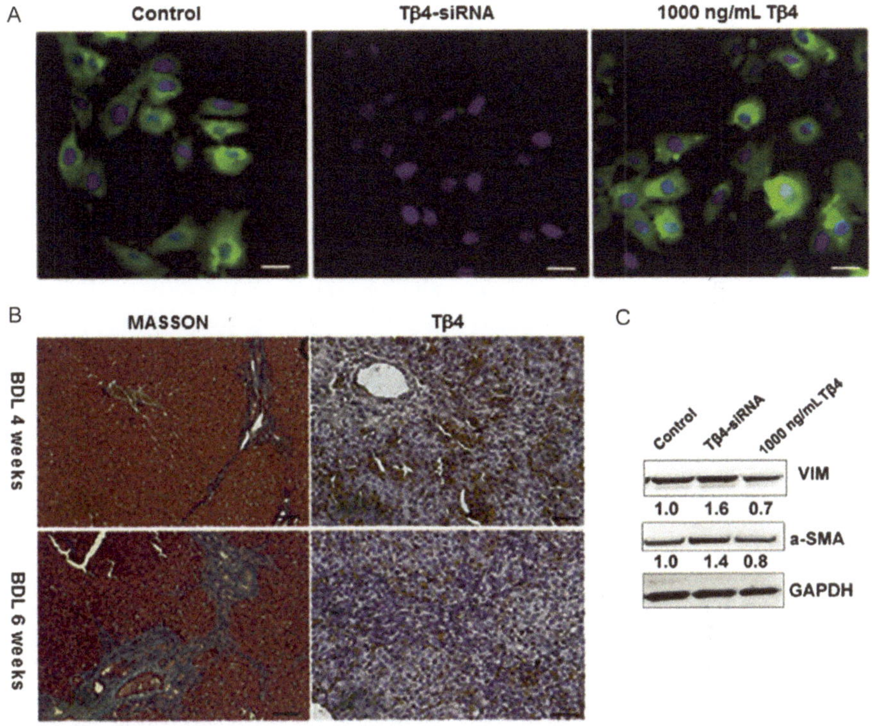

J. Kim and Y. Jung, Fig. 2 Tβ4 is upregulated in LX-2 cells but downregulated in fibrotic liver. (A) Immunofluorescence analysis shows that Tβ4 is expressed in LX-2 cells treated with or without Tβ4 peptide. Tβ4-siRNA downregulates the endogenous expression of Tβ4 in these cells (×20). (B) Immunohistochemical staining for Tβ4 in liver of rats at 4 and 6 weeks after bile duct ligation. (C) Expression of α-SMA and vimentin in Tβ4-siRNA or Tβ4 peptide-treated LX-2 cells, as examined by Western blot assay. The expression of GAPDH was used as an internal control. *Figure modified from Xiao, Y., Qu, C., Ge, W., Wang, B., Wu, J., Xu, L., & Chen, Y. (2014). Depletion of thymosin β4 promotes the proliferation, migration, and activation of human hepatic stellate cells.* Cellular Physiology and Biochemistry: International Journal of Experimental Cellular Physiology, Biochemistry, and Pharmacology, 34*(2), 356–367, ©2014 S. Karger AG, Basel.*

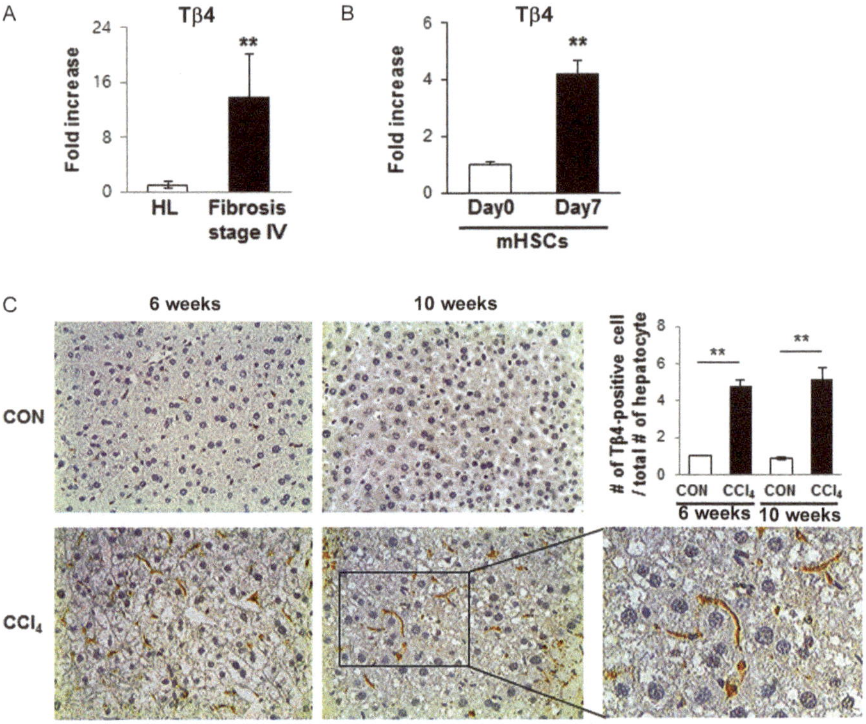

J. Kim and Y. Jung, Fig. 3 Tβ4 increases during hepatic fibrogenesis and HSC activation. (A) QRT-PCR analysis for Tβ4 in human liver obtained from healthy (HL) and fibrotic livers with four stage of fibrosis (fibrosis stage IV). Mean $\pm$ SD results are graphed and data represent the mean $\pm$ SD of three independent experiments ($^{**}p < 0.005$). (B) Tβ4 is upregulated during HSC culture, as assessed by QRT-PCR (Day 0: freshly isolated and not being cultured primary HSCs/Day 7: cultured primary HSCs for 7 days). Mean $\pm$ SD results are graphed. Data represent the mean $\pm$ SD of three independent experiments ($^{**}p < 0.005$). (C) Immunohistochemical staining for Tβ4 in liver sections from the representative mice at 6 and 10 weeks after CCl_4 injection ($\times 40$). Magnified image from CCl_4-treated livers at 10 weeks ($\times 63$) shows Tβ4-positive HSC-looking cells. The number of Tβ4-positive cells are counted and graphed ($^{**}p < 0.005$ vs own control group). *CON*: corn oil-treated mice, *CCl_4*: CCl_4-treated mice. *Figure modified from Kim, J., Wang, S., Hyun, J., Choi, S. S., Cha, H., Ock, M., & Jung, Y. (2015). Hepatic stellate cells express thymosin Beta 4 in chronically damaged liver.* PLoS One, *10(3), e0122758, ©2015 Kim et al.*

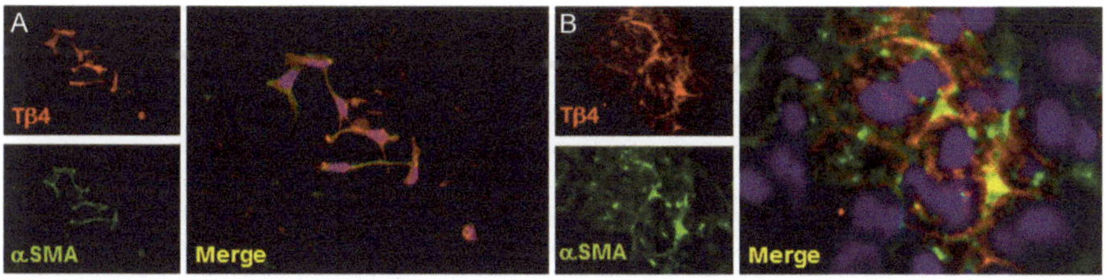

J. Kim and Y. Jung, Fig. 4 Activated HSCs express Tβ4 in the damaged livers. (A) Double immunofluorescent staining for Tβ4 (*red*) and α-SMA (*green*) in acitvated LX-2 cells (×40). (B) Double immunofluorescent staining for Tβ4 (*red*) and α-SMA (*green*) in liver section from representative CCl_4-treated mice (×63). Coexpressing cells are shown as *yellow* to *orange*. DAPI nuclear staining is shown as *blue*. *Figure modified from Kim, J., Wang, S., Hyun, J., Choi, S. S., Cha, H., Ock, M., & Jung, Y. (2015). Hepatic stellate cells express thymosin Beta 4 in chronically damaged liver.* PLoS One, 10*(3), e0122758, ©2015 Kim et al.*

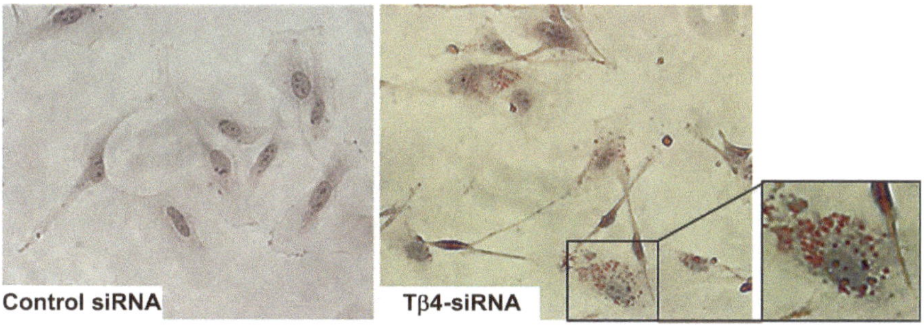

J. Kim and Y. Jung, Fig. 5 Suppression of Tβ4 induces the inactivation of HSCs. Oil Red O (ORO) staining in control siRNA- or Tβ4 siRNA-treated LX-2 cells (X40). The lipid droplets appear as *red*. Magnified image (X63) from Tβ4 siRNA-treated cells show *red* lipid-like droplets around the nuclei (X63). *control siRNA*: control siRNA-treated LX-2 cells, *Tβ4 siRNA*: Tβ4 siRNA-treated LX-2 cells. *Figure modified from Kim, J., Wang, S., Hyun, J., Choi, S. S., Cha, H., Ock, M., & Jung, Y. (2015). Hepatic stellate cells express thymosin beta 4 in chronically damaged liver.* PLoS One, 10*(3), e0122758, ©2015 Kim et al.*

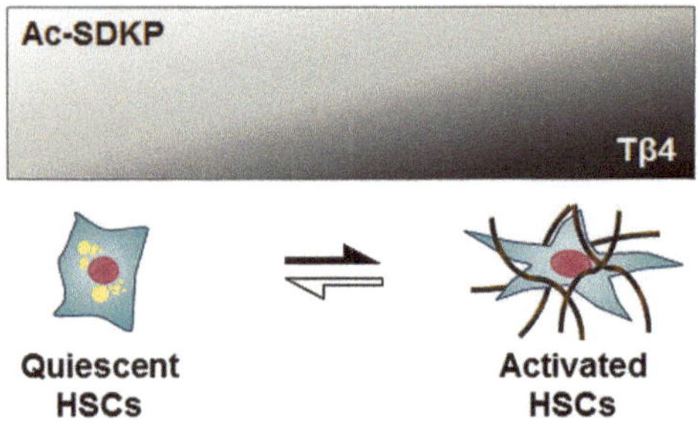

J. Kim and Y. Jung, Fig. 6 Proposed model for role of Tβ4 and Ac-SDKP on activation of HSCs. Quiescent HSCs transdifferentiate into the activated HSCs following liver injury. During activation of HSCs, the endogenous level of Ac-SDKP (*white*) decreases, whereas the endogenous level of Tβ4 (*black*) increases.

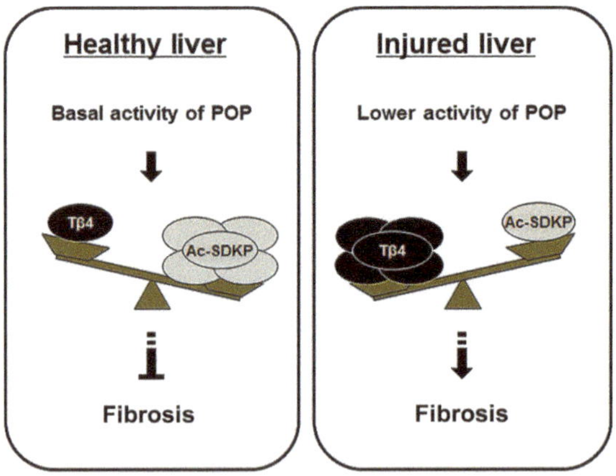

J. Kim and Y. Jung, Fig. 7 Schematic illustration of the effect of balance between Tβ4 and Ac-SDKP level on liver fibrosis. It is depicted how the amount of Tβ4 and Ac-SDKP influences liver fibrosis. Tβ4 is cleaved into Ac-SDKP by prolyl oligopeptidase (POP). In healthy liver, POP with basal activity cleaves Tβ4 into Ac-SDKP normally. However, when activity of POP is reduced in damaged liver, the cleavage of Tβ4 into Ac-SDKP is suppressed. Hence, accumulation of Tβ4 with lower level of Ac-SDKP promotes liver fibrosis.

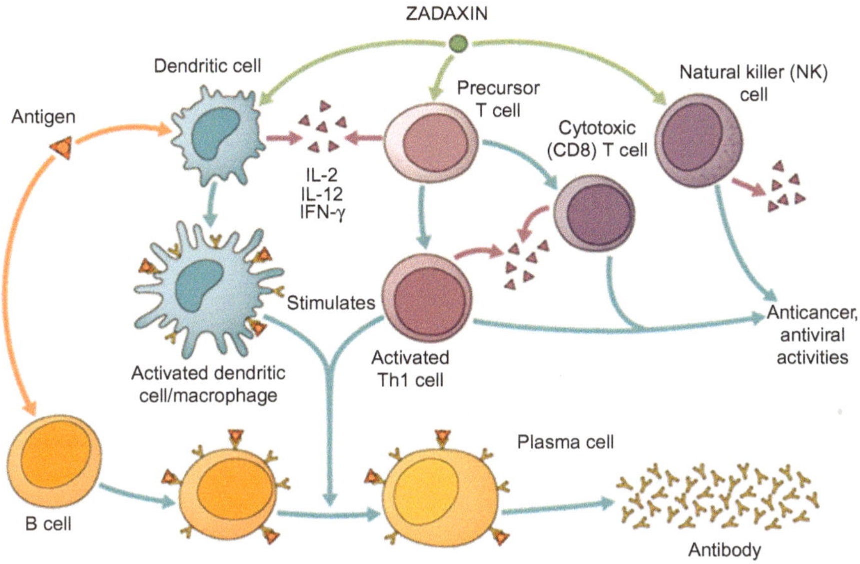

R. King and C. Tuthill, Fig. 1 Immune-stimulating mechanism of action of Ta1.

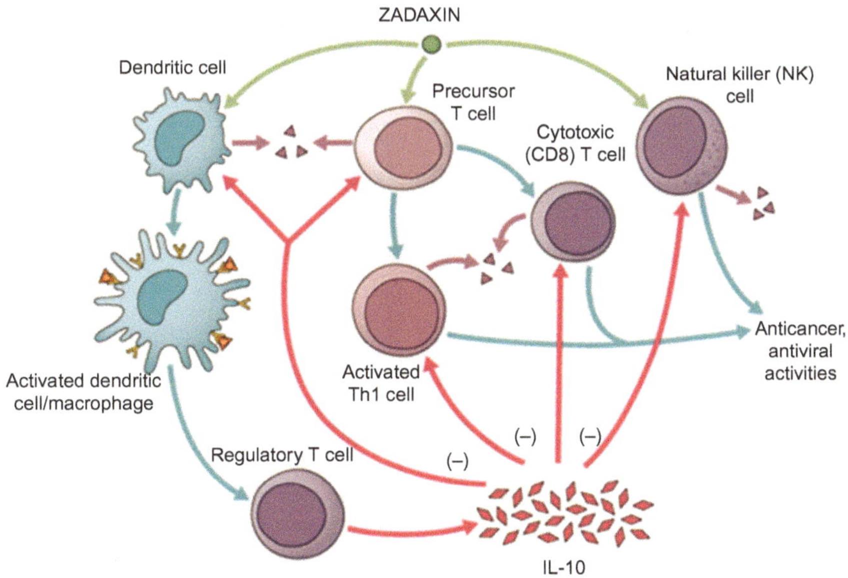

R. King and C. Tuthill, Fig. 2 Immune-modulating mechanism of action of Ta1.

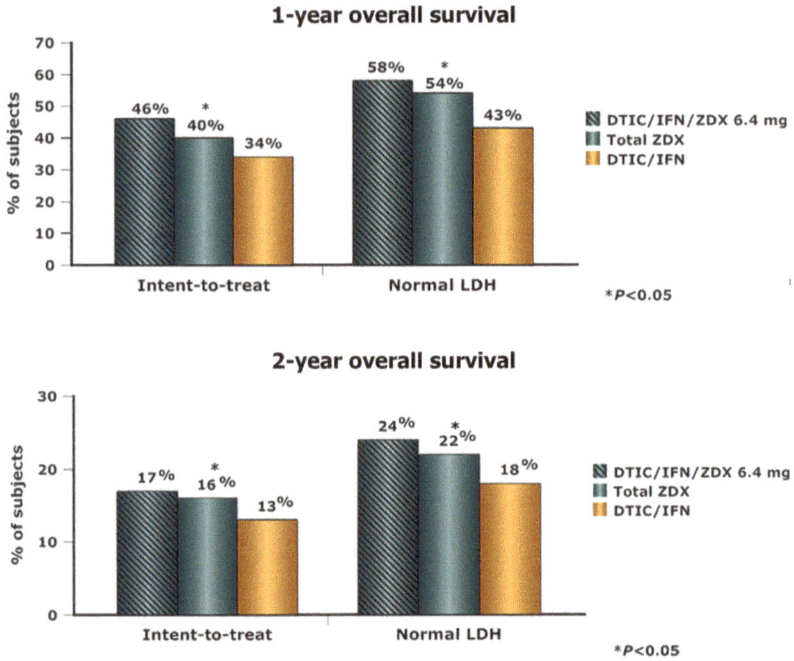

R. King and C. Tuthill, Fig. 4 Ta1 improves survival in Stage IV melanoma.

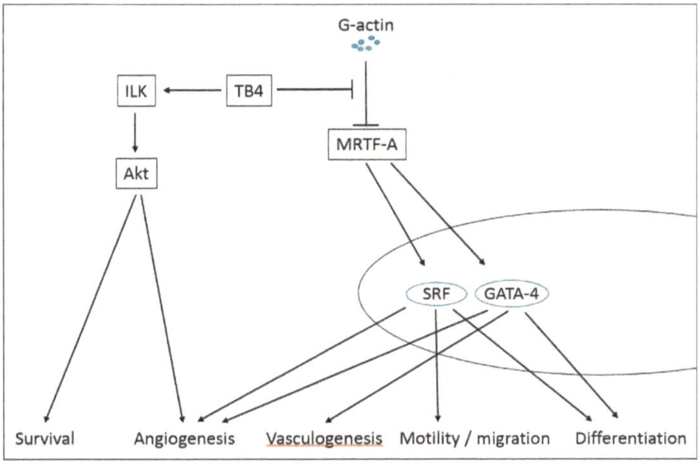

G.T. Pipes and J. Yang, Fig. 1 Simplified diagram depicting linkage between the MRTF-actin pathway and the ILK-Akt pathway via Tβ4, and subsequent effects at the cellular and tissue levels.

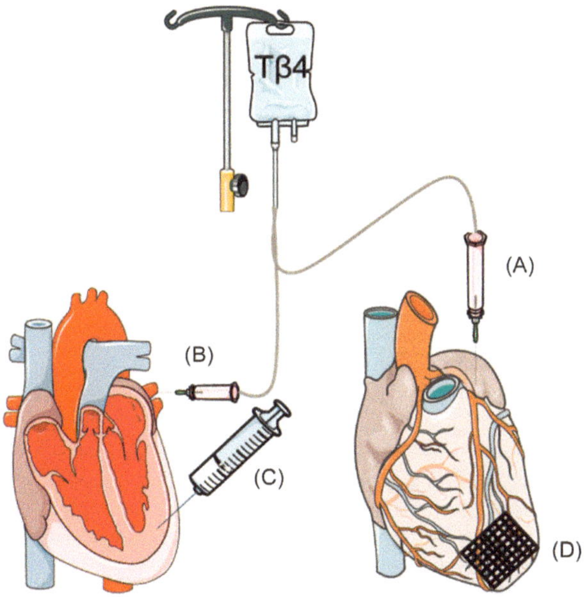

E.D. Marks and A. Kumar, Fig. 1 Schematic of administration routes for Tβ4. (A) Systemic administration via IV delivers peptide to the damaged left ventricle through the left anterior descending (LAD) artery. (B) Catheterization to the left ventricle allows for more specific delivery. (C) Direct intramuscular injections. (D) Drug-eluting patches can be applied to the epicardium or directly within the myocardium. Direct injections and patches require less overall peptide for therapeutic efficiency, but can be more invasive than catheterization or systemic administration through IV.

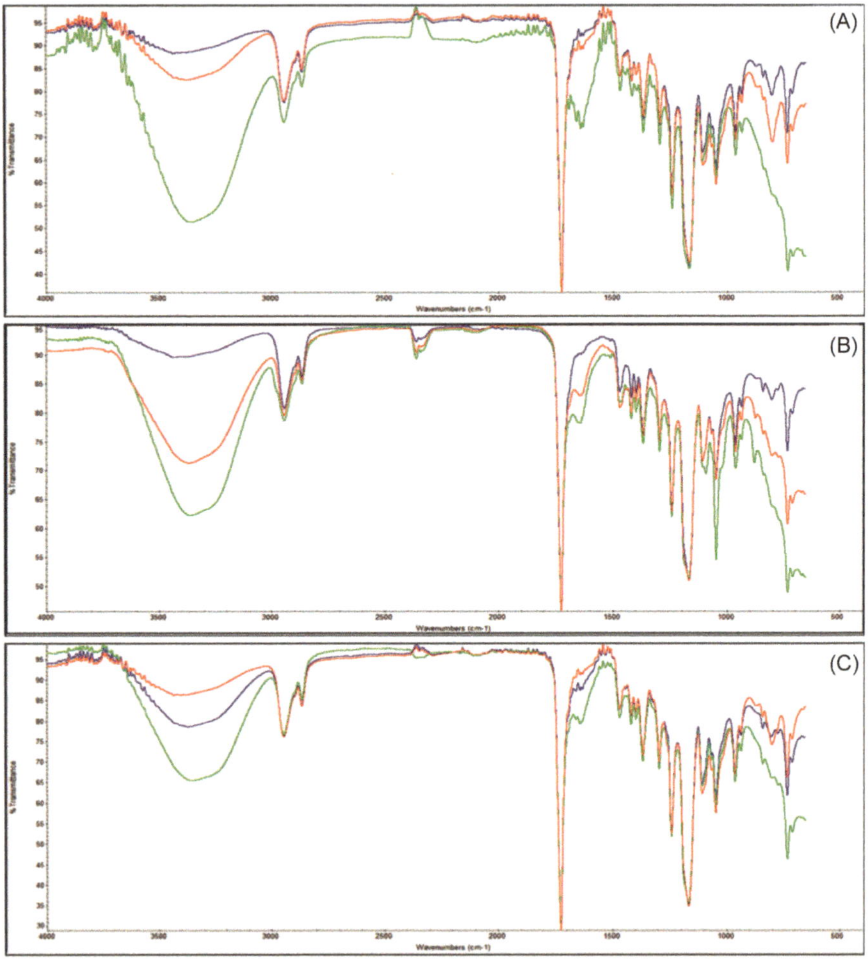

E.D. Marks and A. Kumar, Fig. 2 Drug-eluting nanoscaffolds release peptide and small molecules in a concentration-dependent manner. (A) Large quantities are detected immediately after coating with high concentration coatings, but quickly degrade. (B) Mid-range concentrations are encapsulated with high efficiency and release peptide more slowly. (C) Low concentrations do not stay within the scaffold to any appreciable degree.

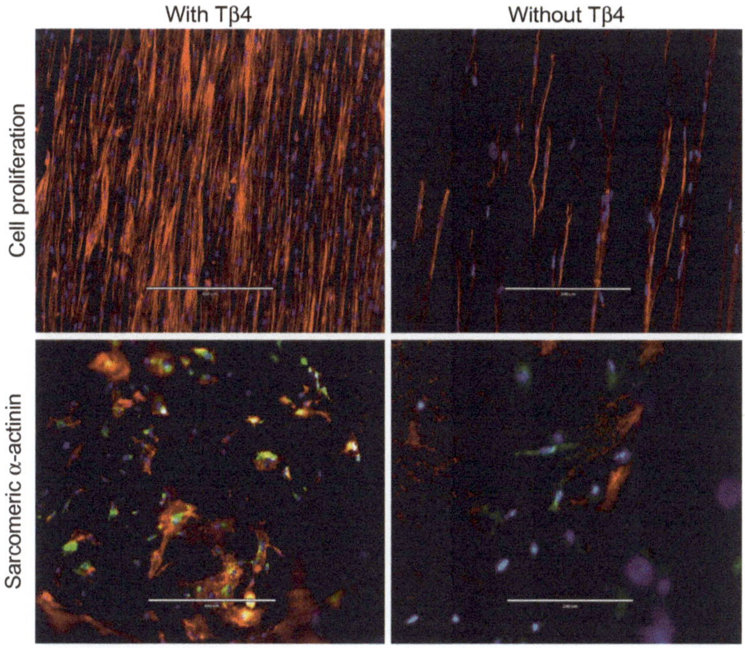

E.D. Marks and A. Kumar, Fig. 3 Tβ4 coatings promote proliferation and differentiation of stem cells toward the cardiac lineage. On coated nanoscaffolds, stem cells proliferate to produce a phenotype similar to myocardial tissue. Immunostaining for sarcomeric α-actinin, a biomarker for cardiomyocyte differentiation, is significantly increased in cells grown on coated nanoscaffolds.

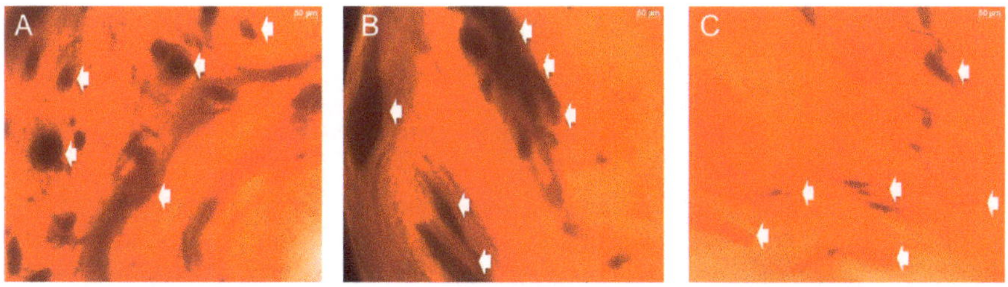

E.D. Marks and A. Kumar, Fig. 4 Nanoscaffolds successfully engraft to myocardium after injection. (A and B) Large and small pieces of scaffold system engraft from epi- to endocardium. (C) Smaller scaffold pieces assist in diffuse spread of cells and factors and are more likely to align to the native myocardial tissue structure. All images taken 48 h after MI induction and 1 h after scaffold injection.

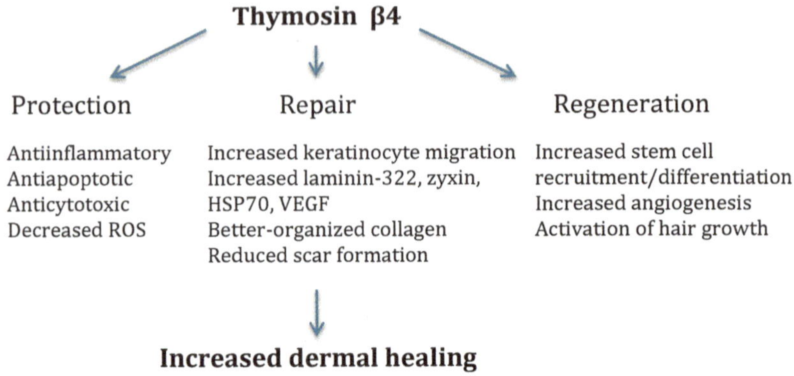

H.K. Kleinman and G. Sosne, Fig. 2 Activity in tissue protection, repair, and regeneration.

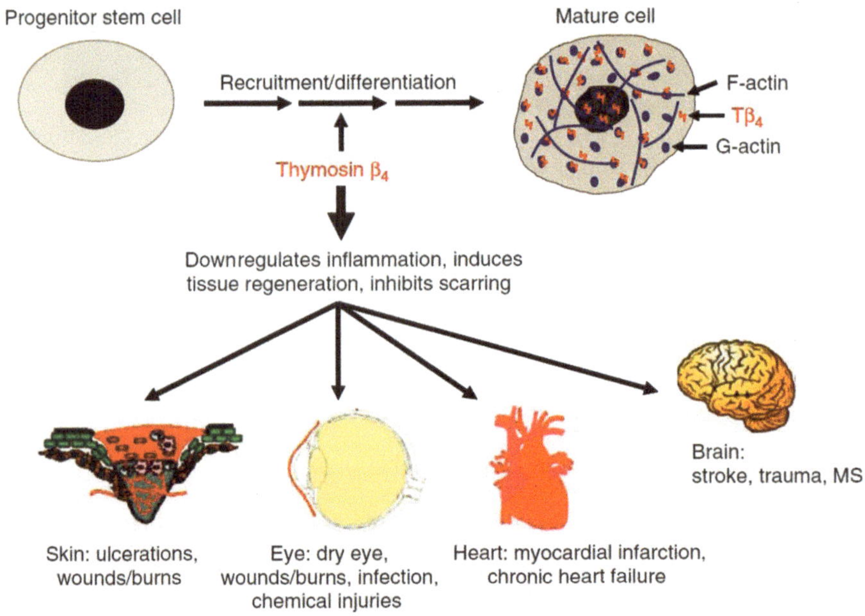

G. Sosne *et al.*, Fig. 2 Known and newly emerging clinical applications of Tβ4. Shown schematically are various organs and the injuries that have been found to be repaired or affected by Tβ4. *Copyright Permission: Goldstein, A. L., Hannappel, E., Sosne, G., & Kleinman, H. K. (2012). Thymosin β4: A multifunctional regenerative peptide. Basic properties and clinical applications.* Expert Opinion on Biological Therapy, 12*(1), 37–51. doi:10.1517/14712598.2012.634793. Used with permission: © Taylor & Francis LLC. http://www.tandfonline.com. Previous work for noncommercial use (doi:10.1517/14712598.2012.634793).*

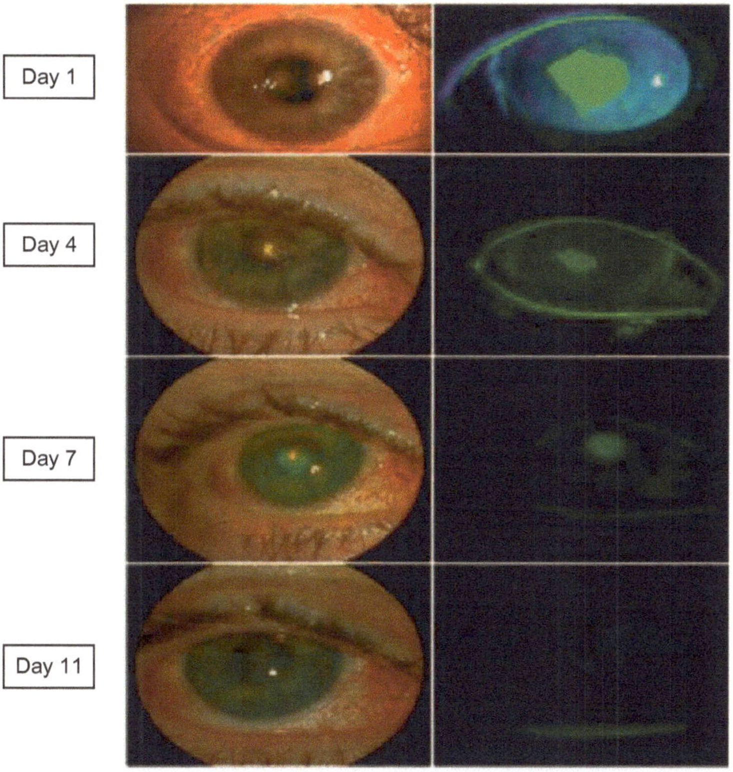

G. Sosne *et al.*, Fig. 3 Slit lamp photographs (*left column*) show the clinical appearance of the right eye at days 1, 4, 7, and 11 after Tβ4 therapy was initiated. The central corneal epithelial defect can be seen (*arrows*) as it decreases in size with the Tβ4 treatment. The fluorescein pictures (*right column*) show areas stained that represent areas absent of epithelium. The defect decreases to almost full closure by day 11 of Tβ4 treatment. *Copyright Permission: Originally published in Sosne, G., Qiu, P., Kurpakus-Wheater, M., & Matthew, H. (2010). Thymosin beta4 and corneal wound healing: Visions of the future.* The Annals of the New York Academy Sciences, 1194, *190–198. © The New York Academy of Sciences.*

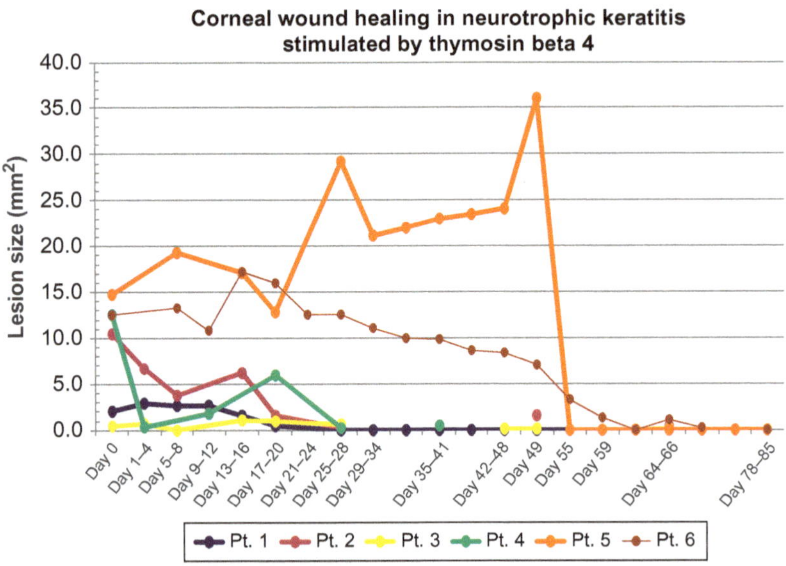

G. Sosne *et al.*, Fig. 4 Corneal wound healing in neurotrophic keratitis stimulated by Tβ4. *Copyright Permission: Originally published in Dunn, S. P., Heidemann, D. G., Chow, C. Y., Crockford, D., Turjman, N., Angel, J., …, Sosne, G. (2010). Treatment of chronic nonhealing neurotrophic corneal epithelial defects with thymosin beta4.* The Annals of the New York Academy of Sciences, 1194, *199–206. © The New York Academy of Sciences.*

FSC
www.fsc.org
MIX
Papier | Fördert
gute Waldnutzung
FSC® C083411

Kolibri 360 GmbH
Ferdinand-Jühlke-Straße 7
99095 Erfurt, Deutschland
produktsicherheit@kolibri360.de